A RELATIVIST'S TOOLKIT
The Mathematics of Black-Hole Mechanics

This textbook fills a gap in the existing literature on general relativity by providing the advanced student with practical tools for the computation of many physically interesting quantities. The context is provided by the mathematical theory of black holes, one of the most elegant, successful and relevant applications of general relativity.

Among the topics discussed are congruences of timelike and null geodesics, the embedding of spacelike, timelike, and null hypersurfaces in spacetime, and the Lagrangian and Hamiltonian formulations of general relativity. The book also describes the application of null congruences to the description of the event horizon, how integration over a null hypersurface relates to black-hole mechanics, and the relationship between the gravitational Hamiltonian and a black hole's mass and angular momentum.

Although the book is self-contained, it is not meant to serve as an introduction to general relativity. Instead, it is meant to help the reader acquire advanced skills and become a competent researcher in relativity and gravitational physics. The primary readership consists of graduate students in gravitational physics. The book will also be a useful reference for more seasoned researchers working in this field.

ERIC POISSON has been a faculty member at the University of Guelph since 1995. He has taught a large number of courses, including an advanced graduate course in general relativity for which this book was written. He obtained his B.Sc. from Laval University in Quebec City, and went to graduate school at the University of Alberta, in Edmonton. Poisson obtained his Ph.D. in 1991, under the supervision of Werner Israel. He then spent three years as a postdoctoral fellow in Kip Thorne's research group at the California Institute of Technology, in Pasadena. Before going to Guelph, he also spent a year with Clifford Will at Washington University in St Louis.

A RELATIVIST'S TOOLKIT

The Mathematics of Black-Hole Mechanics

ERIC POISSON

Department of Physics, University of Guelph

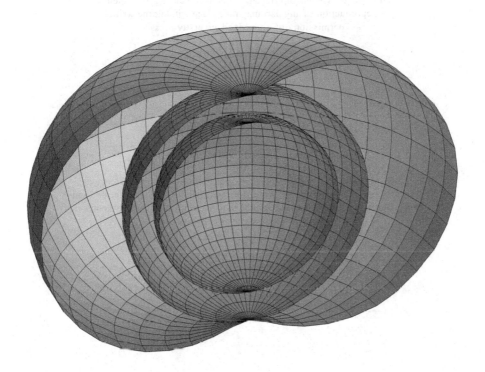

CAMBRIDGE

UNIVERSITY PRESS

CAMBRIDGE UNIVERSITY PRESS
Cambridge, New York, Melbourne, Madrid, Cape Town, Singapore, São Paulo

Cambridge University Press
The Edinburgh Building, Cambridge CB2 8RU, UK

Published in the United States of America by Cambridge University Press, New York

www.cambridge.org
Information on this title: www.cambridge.org/9780521830911

First published 2004
This digitally printed version 2007

A catalogue record for this publication is available from the British Library

ISBN 978-0-521-83091-1 hardback
ISBN 978-0-521-53780-3 paperback

Contents

Preface		*page* xi
Notation and conventions		xvi
1	**Fundamentals**	1
1.1	Vectors, dual vectors, and tensors	2
1.2	Covariant differentiation	4
1.3	Geodesics	6
1.4	Lie differentiation	8
1.5	Killing vectors	10
1.6	Local flatness	11
1.7	Metric determinant	12
1.8	Levi-Civita tensor	13
1.9	Curvature	15
1.10	Geodesic deviation	16
1.11	Fermi normal coordinates	18
	1.11.1 Geometric construction	19
	1.11.2 Coordinate transformation	20
	1.11.3 Deviation vectors	21
	1.11.4 Metric on γ	22
	1.11.5 First derivatives of the metric on γ	22
	1.11.6 Second derivatives of the metric on γ	22
	1.11.7 Riemann tensor in Fermi normal coordinates	24
1.12	Bibliographical notes	24
1.13	Problems	25
2	**Geodesic congruences**	28
2.1	Energy conditions	29
	2.1.1 Introduction and summary	29
	2.1.2 Weak energy condition	30
	2.1.3 Null energy condition	31

	2.1.4	Strong energy condition	31
	2.1.5	Dominant energy condition	32
	2.1.6	Violations of the energy conditions	32
2.2	Kinematics of a deformable medium		33
	2.2.1	Two-dimensional medium	33
	2.2.2	Expansion	34
	2.2.3	Shear	34
	2.2.4	Rotation	35
	2.2.5	General case	35
	2.2.6	Three-dimensional medium	35
2.3	Congruence of timelike geodesics		36
	2.3.1	Transverse metric	37
	2.3.2	Kinematics	37
	2.3.3	Frobenius' theorem	38
	2.3.4	Raychaudhuri's equation	40
	2.3.5	Focusing theorem	40
	2.3.6	Example	41
	2.3.7	Another example	42
	2.3.8	Interpretation of θ	43
2.4	Congruence of null geodesics		45
	2.4.1	Transverse metric	46
	2.4.2	Kinematics	47
	2.4.3	Frobenius' theorem	48
	2.4.4	Raychaudhuri's equation	50
	2.4.5	Focusing theorem	50
	2.4.6	Example	51
	2.4.7	Another example	52
	2.4.8	Interpretation of θ	52
2.5	Bibliographical notes		54
2.6	Problems		54
3	**Hypersurfaces**		**59**
3.1	Description of hypersurfaces		60
	3.1.1	Defining equations	60
	3.1.2	Normal vector	60
	3.1.3	Induced metric	62
	3.1.4	Light cone in flat spacetime	63
3.2	Integration on hypersurfaces		64
	3.2.1	Surface element (non-null case)	64
	3.2.2	Surface element (null case)	65
	3.2.3	Element of two-surface	67

3.3	Gauss–Stokes theorem	69
	3.3.1 First version	69
	3.3.2 Conservation	71
	3.3.3 Second version	72
3.4	Differentiation of tangent vector fields	73
	3.4.1 Tangent tensor fields	73
	3.4.2 Intrinsic covariant derivative	73
	3.4.3 Extrinsic curvature	75
3.5	Gauss–Codazzi equations	76
	3.5.1 General form	76
	3.5.2 Contracted form	78
	3.5.3 Ricci scalar	78
3.6	Initial-value problem	79
	3.6.1 Constraints	79
	3.6.2 Cosmological initial values	80
	3.6.3 Moment of time symmetry	81
	3.6.4 Stationary and static spacetimes	82
	3.6.5 Spherical space, moment of time symmetry	82
	3.6.6 Spherical space, empty and flat	82
	3.6.7 Conformally-flat space	84
3.7	Junction conditions and thin shells	84
	3.7.1 Notation and assumptions	85
	3.7.2 First junction condition	86
	3.7.3 Riemann tensor	87
	3.7.4 Surface stress-energy tensor	87
	3.7.5 Second junction condition	88
	3.7.6 Summary	89
3.8	Oppenheimer–Snyder collapse	90
3.9	Thin-shell collapse	93
3.10	Slowly rotating shell	94
3.11	Null shells	98
	3.11.1 Geometry	98
	3.11.2 Surface stress-energy tensor	100
	3.11.3 Intrinsic formulation	102
	3.11.4 Summary	104
	3.11.5 Parameterization of the null generators	104
	3.11.6 Imploding spherical shell	107
	3.11.7 Accreting black hole	109
	3.11.8 Cosmological phase transition	112
3.12	Bibliographical notes	114

3.13   Problems   114
4   Lagrangian and Hamiltonian formulations of general relativity   118
4.1   Lagrangian formulation   119
    4.1.1   Mechanics   119
    4.1.2   Field theory   120
    4.1.3   General relativity   121
    4.1.4   Variation of the Hilbert term   122
    4.1.5   Variation of the boundary term   124
    4.1.6   Variation of the matter action   125
    4.1.7   Nondynamical term   126
    4.1.8   Bianchi identities   127
4.2   Hamiltonian formulation   128
    4.2.1   Mechanics   128
    4.2.2 $3 + 1$ decomposition   129
    4.2.3   Field theory   131
    4.2.4   Foliation of the boundary   134
    4.2.5   Gravitational action   136
    4.2.6   Gravitational Hamiltonian   139
    4.2.7   Variation of the Hamiltonian   141
    4.2.8   Hamilton's equations   145
    4.2.9   Value of the Hamiltonian for solutions   146
4.3   Mass and angular momentum   146
    4.3.1   Hamiltonian definitions   146
    4.3.2   Mass and angular momentum for stationary, axially
        symmetric spacetimes   148
    4.3.3   Komar formulae   149
    4.3.4   Bondi–Sachs mass   151
    4.3.5   Distinction between ADM and Bondi–Sachs masses:
        Vaidya spacetime   152
    4.3.6   Transfer of mass and angular momentum   155
4.4   Bibliographical notes   156
4.5   Problems   157
5   Black holes   163
5.1   Schwarzschild black hole   163
    5.1.1   Birkhoff's theorem   163
    5.1.2   Kruskal coordinates   164
    5.1.3   Eddington–Finkelstein coordinates   167
    5.1.4   Painlevé–Gullstrand coordinates   168
    5.1.5   Penrose–Carter diagram   168
    5.1.6   Event horizon   170

	5.1.7	Apparent horizon	171
	5.1.8	Distinction between event and apparent horizons: Vaidya spacetime	172
	5.1.9	Killing horizon	175
	5.1.10	Bifurcation two-sphere	176
5.2	Reissner–Nordström black hole		176
	5.2.1	Derivation of the Reissner–Nordström solution	176
	5.2.2	Kruskal coordinates	178
	5.2.3	Radial observers in Reissner–Nordström spacetime	181
	5.2.4	Surface gravity	184
5.3	Kerr black hole		187
	5.3.1	The Kerr metric	187
	5.3.2	Dragging of inertial frames: ZAMOs	188
	5.3.3	Static limit: static observers	188
	5.3.4	Event horizon: stationary observers	189
	5.3.5	The Penrose process	191
	5.3.6	Principal null congruences	192
	5.3.7	Kerr–Schild coordinates	194
	5.3.8	The nature of the singularity	195
	5.3.9	Maximal extension of the Kerr spacetime	196
	5.3.10	Surface gravity	198
	5.3.11	Bifurcation two-sphere	199
	5.3.12	Smarr's formula	200
	5.3.13	Variation law	201
5.4	General properties of black holes		201
	5.4.1	General black holes	202
	5.4.2	Stationary black holes	204
	5.4.3	Stationary black holes in vacuum	205
5.5	The laws of black-hole mechanics		206
	5.5.1	Preliminaries	207
	5.5.2	Zeroth law	208
	5.5.3	Generalized Smarr formula	209
	5.5.4	First law	211
	5.5.5	Second law	212
	5.5.6	Third law	213
	5.5.7	Black-hole thermodynamics	214
5.6	Bibliographical notes		215
5.7	Problems		216
References			224
Index			229

Preface

Does the world really need a new textbook on general relativity? I feel that my first duty in presenting this book should be to provide a convincing affirmative answer to this question.

There already exists a vast array of available books. I will not attempt here to make an exhaustive list, but I will mention three of my favourites. For its unsurpassed pedagogical presentation of the elementary aspects of general relativity, I like Schutz's *A first course in general relativity*. For its unsurpassed completeness, I like *Gravitation* by Misner, Thorne, and Wheeler. And for its unsurpassed elegance and rigour, I like Wald's *General Relativity*. In my view, a serious student could do no better than start with Schutz for an outstanding introductory course, then move on to Misner, Thorne, and Wheeler to get a broad coverage of many different topics and techniques, and then finish off with Wald to gain access to the more modern topics and the mathematical standard that Wald has since imposed on this field. This is a long route, but with this book I hope to help the student along. I see my place as being somewhere between Schutz and Wald – more advanced than Schutz but less sophisticated than Wald – and I cover some of the few topics that are not handled by Misner, Thorne, and Wheeler.

In the winter of 1998 I was given the responsibility of creating an advanced course in general relativity. The course was intended for graduate students working in the Gravitation Group of the Guelph-Waterloo Physics Institute, a joint graduate programme in Physics shared by the Universities of Guelph and Waterloo. I thought long and hard before giving the first offering of this course, in an effort to round up the most useful and interesting topics, and to create the best possible course. I came up with a few guiding principles. First, I wanted to let the students in on a number of results and techniques that are part of every relativist's arsenal, but are not adequately covered in the popular texts. Second, I wanted the course to be practical, in the sense that the students would learn how to compute things instead of being subjected to a bunch of abstract concepts. And third, I wanted to put these techniques to work in a really cool application of the theory, so that this whole enterprise would seem to have purpose.

As I developed the course it became clear that it would not match the material covered in any of the existing textbooks; to meet my requirements I would have to form a synthesis of many texts, I would have to consult review articles, and I would have to go to the technical literature. This was a long but enjoyable undertaking, and I learned a lot. It gave me the opportunity to homogenize the various separate treatments, consolidate the various different notations, and present this synthesis as a unified whole. During this process I started to type up lecture notes that would be distributed to the students. These have evolved into this book.

In the end, the course was designed around my choice of 'really cool application'. There was no contest: the immediate winner was the mathematical theory of black holes, surely one of the most elegant, successful, and relevant applications of general relativity. This is covered in Chapter 5 of this book, which offers a thorough review of the solutions to the Einstein field equations that describe isolated black holes, a description of the fundamental properties of black holes that are independent of the details of any particular solution, and an introduction to the four laws of black-hole mechanics. In the next paragraphs I outline the material covered in the other chapters, and describe the connections with the theory of black holes.

The most important aspect of black-hole spacetimes is that they contain an event horizon, a null hypersurface that marks the boundary of the black hole and shields external observers from events going on inside. On this hypersurface there runs a network (or congruence) of non-intersecting null geodesics; these are called the null generators of the event horizon. To understand the behaviour of the horizon as a whole it proves necessary to understand how the generators themselves behave, and in Chapter 2 of this book we develop the relevant techniques. The description of congruences is concerned with the motion of nearby geodesics relative to a given reference geodesic; this motion is described by a deviation vector that lives in a space orthogonal to the tangent vector of the reference geodesic. This transverse space is easy to construct when the geodesics are timelike, but the case of null geodesics is subtle. This has to do with the fact that the transverse space is then two-dimensional – the null vector tangent to the generators is orthogonal to itself and this direction must be explicitly removed from the transverse space. I show how this is done in Chapter 2. While null congruences are treated in other textbooks (most notably in Wald), the student is likely to find my presentation (which I have adapted from Carter (1979)) better suited for practical computations. While Chapter 2 is concerned mostly with congruences of null geodesics, I present also a complete treatment of the timelike case. There are two reasons for this. First, this forms a necessary basis to understand the subtleties associated with the null case. Second, and more importantly, the mathematical techniques involved in the study of congruences of timelike geodesics are used widely in the general relativity literature, most notably in the field of mathematical cosmology. Another topic covered in Chapter 2 is the standard energy conditions of general relativity; these constraints on the stress-energy tensor ensure that under normal circumstances, gravity acts as an attractive force – it tends to focus geodesics. Energy conditions appear in most theorems governing the behaviour of black holes.

Many quantities of interest in black-hole physics are defined by integration over the event horizon. An obvious example is the hole's surface area. Another example is the gain in mass of an accreting black hole; this is obtained by integrating a certain component of the accreting material's stress-energy tensor over the event horizon. These integrations require techniques that are introduced in Chapter 3 of this book. In particular, we shall need a notion of surface element on the event horizon. If the horizon were a timelike or a spacelike hypersurface, the construction of a surface element would pose no particular challenge, but once again there are interesting subtleties associated with the null case. I provide a complete treatment of these issues in Chapter 3. I believe that my presentation is more systematic, and more practical, than what can be found in the popular textbooks. Other topics covered in Chapter 3 include the initial-value problem of general relativity (which involves the induced metric and extrinsic curvature of a spacelike hypersurface) and the Darmois–Lanczos–Israel–Barrabès formalism for junction conditions and thin shells (which constrains the possible discontinuities in the induced metric and extrinsic curvature). The initial-value problem is discussed at a much deeper level in Wald, but I felt it was important to include this material here: it provides a useful illustration of the physical meaning of the extrinsic curvature, an object that plays an important role in Chapter 4 of this book. Junction conditions and thin shells, on the other hand, are not covered adequately in any textbook, in spite of the fact that the Darmois–Lanczos–Israel–Barrabès formalism is used very widely in the literature. (Junction conditions and thin shells are touched upon in Misner, Thorne, and Wheeler, but I find that their treatment is too brief to do justice to the formalism.)

Among the most important quantities characterizing black holes are their mass and angular momentum, and the question arises as to how the mass and angular momentum of an isolated body are to be defined in general relativity. I find that the most compelling definitions come from the gravitational Hamiltonian, whose value for a given solution to the Einstein field equations depends on a specifiable vector field. If this vector corresponds to a time translation at spatial infinity, then the Hamiltonian gives the total mass of the spacetime; if, on the other hand, the vector corresponds to an asymptotic rotation about an axis, then the Hamiltonian gives the spacetime's total angular momentum in the direction of this axis. This connection is both deep and beautiful, and in this book it forms the starting point for defining black-hole mass and angular momentum. Chapter 4 is devoted to a systematic treatment of the Lagrangian and Hamiltonian formulations of general relativity, with the goal in mind of arriving at well-motivated notions of mass and angular momentum. What sets my presentation apart from what can be found in other texts, including Misner, Thorne, and Wheeler and Wald, is that I pay careful attention to the 'boundary terms' that must be included in the gravitational action to produce a well-posed variational principle. These boundary terms have been around for a very long time, but it is only fairly recently that their importance has been fully recognized. In particular, they are directly involved in defining the mass and angular momentum of an asymptotically-flat spacetime.

To set the stage, I review the fundamentals of differential geometry in Chapter 1 of this book. The collection of topics is standard: vectors and tensors, covariant differentiation, geodesics, Lie differentiation, Killing vectors, curvature tensors, geodesic deviation, and some others. The goal here is not to provide an introduction to these topics; although some may be new, I assume that for the most part, the student will have encountered them before (in an introductory course at the level of Schutz, for example). Instead, my objective with this chapter is to refresh the student's memory and establish the style and notation that I adopt throughout the book.

As I have indicated, I have tried to present this material as a unified whole, using a consistent notation and maintaining a fairly uniform level of precision and rigour. While I have tried to be somewhat precise and rigourous, I have deliberately avoided putting too much emphasis on this. My attitude is that it is more important to illustrate how a theorem works and can be used in a practical situation, than it is to provide all the fine print that goes into a rigourous proof. The proofs that I do provide are informal; they may sometimes be incomplete, but they should suffice to convince the student that the theorems are true. They may, however, leave the student wanting for more; in this case I shall have to refer her to a more authoritative text such as Wald.

I have also indicated that I wanted this book to be practical – I hope that after studying this book, the student will be able to use what she has learned to compute things of direct relevance to her. To encourage this I have inserted a large number of examples within the text. I also provide problem sets at the end of each chapter; here the student's understanding will be put to the test. The problems vary in difficulty, from the plug-and-grind type designed to increase the student's familiarity with a new technique, to the more challenging type that is supposed to make the student think. Some of the problems require a large amount of tensor algebra, and I strongly advise the student to let the computer perform the most routine operations. (My favourite package for tensor manipulations is GRTensorII, developed by Peter Musgrave, Denis Pollney and Kayll Lake. It is available free of charge at http://grtensor.phy.queensu.ca/.)

Early versions of this book have been used by graduate students who took my course over the years. A number of them have expressed great praise by involving some of the techniques covered here in their own research. This is extraordinarily gratifying, and it has convinced me that a wider release of this book might do more than just service my vanity. A number of students have carefully checked through the manuscript for errors (typographical or otherwise), and some have made useful suggestions for improvements. For this I thank Daniel Bruni, Sean Crowe, Luis de Menezes, Paul Kobak, Karl Martel, Peter Martin, Sanjeev Seahra and Katrin Rohlf. Of course, I accept full responsibility for whatever errors remain. The reader is invited to report any error she may find (poisson@physics.uoguelph.ca), and can look up those already reported at http://www.physics.uoguelph.ca/poisson/toolkit/.

This book is dedicated to Werner Israel, my teacher, mentor, and friend, whose influence on me, both as a relativist and as a human being, runs deep. His influence, I trust, will be felt throughout the book. Each time I started the elaboration of a new topic I would ask myself: 'How would Werner approach this?' I do not believe that the answers I came up with would come even close to his level of pedagogical excellence, but there is no doubt that to ask the question has made me try harder to reach that level.

Notation and conventions

I adopt the sign conventions of Misner, Thorne, and Wheeler (1973), with a metric of signature $(-1, 1, 1, 1)$, a Riemann tensor defined by $R^{\alpha}{}_{\beta\gamma\delta} = \Gamma^{\alpha}{}_{\beta\delta,\gamma} + \cdots$, and a Ricci tensor defined by $R_{\alpha\beta} = R^{\mu}{}_{\alpha\mu\beta}$. Greek indices $(\alpha, \beta, \dots)$ run from 0 to 3, lower-case Latin indices $(a, b, \dots)$ run from 1 to 3, and upper-case Latin indices $(A, B, \dots)$ run from 2 to 3. Geometrized units, in which $G = c = 1$, are employed.

    Here's a list of frequently occurring symbols:

Symbol	Description
x^{α}	Arbitrary coordinates on manifold $\mathscr{M}$
y^{a}	Arbitrary coordinates on hypersurface Σ
θ^{A}	Arbitrary coordinates on two-surface S
$\overset{*}{=}$	Equals in specified coordinates
$e^{\alpha}_{a} = \partial x^{\alpha}/\partial y^{a}, \quad e^{\alpha}_{A} = \partial x^{\alpha}/\partial \theta^{A}$	Holonomic basis vectors
$\hat{e}^{\alpha}_{\mu}, \quad \hat{e}^{\alpha}_{A}$	Orthonormal basis vectors
$g_{\alpha\beta}$	Metric on $\mathscr{M}$
$h_{ab} = g_{\alpha\beta} e^{\alpha}_{a} e^{\beta}_{b}$	Induced metric on Σ
$\sigma_{AB} = g_{\alpha\beta} e^{\alpha}_{A} e^{\beta}_{B}$	Induced metric on S
$g, \quad h, \quad \sigma$	Metric determinants
$A_{(\alpha\beta)} = \frac{1}{2}(A_{\alpha\beta} + A_{\beta\alpha})$	Symmetrization
$A_{[\alpha\beta]} = \frac{1}{2}(A_{\alpha\beta} - A_{\beta\alpha})$	Antisymmetrization
$\Gamma^{\alpha}{}_{\beta\gamma}$	Christoffel symbols constructed from $g_{\alpha\beta}$
$\Gamma^{a}{}_{bc}$	Christoffel symbols constructed from h_{ab}
$R_{\alpha\beta\gamma\delta}, \quad R_{\alpha\beta}, \quad R$	As constructed from $g_{\alpha\beta}$
$R_{abcd}, \quad R_{ab}, \quad {}^{3}R$	As constructed from h_{ab}
$\psi_{,\alpha} = \partial_{\alpha}\psi$	Partial differentiation with respect to x^{α}
$\psi_{,a} = \partial_{a}\psi$	Partial differentiation with respect to y^{a}
$A^{\alpha}{}_{;\beta} = \nabla_{\beta} A^{\alpha}$	Covariant differentiation ($g_{\alpha\beta}$-compatible)
$A^{a}{}_{|b} = D_{b} A^{a}$	Covariant differentiation (h_{ab}-compatible)
$\pounds_{u} A^{\alpha}$	Lie derivative of A^{α} along u^{α}
$[\alpha\ \beta\ \gamma\ \delta]$	Permutation symbol
$\varepsilon_{\alpha\beta\gamma\delta} = \sqrt{-g}\,[\alpha\ \beta\ \gamma\ \delta]$	Levi-Civita tensor
$d\Sigma_{\mu} = \varepsilon_{\mu\alpha\beta\gamma}\, e^{\alpha}_{1} e^{\beta}_{2} e^{\gamma}_{3}\, d^{3}y$	Directed surface element on Σ
$dS_{\mu\nu} = \varepsilon_{\mu\nu\alpha\beta}\, e^{\alpha}_{2} e^{\beta}_{3}\, d^{2}\theta$	Directed surface element on S
n^{α}	Unit normal on Σ (if timelike or spacelike)
$\varepsilon = n^{\alpha} n_{\alpha}$	$+1$ if Σ is timelike, -1 if Σ is spacelike
$K_{ab} = n_{\alpha;\beta}\, e^{\alpha}_{a} e^{\beta}_{b}$	Extrinsic curvature of Σ
$\theta, \quad \sigma_{\alpha\beta}, \quad \omega_{\alpha\beta}$	Expansion, shear, and rotation
$d\Omega^{2} = d\theta^{2} + \sin^{2}\theta\, d\phi^{2}$	Line element on unit two-sphere

1

Fundamentals

This first chapter is devoted to a brisk review of the fundamentals of differential geometry. The collection of topics presented here is fairly standard, and most of these topics should have been encountered in a previous introductory course on general relativity. Some, however, may be new, or may be treated here from a different point of view, or with an increased degree of completeness.

We begin in Section 1.1 by providing definitions for tensors on a differentiable manifold. The point of view adopted here, and throughout the text, is entirely unsophisticated: We do without the abstract formulation of differential geometry and define tensors in the old-fashioned way, in terms of how their components transform under a coordinate transformation. While the abstract formulation (in which tensors are defined as multilinear mappings of vectors and dual vectors into real numbers) is decidedly more elegant and beautiful, and should be an integral part of an education in general relativity, the old approach has the advantage of economy, and this motivated its adoption here. Also, the old-fashioned way of defining tensors produces an immediate distinction between tensor fields in spacetime (four-tensors) and tensor fields on a hypersurface (three-tensors); this distinction will be important in later chapters of this book.

Covariant differentiation is reviewed in Section 1.2, Lie differentiation in Section 1.4, and Killing vectors are introduced in Section 1.5. In Section 1.3 we develop the mathematical theory of geodesics. The theory is based on a variational principle and employs an arbitrary parameterization of the curve. The advantage of this approach (over one in which geodesics are defined by parallel transport of the tangent vector) is that the limiting case of null geodesics can be treated more naturally. Also, it is often convenient, especially with null geodesics, to use a parameterization that is not affine; we will do so in later portions of this book.

In Section 1.6 we review a fundamental theorem of differential geometry, the local flatness theorem. Here we prove the theorem in the standard way, by counting the number of functions required to go from an arbitrary coordinate system

1

to a locally Lorentzian frame. In Section 1.11 we extend the theorem to an entire geodesic, and we prove it by constructing Fermi normal coordinates in a neighbourhood of this geodesic.

Useful results involving the determinant of the metric tensor are derived in Section 1.7. The metric determinant is used in Section 1.8 to define the Levi-Civita tensor, which will be put to use in later parts of this book (most notably in Chapter 3). The Riemann curvature tensor and its contractions are introduced in Section 1.9, along with the Einstein field equations. The geometrical meaning of the Riemann tensor is explored in Section 1.10, in which we derive the equation of geodesic deviation.

1.1 Vectors, dual vectors, and tensors

Consider a curve γ on a manifold. The curve is parameterized by λ and is described in an arbitrary coordinate system by the relations $x^\alpha(\lambda)$. We wish to calculate the rate of change of a scalar function $f(x^\alpha)$ along this curve:

$$\frac{\mathrm{d}f}{\mathrm{d}\lambda} = \frac{\partial f}{\partial x^\alpha}\frac{\mathrm{d}x^\alpha}{\mathrm{d}\lambda} = f_{,\alpha}u^\alpha.$$

This procedure allows us to introduce two types of objects on the manifold: $u^\alpha = \mathrm{d}x^\alpha/\mathrm{d}\lambda$ is a vector that is everywhere tangent to γ, and $f_{,\alpha} = \partial f/\partial x^\alpha$ is a dual vector, the gradient of the function f. These objects transform as follows under an arbitrary coordinate transformation from x^α to $x^{\alpha'}$:

$$f_{,\alpha'} = \frac{\partial f}{\partial x^{\alpha'}} = \frac{\partial f}{\partial x^\alpha}\frac{\partial x^\alpha}{\partial x^{\alpha'}} = \frac{\partial x^\alpha}{\partial x^{\alpha'}}f_{,\alpha}$$

and

$$u^{\alpha'} = \frac{\mathrm{d}x^{\alpha'}}{\mathrm{d}\lambda} = \frac{\partial x^{\alpha'}}{\partial x^\alpha}\frac{\mathrm{d}x^\alpha}{\mathrm{d}\lambda} = \frac{\partial x^{\alpha'}}{\partial x^\alpha}u^\alpha.$$

From these equations we recover the fact that $\mathrm{d}f/\mathrm{d}\lambda$ is an invariant: $f_{,\alpha'}u^{\alpha'} = f_{,\alpha}u^\alpha$.

Any object A^α which transforms as

$$A^{\alpha'} = \frac{\partial x^{\alpha'}}{\partial x^\alpha}A^\alpha \tag{1.1}$$

under a coordinate transformation will be called a *vector*. On the other hand, any object p_α which transforms as

$$p_{\alpha'} = \frac{\partial x^\alpha}{\partial x^{\alpha'}}p_\alpha \tag{1.2}$$

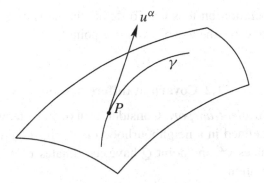

Figure 1.1   A tensor at P lives in the manifold's tangent plane at P.

under the same coordinate transformation will be called a *dual vector*. The contraction $A^\alpha p_\alpha$ between a vector and a dual vector is invariant under the coordinate transformation, and is therefore a scalar.

Generalizing these definitions, a *tensor* of type (n, m) is an object $T^{\alpha \cdots \beta}{}_{\gamma \cdots \delta}$ which transforms as

$$T^{\alpha' \cdots \beta'}{}_{\gamma' \cdots \delta'} = \frac{\partial x^{\alpha'}}{\partial x^\alpha} \cdots \frac{\partial x^{\beta'}}{\partial x^\beta} \frac{\partial x^\gamma}{\partial x^{\gamma'}} \cdots \frac{\partial x^\delta}{\partial x^{\delta'}} T^{\alpha \cdots \beta}{}_{\gamma \cdots \delta} \qquad (1.3)$$

under a coordinate transformation. The integer n is equal to the number of superscripts, while m is equal to the number of subscripts. It should be noted that the order of the indices is important; in general, $T^{\beta \cdots \alpha}{}_{\gamma \cdots \delta} \neq T^{\alpha \cdots \beta}{}_{\gamma \cdots \delta}$. By definition, vectors are tensors of type $(1, 0)$, and dual vectors are tensors of type $(0, 1)$.

A very special tensor is the *metric tensor* $g_{\alpha\beta}$, which is used to define the inner product between two vectors. It is also the quantity that represents the gravitational field in general relativity. The metric or its inverse $g^{\alpha\beta}$ can be used to lower or raise indices. For example, $A_\alpha \equiv g_{\alpha\beta} A^\beta$ and $p^\alpha \equiv g^{\alpha\beta} p_\beta$. The inverse metric is defined by the relations $g^{\alpha\mu} g_{\mu\beta} = \delta^\alpha{}_\beta$. The metric and its inverse are symmetric tensors.

Tensors are not actually defined on the manifold itself. To illustrate this, consider the vector u^α tangent to the curve γ, as represented in Fig. 1.1. The diagram makes it clear that the tangent vector actually 'sticks out' of the manifold. In fact, a vector at a point P on the manifold is defined in a plane tangent to the manifold at that point; this plane is called the *tangent plane* at P. Similarly, tensors at a point P can be thought of as living in this tangent plane. Tensors at P can be added and contracted, and the result is also a tensor. However, a tensor at P and another tensor at Q cannot be combined in a tensorial way, because these tensors belong to different tangent planes. For example, the operations $A^\alpha(P) B^\beta(Q)$ and $A^\alpha(Q) - A^\alpha(P)$ are not defined as tensorial operations. This implies that differentiation is not a straightforward operation on tensors. To define the derivative of a tensor, a *rule* must be provided to carry the tensor from one point to another.

1.2 Covariant differentiation

One such rule is *parallel transport*. Consider a curve γ, its tangent vector u^α, and a vector field A^α defined in a neighbourhood of γ (Fig. 1.2). Let point P on the curve have coordinates x^α, and point Q have coordinates $x^\alpha + dx^\alpha$. As was stated previously, the operation

$$
\begin{aligned}
dA^\alpha &\equiv A^\alpha(Q) - A^\alpha(P) \\
&= A^\alpha(x^\beta + dx^\beta) - A^\alpha(x^\beta) \\
&= A^\alpha{}_{,\beta}\, dx^\beta
\end{aligned}
$$

is not tensorial. This is easily checked: under a coordinate transformation,

$$
A^{\alpha'}{}_{,\beta'} = \frac{\partial}{\partial x^{\beta'}} \frac{\partial x^{\alpha'}}{\partial x^\alpha} A^\alpha = \frac{\partial x^{\alpha'}}{\partial x^\alpha} \frac{\partial x^\beta}{\partial x^{\beta'}} A^\alpha{}_{,\beta} + \frac{\partial^2 x^{\alpha'}}{\partial x^\alpha \partial x^\beta} \frac{\partial x^\beta}{\partial x^{\beta'}} A^\alpha,
$$

and this is not a tensorial transformation. To be properly tensorial the derivative operator should have the form $DA^\alpha = A_T^\alpha(P) - A^\alpha(P)$, where $A_T^\alpha(P)$ is the vector that is obtained by 'transporting' A^α from Q to P. We may write this as $DA^\alpha = dA^\alpha + \delta A^\alpha$, where $\delta A^\alpha \equiv A_T^\alpha(P) - A^\alpha(Q)$ is also not a tensorial operation. The precise rule for parallel transport must now be specified. We demand that δA^α be linear in both A^μ and dx^β, so that $\delta A^\alpha = \Gamma^\alpha{}_{\mu\beta} A^\mu dx^\beta$ for some (non-tensorial) field $\Gamma^\alpha{}_{\mu\beta}$ called the *connection*. A priori, this field is freely specifiable.

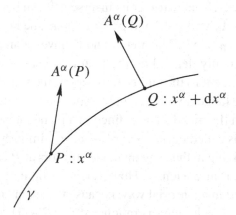

Figure 1.2   Differentiation of a tensor.

We now have $DA^\alpha = A^\alpha{}_{,\beta}\, dx^\beta + \Gamma^\alpha{}_{\mu\beta}\, A^\mu\, dx^\beta$, and dividing through by $d\lambda$, the increment in the curve's parameter, we obtain

$$\frac{DA^\alpha}{d\lambda} = A^\alpha{}_{;\beta} u^\beta, \tag{1.4}$$

where $u^\beta = dx^\beta/d\lambda$ is the tangent vector, and

$$A^\alpha{}_{;\beta} \equiv A^\alpha{}_{,\beta} + \Gamma^\alpha{}_{\mu\beta}\, A^\mu. \tag{1.5}$$

This is the *covariant derivative* of the vector A^α. Other standard notations are $A^\alpha{}_{;\beta} \equiv \nabla_\beta A^\alpha$ and $DA^\alpha/d\lambda \equiv \nabla_u A^\alpha$.

The fact that $A^\alpha{}_{;\beta}$ is a tensor allows us to deduce the transformation property of the connection. Starting from $\Gamma^\alpha{}_{\mu\beta} A^\mu = A^\alpha{}_{;\beta} - A^\alpha{}_{,\beta}$ it is easy to show that

$$\Gamma^{\alpha'}{}_{\mu'\beta'} A^{\mu'} = \frac{\partial x^{\alpha'}}{\partial x^\alpha}\frac{\partial x^\beta}{\partial x^{\beta'}}\, \Gamma^\alpha{}_{\mu\beta} A^\mu - \frac{\partial^2 x^{\alpha'}}{\partial x^\mu \partial x^\beta}\frac{\partial x^\beta}{\partial x^{\beta'}}\, A^\mu.$$

Expressing $A^{\mu'}$ in terms of A^μ on the left-hand side and using the fact that A^μ is an arbitrary vector field, we obtain

$$\Gamma^{\alpha'}{}_{\mu'\beta'} \frac{\partial x^{\mu'}}{\partial x^\mu} = \frac{\partial x^{\alpha'}}{\partial x^\alpha}\frac{\partial x^\beta}{\partial x^{\beta'}}\, \Gamma^\alpha{}_{\mu\beta} - \frac{\partial^2 x^{\alpha'}}{\partial x^\mu \partial x^\beta}\frac{\partial x^\beta}{\partial x^{\beta'}}.$$

Multiplying through by $\partial x^\mu/\partial x^{\gamma'}$ and rearranging the indices, we arrive at

$$\Gamma^{\alpha'}{}_{\mu'\beta'} = \frac{\partial x^{\alpha'}}{\partial x^\alpha}\frac{\partial x^\beta}{\partial x^{\beta'}}\frac{\partial x^\mu}{\partial x^{\mu'}}\, \Gamma^\alpha{}_{\mu\beta} - \frac{\partial^2 x^{\alpha'}}{\partial x^\mu \partial x^\beta}\frac{\partial x^\beta}{\partial x^{\beta'}}\frac{\partial x^\mu}{\partial x^{\mu'}}. \tag{1.6}$$

This is the transformation law for the connection; the second term prevents it from transforming as a tensor.

Covariant differentiation can be extended to other types of tensors by demanding that the operator D obeys the product rule of differential calculus. (For scalars, it is understood that $D \equiv d$.) For example, we may derive an expression for the covariant derivative of a dual vector from the requirement

$$d(A^\alpha p_\alpha) \equiv D(A^\alpha p_\alpha) = (DA^\alpha)p_\alpha + A^\alpha D(p_\alpha).$$

Writing the left-hand side as $A^\alpha{}_{,\beta}\, p_\alpha\, dx^\beta + A^\alpha p_{\alpha,\beta}\, dx^\beta$ and using Eqs. (1.4) and (1.5), we obtain

$$\frac{Dp_\alpha}{d\lambda} = p_{\alpha;\beta} u^\beta, \tag{1.7}$$

where

$$p_{\alpha;\beta} \equiv p_{\alpha,\beta} - \Gamma^\mu{}_{\alpha\beta} p_\mu. \tag{1.8}$$

This procedure generalizes to tensors of arbitrary type. For example, the covariant derivative of a type-(1, 1) tensor is given by

$$T^\alpha_{\ \beta;\gamma} = T^\alpha_{\ \beta,\gamma} + \Gamma^\alpha_{\ \mu\gamma} T^\mu_{\ \beta} - \Gamma^\mu_{\ \beta\gamma} T^\alpha_{\ \mu}. \tag{1.9}$$

The rule is that there is a connection term for each tensorial index; it comes with a plus sign if the index is a superscript, and with a minus sign if the index is a subscript.

Up to now the connection has been left completely arbitrary. A specific choice is made by demanding that it be *symmetric* and *metric compatible*,

$$\Gamma^\alpha_{\ \gamma\beta} = \Gamma^\alpha_{\ \beta\gamma}, \qquad g_{\alpha\beta;\gamma} = 0. \tag{1.10}$$

In general relativity, these properties come as a consequence of Einstein's principle of equivalence. It is easy to show that Eqs. (1.10) imply

$$\Gamma^\alpha_{\ \beta\gamma} = \frac{1}{2} g^{\alpha\mu} \left(g_{\mu\beta,\gamma} + g_{\mu\gamma,\beta} - g_{\beta\gamma,\mu} \right). \tag{1.11}$$

Thus, the connection is fully determined by the metric. In this context $\Gamma^\alpha_{\ \beta\gamma}$ are called the *Christoffel symbols*.

We conclude this section by introducing some terminology. A tensor field $T^{\alpha\cdots}_{\ \ \beta\cdots}$ is said to be *parallel transported* along a curve γ if its covariant derivative along the curve vanishes: $DT^{\alpha\cdots}_{\ \ \beta\cdots}/d\lambda = T^{\alpha\cdots}_{\ \ \beta\cdots;\mu} u^\mu = 0$.

1.3 Geodesics

A curve is a *geodesic* if it extremizes the distance between two fixed points.

Let a curve γ be described by the relations $x^\alpha(\lambda)$, where λ is an arbitrary parameter, and let P and Q be two points on this curve. The distance between P and Q along γ is given by

$$\ell = \int_P^Q \sqrt{\pm g_{\alpha\beta} \dot{x}^\alpha \dot{x}^\beta} \, d\lambda, \tag{1.12}$$

where $\dot{x}^\alpha \equiv dx^\alpha/d\lambda$. In the square root, the positive (negative) sign is chosen if the curve is spacelike (timelike); it is assumed that γ is nowhere null. It should be clear that ℓ is invariant under a reparameterization of the curve, $\lambda \to \lambda'(\lambda)$.

The curve for which ℓ is an extremum is determined by substituting the 'Lagrangian' $L(\dot{x}^\mu, x^\mu) = (\pm g_{\mu\nu} \dot{x}^\mu \dot{x}^\nu)^{1/2}$ into the Euler–Lagrange equations,

$$\frac{d}{d\lambda} \frac{\partial L}{\partial \dot{x}^\alpha} - \frac{\partial L}{\partial x^\alpha} = 0.$$

A straightforward calculation shows that $x^\alpha(\lambda)$ must satisfy the differential equation

$$\ddot{x}^\alpha + \Gamma^\alpha_{\beta\gamma}\dot{x}^\beta\dot{x}^\gamma = \kappa(\lambda)\dot{x}^\alpha \qquad \text{(arbitrary parameter)}, \qquad (1.13)$$

where $\kappa \equiv \mathrm{d}\ln L/\mathrm{d}\lambda$. The *geodesic equation* can also be written as $u^\alpha_{\ ;\beta}u^\beta = \kappa u^\alpha$, in which $u^\alpha = \dot{x}^\alpha$ is tangent to the geodesic.

A particularly useful choice of parameter is proper time τ when the geodesic is timelike, or proper distance s when the geodesic is spacelike. (It is important that this choice be made *after* extremization, and not before.) Because $\mathrm{d}\tau^2 = -g_{\alpha\beta}\,\mathrm{d}x^\alpha\,\mathrm{d}x^\beta$ for timelike geodesics and $\mathrm{d}s^2 = g_{\alpha\beta}\,\mathrm{d}x^\alpha\,\mathrm{d}x^\beta$ for spacelike geodesics, we have that $L = 1$ in either case, and this implies $\kappa = 0$. The geodesic equation becomes

$$\ddot{x}^\alpha + \Gamma^\alpha_{\beta\gamma}\dot{x}^\beta\dot{x}^\gamma = 0 \qquad \text{(affine parameter)}, \qquad (1.14)$$

or $u^\alpha_{\ ;\beta}u^\beta = 0$, which states that the tangent vector is parallel transported along the geodesic. These equations are invariant under reparameterizations of the form $\lambda \to \lambda' = a\lambda + b$, where a and b are constants. Parameters related to s and τ by such transformations are called *affine parameters*. It is useful to note that Eq. (1.14) can be recovered by substituting $L' = \frac{1}{2}g_{\alpha\beta}\dot{x}^\alpha\dot{x}^\beta$ into the Euler-Lagrange equations; this gives rise to practical method of computing the Christoffel symbols.

By continuity, the general form $u^\alpha_{\ ;\beta}u^\beta = \kappa u^\alpha$ for the geodesic equation must be valid also for null geodesics. For this to be true, the parameter λ cannot be affine, because $\mathrm{d}s = \mathrm{d}\tau = 0$ along a null geodesic, and the limit is then singular. However, affine parameters can nevertheless be found for null geodesics. Starting from Eq. (1.13) it is always possible to introduce a new parameter λ^* such that the geodesic equation will take the form of Eq. (1.14). It is easy to check that the appropriate transformation is

$$\frac{\mathrm{d}\lambda^*}{\mathrm{d}\lambda} = \exp\left[\int^\lambda \kappa(\lambda')\,\mathrm{d}\lambda'\right]. \qquad (1.15)$$

(You will be asked to provide a proof of this statement in Section 1.13, Problem 2.) It should be noted that while the null version of Eq. (1.13) was obtained by a limiting procedure, the null version of Eq. (1.14) cannot be considered to be a limit of the same equation for timelike or spacelike geodesics: the parameterization is highly discontinuous.

We conclude this section with the following remark: Along an affinely parameterized geodesic (timelike, spacelike, or null), the scalar quantity $\varepsilon = u^\alpha u_\alpha$ is a constant. The proof requires a single line:

$$\frac{\mathrm{d}\varepsilon}{\mathrm{d}\lambda} = (u^\alpha u_\alpha)_{;\beta}u^\beta = (u^\alpha_{\ ;\beta}u^\beta)u_\alpha + u^\alpha(u_{\alpha;\beta}u^\beta) = 0.$$

If proper time or proper distance is chosen for λ, then $\varepsilon = \mp 1$, respectively. For a null geodesic, $\varepsilon = 0$.

1.4 Lie differentiation

In Section 1.2, covariant differentiation was defined by introducing a rule to transport a tensor from a point Q to a neighbouring point P, at which the derivative was to be evaluated. This rule involved the introduction of a new structure on the manifold, the connection. In this section we define another type of derivative – the Lie derivative – without introducing any additional structure.

Consider a curve γ, its tangent vector $u^\alpha = dx^\alpha/d\lambda$, and a vector field A^α defined in a neighbourhood of γ (Fig. 1.2). As before, the point P shall have the coordinates x^α, while the point Q shall be at $x^\alpha + dx^\alpha$. The equation

$$x'^\alpha \equiv x^\alpha + dx^\alpha = x^\alpha + u^\alpha \, d\lambda$$

can be interpreted as an infinitesimal coordinate transformation from the system x to the system x'. Under this transformation, the vector A^α becomes

$$A'^\alpha(x') = \frac{\partial x'^\alpha}{\partial x^\beta} A^\beta(x)$$
$$= (\delta^\alpha_\beta + u^\alpha_{,\beta} \, d\lambda) A^\beta(x)$$
$$= A^\alpha(x) + u^\alpha_{,\beta} A^\beta(x) \, d\lambda.$$

In other words,

$$A'^\alpha(Q) = A^\alpha(P) + u^\alpha_{,\beta} A^\beta(P) \, d\lambda.$$

On the other hand, $A^\alpha(Q)$, the value of the original vector field at the point Q, can be expressed as

$$A^\alpha(Q) = A^\alpha(x + dx)$$
$$= A^\alpha(x) + A^\alpha_{,\beta}(x) \, dx^\beta$$
$$= A^\alpha(P) + u^\beta A^\alpha_{,\beta}(P) \, d\lambda.$$

In general, $A'^\alpha(Q)$ and $A^\alpha(Q)$ will not be equal. Their difference defines the *Lie derivative* of the vector A^α along the curve γ:

$$\pounds_u A^\alpha(P) \equiv \frac{A^\alpha(Q) - A'^\alpha(Q)}{d\lambda}.$$

Combining the previous three equations yields

$$\pounds_u A^\alpha = A^\alpha_{,\beta} u^\beta - u^\alpha_{,\beta} A^\beta. \qquad (1.16)$$

Despite an appearance to the contrary, $\pounds_u A^\alpha$ is a tensor: It is easy to check that Eq. (1.16) is equivalent to

$$\pounds_u A^\alpha = A^\alpha_{\;;\beta} u^\beta - u^\alpha_{\;;\beta} A^\beta, \tag{1.17}$$

whose tensorial nature is evident. The fact that $\pounds_u A^\alpha$ can be defined without a connection means that Lie differentiation is a more primitive operation than covariant differentiation.

The definition of the Lie derivative extends to all types of tensors. For scalars, $\pounds_u f \equiv df/d\lambda = f_{,\alpha} u^\alpha$. For dual vectors, the same steps reveal that

$$\begin{aligned} \pounds_u p_\alpha &= p_{\alpha,\beta} u^\beta + u^\beta_{\;,\alpha} p_\beta \\ &= p_{\alpha;\beta} u^\beta + u^\beta_{\;;\alpha} p_\beta. \end{aligned} \tag{1.18}$$

As another example, the Lie derivative of a type-$(1, 1)$ tensor is given by

$$\begin{aligned} \pounds_u T^\alpha_{\;\beta} &= T^\alpha_{\;\beta,\mu} u^\mu - u^\alpha_{\;,\mu} T^\mu_{\;\beta} + u^\mu_{\;,\beta} T^\alpha_{\;\mu} \\ &= T^\alpha_{\;\beta;\mu} u^\mu - u^\alpha_{\;;\mu} T^\mu_{\;\beta} + u^\mu_{\;;\beta} T^\alpha_{\;\mu}. \end{aligned} \tag{1.19}$$

Further generalizations are obvious. It may be verified that the Lie derivative obeys the product rule of differential calculus. For example, the relation

$$\pounds_u (A^\alpha p_\beta) = (\pounds_u A^\alpha) p_\beta + A^\alpha (\pounds_u p_\beta) \tag{1.20}$$

is easily established.

A tensor field $T^{\alpha\cdots}_{\;\;\beta\cdots}$ is said to be *Lie transported* along a curve γ if its Lie derivative along the curve vanishes: $\pounds_u T^{\alpha\cdots}_{\;\;\beta\cdots} = 0$, where u^α is the curve's tangent vector. Suppose that the coordinates are chosen so that x^1, x^2, and x^3 are all constant on γ, while $x^0 \equiv \lambda$ varies on γ. In such a coordinate system,

$$u^\alpha = \frac{dx^\alpha}{d\lambda} \stackrel{*}{=} \delta^\alpha_{\;0},$$

where the symbol '$\stackrel{*}{=}$' means 'equals in the specified coordinate system'. It follows that $u^\alpha_{\;,\beta} \stackrel{*}{=} 0$, so that

$$\pounds_u T^{\alpha\cdots}_{\;\;\beta\cdots} \stackrel{*}{=} T^{\alpha\cdots}_{\;\;\beta\cdots,\mu} u^\mu \stackrel{*}{=} \frac{\partial}{\partial x^0} T^{\alpha\cdots}_{\;\;\beta\cdots}.$$

If the tensor is Lie transported along γ, then the tensor's components are all independent of x^0 in the specified coordinate system.

We have formulated the following theorem:

If $\pounds_u T^{\alpha\cdots}_{\;\;\beta\cdots} = 0$, that is, if a tensor is Lie transported along a curve γ with tangent vector u^α, then a coordinate system can be constructed such that $u^\alpha \stackrel{*}{=} \delta^\alpha_{\;0}$ and $T^{\alpha\cdots}_{\;\;\beta\cdots,0} \stackrel{*}{=} 0$. Conversely, if in a given coordinate system the components of a tensor do not depend on a particular coordinate x^0, then the Lie derivative of the tensor in the direction of u^α vanishes.

The Lie derivative is therefore the natural construct to express, covariantly, the invariance of a tensor under a change of position.

1.5 Killing vectors

If, in a given coordinate system, the components of the metric do not depend on x^0, then by the preceding theorem $\pounds_\xi g_{\alpha\beta} = 0$, where $\xi^\alpha \stackrel{*}{=} \delta^\alpha_0$. The vector ξ^α is then called a *Killing vector*. The condition for ξ^α to be a Killing vector is that

$$0 = \pounds_\xi g_{\alpha\beta} = \xi_{\alpha;\beta} + \xi_{\beta;\alpha}. \tag{1.21}$$

Thus, the tensor $\xi_{\alpha;\beta}$ is antisymmetric if ξ^α is a Killing vector.

Killing vectors can be used to find constants associated with the motion along a geodesic. Suppose that u^α is tangent to a geodesic affinely parameterized by λ. Then

$$\frac{d}{d\lambda}\left(u^\alpha \xi_\alpha\right) = (u^\alpha \xi_\alpha)_{;\beta} u^\beta$$
$$= u^\alpha_{;\beta} u^\beta \xi_\alpha + \xi_{\alpha;\beta} u^\alpha u^\beta$$
$$= 0.$$

In the second line, the first term vanishes by virtue of the geodesic equation, and the second term vanishes because $\xi_{\alpha;\beta}$ is an antisymmetric tensor while $u^\alpha u^\beta$ is symmetric. Thus, $u^\alpha \xi_\alpha$ is constant along the geodesic.

As an example, consider a static, spherically symmetric spacetime with metric

$$ds^2 = -A(r)\,dt^2 + B(r)\,dr^2 + r^2\,d\Omega^2,$$

where $d\Omega^2 = d\theta^2 + \sin^2\theta\,d\phi^2$. Because the metric does not depend on t nor ϕ, the vectors

$$\xi^\alpha_{(t)} = \frac{\partial x^\alpha}{\partial t}, \qquad \xi^\alpha_{(\phi)} = \frac{\partial x^\alpha}{\partial \phi}$$

are Killing vectors. The quantities

$$\tilde{E} = -u_\alpha \xi^\alpha_{(t)}, \qquad \tilde{L} = u_\alpha \xi^\alpha_{(\phi)}$$

are then constant along a geodesic to which u^α is tangent. If the geodesic is time-like and u^α is the four-velocity of a particle moving on that geodesic, then $\tilde{E}$ and $\tilde{L}$ can be interpreted as energy and angular momentum per unit mass, respectively. It should also be noted that spherical symmetry implies the existence of two additional Killing vectors,

$$\xi^\alpha_{(1)} \partial_\alpha = \sin\phi\, \partial_\theta + \cot\theta \cos\phi\, \partial_\phi, \qquad \xi^\alpha_{(2)} \partial_\alpha = -\cos\phi\, \partial_\theta + \cot\theta \sin\phi\, \partial_\phi.$$

It is straightforward to show that these do indeed satisfy Killing's equation (1.21). (To prove this is the purpose of Section 1.13, Problem 5.)

1.6 Local flatness

For a given point P in spacetime, it is always possible to find a coordinate system $x^{\alpha'}$ such that

$$g_{\alpha'\beta'}(P) = \eta_{\alpha'\beta'}, \qquad \Gamma^{\alpha'}_{\beta'\gamma'}(P) = 0, \qquad (1.22)$$

where $\eta_{\alpha'\beta'} = \mathrm{diag}(-1, 1, 1, 1)$ is the Minkowski metric. Such a coordinate system will be called a *local Lorentz frame* at P. We note that it is not possible to also set the derivatives of the connection to zero when the spacetime is curved. The physical interpretation of the *local-flatness theorem* is that free-falling observers see no effect of gravity in their immediate vicinity, as required by Einstein's principle of equivalence.

We now prove the theorem. Let x^α be an arbitrary coordinate system, and let us assume, with no loss of generality, that P is at the origin of both coordinate systems. Then the coordinates of a point near P are related by

$$x^{\alpha'} = A^{\alpha'}_\beta x^\beta + O(x^2), \qquad x^\alpha = A^\alpha_{\beta'} x^{\beta'} + O(x'^2),$$

where $A^{\alpha'}_\beta$ and $A^\alpha_{\beta'}$ are constant matrices. It is easy to check that one is in fact the inverse of the other:

$$A^{\alpha'}_\mu A^\mu_{\beta'} = \delta^{\alpha'}_{\beta'}, \qquad A^\alpha_{\mu'} A^{\mu'}_\beta = \delta^\alpha_\beta.$$

Under this transformation, the metric becomes

$$g_{\alpha'\beta'}(P) = A^\alpha_{\alpha'} A^\beta_{\beta'} g_{\alpha\beta}(P).$$

We demand that the left-hand side be equal to $\eta_{\alpha'\beta'}$. This gives us 10 equations for the 16 unknown components of the matrix $A^\alpha_{\alpha'}$. A solution can always be found, with 6 undetermined components. This corresponds to the freedom of performing a Lorentz transformation (3 rotation parameters and 3 boost parameters) which does not alter the form of the Minkowski metric.

Suppose that a particular choice has been made for $A^\alpha_{\alpha'}$. Then $A^{\alpha'}_\alpha$ is found by inverting the matrix, and the coordinate transformation is known to first order. Let us proceed to second order:

$$x^{\alpha'} = A^{\alpha'}_{\ \beta} x^{\beta} + \frac{1}{2} B^{\alpha'}_{\ \beta\gamma} x^{\beta} x^{\gamma} + O(x^3),$$

where the constant coefficients $B^{\alpha'}_{\ \beta\gamma}$ are symmetric in the lower indices. Recalling Eq. (1.6), we have that the connection transforms as

$$\Gamma^{\alpha'}_{\ \beta'\gamma'}(P) = A^{\alpha'}_{\ \alpha} A^{\beta}_{\ \beta'} A^{\gamma}_{\ \gamma'} \Gamma^{\alpha}_{\ \beta\gamma}(P) - B^{\alpha'}_{\ \beta\gamma} A^{\beta}_{\ \beta'} A^{\gamma}_{\ \gamma'}.$$

To put the left-hand side to zero, it is sufficient to impose

$$B^{\alpha'}_{\ \beta\gamma} = A^{\alpha'}_{\ \alpha} \Gamma^{\alpha}_{\ \beta\gamma}(P).$$

These equations determine $B^{\alpha'}_{\ \beta\gamma}$ uniquely, and the coordinate transformation is now known to second order. Irrespective of the higher-order terms, it enforces Eqs. (1.22).

We shall return in Section 1.11 with a more geometric proof of the local-flatness theorem, and its extension from a single point P to an entire geodesic γ.

1.7 Metric determinant

The quantity $\sqrt{-g}$, where $g \equiv \det[g_{\alpha\beta}]$, occurs frequently in differential geometry. We first note that $\sqrt{g'/g}$, where $g' = \det[g_{\alpha'\beta'}]$, is the Jacobian of the transformation $x^{\alpha} \to x^{\alpha'}(x^{\alpha})$. To see this, recall from ordinary differential calculus that under such a transformation, $d^4x = J d^4x'$, where $J = \det[\partial x^{\alpha}/\partial x^{\alpha'}]$ is the Jacobian. Now consider the transformation of the metric,

$$g_{\alpha'\beta'} = \frac{\partial x^{\alpha}}{\partial x^{\alpha'}} \frac{\partial x^{\beta}}{\partial x^{\beta'}} g_{\alpha\beta}.$$

Because the determinant of a product of matrices is equal to the product of their determinants, this equation implies $g' = g J^2$, which proves the assertion.

As an important application, consider the transformation from $x^{\alpha'}$, a local Lorentz frame at P, to x^{α}, an arbitrary coordinate system. The four-dimensional volume element around P is $d^4x' = J^{-1} d^4x = \sqrt{g/g'}\, d^4x$. But since $g' = -1$ we have that

$$\sqrt{-g}\, d^4x \qquad\qquad\qquad (1.23)$$

is an invariant volume element around the arbitrary point P. This result generalizes to a manifold of any dimension with a metric of any signature; in this case, $|g|^{1/2} d^n x$ is the invariant volume element, where n is the dimension of the manifold.

We shall now derive another useful result,

$$\Gamma^{\mu}_{\ \mu\alpha} = \frac{1}{2} g^{\mu\nu} g_{\mu\nu,\alpha} = \frac{1}{\sqrt{-g}} \left(\sqrt{-g}\right)_{,\alpha}. \tag{1.24}$$

Consider, for any matrix M, the variation of $\ln |\det M|$ induced by a variation of M's elements. Using the product rule for determinants we have

$$\begin{aligned}
\delta \ln |\det M| &\equiv \ln |\det(M + \delta M)| - \ln |\det M| \\
&= \ln \frac{\det(M + \delta M)}{\det M} \\
&= \ln \det M^{-1}(M + \delta M) \\
&= \ln \det(1 + M^{-1}\delta M).
\end{aligned}$$

We now use the identity $\det(1 + \epsilon) = 1 + \mathrm{Tr}\,\epsilon + O(\epsilon^2)$, valid for any 'small' matrix ϵ. (Try proving this for 3×3 matrices.) This gives

$$\begin{aligned}
\delta \ln |\det M| &= \ln(1 + \mathrm{Tr}\, M^{-1}\delta M) \\
&= \mathrm{Tr}\, M^{-1}\delta M.
\end{aligned}$$

Substituting the metric tensor in place of M gives $\delta \ln |g| = g^{\alpha\beta}\delta g_{\alpha\beta}$, or

$$\frac{\partial}{\partial x^{\mu}} \ln |g| = g^{\alpha\beta} g_{\alpha\beta,\mu}.$$

This establishes Eq. (1.24).

Equation (1.24) gives rise to the *divergence formula*: For any vector field A^{α},

$$A^{\alpha}_{\ ,\alpha} = \frac{1}{\sqrt{-g}} \left(\sqrt{-g}\, A^{\alpha}\right)_{,\alpha}. \tag{1.25}$$

A similar result holds for any *antisymmetric* tensor field $B^{\alpha\beta}$:

$$B^{\alpha\beta}_{\ \ ;\beta} = \frac{1}{\sqrt{-g}} \left(\sqrt{-g}\, B^{\alpha\beta}\right)_{,\beta}. \tag{1.26}$$

These formulae are useful for the efficient computation of covariant divergences.

1.8 Levi-Civita tensor

The *permutation symbol* $[\alpha\,\beta\,\gamma\,\delta]$, defined by

$$[\alpha\,\beta\,\gamma\,\delta] = \begin{cases} +1 & \text{if } \alpha\beta\gamma\delta \text{ is an even permutation of } 0123 \\ -1 & \text{if } \alpha\beta\gamma\delta \text{ is an odd permutation of } 0123, \\ \ \ 0 & \text{if any two indices are equal} \end{cases} \tag{1.27}$$

is a very useful, non-tensorial quantity. For example, it can be used to give a definition for the determinant: For any 4×4 matrix $M_{\alpha\beta}$,

$$\det[M_{\alpha\beta}] = [\alpha\, \beta\, \gamma\, \delta] M_{0\alpha} M_{1\beta} M_{2\gamma} M_{3\delta}$$
$$= [\alpha\, \beta\, \gamma\, \delta] M_{\alpha 0} M_{\beta 1} M_{\gamma 2} M_{\delta 3}. \tag{1.28}$$

Either equality can be established by brute-force computation. The well-known property that $\det[M_{\beta\alpha}] = \det[M_{\alpha\beta}]$ follows directly from Eq. (1.28).

We shall now show that the combination

$$\varepsilon_{\alpha\beta\gamma\delta} = \sqrt{-g}\, [\alpha\, \beta\, \gamma\, \delta] \tag{1.29}$$

is a tensor, called the *Levi-Civita tensor*. Consider the quantity

$$[\alpha\, \beta\, \gamma\, \delta] \frac{\partial x^\alpha}{\partial x^{\alpha'}} \frac{\partial x^\beta}{\partial x^{\beta'}} \frac{\partial x^\gamma}{\partial x^{\gamma'}} \frac{\partial x^\delta}{\partial x^{\delta'}},$$

which is completely antisymmetric in the primed indices. This must therefore be proportional to $[\alpha'\, \beta'\, \gamma'\, \delta']$:

$$[\alpha\, \beta\, \gamma\, \delta] \frac{\partial x^\alpha}{\partial x^{\alpha'}} \frac{\partial x^\beta}{\partial x^{\beta'}} \frac{\partial x^\gamma}{\partial x^{\gamma'}} \frac{\partial x^\delta}{\partial x^{\delta'}} = \lambda [\alpha'\, \beta'\, \gamma'\, \delta'],$$

for some proportionality factor λ. Putting $\alpha'\beta'\gamma'\delta' = 0123$ yields

$$\lambda = [\alpha\, \beta\, \gamma\, \delta] \frac{\partial x^\alpha}{\partial x^{0'}} \frac{\partial x^\beta}{\partial x^{1'}} \frac{\partial x^\gamma}{\partial x^{2'}} \frac{\partial x^\delta}{\partial x^{3'}},$$

which determines λ. But the right-hand side is just the determinant of the matrix $\partial x^\alpha / \partial x^{\alpha'}$, that is, the Jacobian of the transformation $x^{\alpha'}(x^\alpha)$. So $\lambda = \sqrt{g'/g}$, and we have

$$\sqrt{-g}\, [\alpha\, \beta\, \gamma\, \delta] \frac{\partial x^\alpha}{\partial x^{\alpha'}} \frac{\partial x^\beta}{\partial x^{\beta'}} \frac{\partial x^\gamma}{\partial x^{\gamma'}} \frac{\partial x^\delta}{\partial x^{\delta'}} = \sqrt{-g'}\, [\alpha'\, \beta'\, \gamma'\, \delta'].$$

This establishes the fact that $\varepsilon_{\alpha\beta\gamma\delta}$ does indeed transform as a type-(0, 4) tensor.

The proof could have started instead with the relation

$$[\alpha\, \beta\, \gamma\, \delta] \frac{\partial x^{\alpha'}}{\partial x^\alpha} \frac{\partial x^{\beta'}}{\partial x^\beta} \frac{\partial x^{\gamma'}}{\partial x^\gamma} \frac{\partial x^{\delta'}}{\partial x^\delta} = \lambda' [\alpha'\, \beta'\, \gamma'\, \delta'],$$

implying $\lambda' = \sqrt{g/g'}$ and showing that

$$\varepsilon^{\alpha\beta\gamma\delta} = -\frac{1}{\sqrt{-g}}\, [\alpha\, \beta\, \gamma\, \delta] \tag{1.30}$$

transforms as a type-(4, 0) tensor. (The minus sign is important.) It is easy to check that this is also the Levi-Civita tensor, obtained from $\varepsilon_{\alpha\beta\gamma\delta}$ by raising all four indices. Alternatively, we may show that $\varepsilon_{\alpha\beta\gamma\delta} = g_{\alpha\mu} g_{\beta\nu} g_{\gamma\lambda} g_{\delta\rho} \varepsilon^{\mu\nu\lambda\rho}$. This relation implies

$$\varepsilon_{0123} = -\frac{1}{\sqrt{-g}}[\mu\,\nu\,\lambda\,\rho]g_{0\mu}g_{1\nu}g_{2\lambda}g_{3\rho} = -\frac{1}{\sqrt{-g}}g = \sqrt{-g},$$

which is evidently compatible with Eq. (1.29).

The Levi-Civita tensor is used in a variety of contexts in differential geometry. We will meet it again in Chapter 3.

1.9 Curvature

The *Riemann tensor* $R^{\alpha}{}_{\beta\gamma\delta}$ may be defined by the relation

$$A^{\mu}{}_{;\alpha\beta} - A^{\mu}{}_{;\beta\alpha} = -R^{\mu}{}_{\nu\alpha\beta}A^{\nu}, \tag{1.31}$$

which holds for any vector field A^{α}. Evaluating the left-hand side explicitly yields

$$R^{\alpha}{}_{\beta\gamma\delta} = \Gamma^{\alpha}{}_{\beta\delta,\gamma} - \Gamma^{\alpha}{}_{\beta\gamma,\delta} + \Gamma^{\alpha}{}_{\mu\gamma}\Gamma^{\mu}{}_{\beta\delta} - \Gamma^{\alpha}{}_{\mu\delta}\Gamma^{\mu}{}_{\beta\gamma}. \tag{1.32}$$

The Riemann tensor is obviously antisymmetric in the last two indices. Its other symmetry properties can be established by evaluating $R^{\alpha}{}_{\beta\gamma\delta}$ in a local Lorentz frame at some point P. A straightforward computation gives

$$R_{\alpha\beta\gamma\delta} \overset{*}{=} \frac{1}{2}\big(g_{\alpha\delta,\beta\gamma} - g_{\alpha\gamma,\beta\delta} - g_{\beta\delta,\alpha\gamma} + g_{\beta\gamma,\alpha\delta}\big),$$

and this implies the tensorial relations

$$R_{\alpha\beta\gamma\delta} = -R_{\beta\alpha\gamma\delta} = -R_{\alpha\beta\delta\gamma} = R_{\gamma\delta\alpha\beta} \tag{1.33}$$

and

$$R_{\mu\alpha\beta\gamma} + R_{\mu\gamma\alpha\beta} + R_{\mu\beta\gamma\alpha} = 0, \tag{1.34}$$

which are valid in any coordinate system. A little more work along the same lines reveals that the Riemann tensor satisfies the *Bianchi identities*,

$$R_{\mu\nu\alpha\beta;\gamma} + R_{\mu\nu\gamma\alpha;\beta} + R_{\mu\nu\beta\gamma;\alpha} = 0. \tag{1.35}$$

In addition to Eq. (1.31), the Riemann tensor satisfies the relations

$$p_{\mu;\alpha\beta} - p_{\mu;\beta\alpha} = R^{\nu}{}_{\mu\alpha\beta}p_{\nu} \tag{1.36}$$

and

$$T^{\mu}{}_{\nu;\alpha\beta} - T^{\mu}{}_{\nu;\beta\alpha} = -R^{\mu}{}_{\lambda\alpha\beta}T^{\lambda}{}_{\nu} + R^{\lambda}{}_{\nu\alpha\beta}T^{\mu}{}_{\lambda}, \tag{1.37}$$

which hold for arbitrary tensors p_α and T^α_β. Generalization to tensors of higher ranks is obvious: the number of Riemann-tensor terms on the right-hand side is equal to the number of tensorial indices.

Contractions of the Riemann tensor produce the *Ricci tensor* $R_{\alpha\beta}$ and the *Ricci scalar R*. These are defined by

$$R_{\alpha\beta} = R^\mu{}_{\alpha\mu\beta}, \qquad R = R^\alpha{}_\alpha. \tag{1.38}$$

It is easy to show that $R_{\alpha\beta}$ is a symmetric tensor. The *Einstein tensor* is defined by

$$G_{\alpha\beta} = R_{\alpha\beta} - \frac{1}{2} R g_{\alpha\beta}; \tag{1.39}$$

this is also a symmetric tensor. By virtue of Eq. (1.35), the Einstein tensor satisfies

$$G^{\alpha\beta}{}_{;\beta} = 0, \tag{1.40}$$

the contracted Bianchi identities.

The *Einstein field equations*,

$$G^{\alpha\beta} = 8\pi T^{\alpha\beta}, \tag{1.41}$$

relate the spacetime curvature (as represented by the Einstein tensor) to the distribution of matter (as represented by $T^{\alpha\beta}$, the stress–energy tensor). Equation (1.40) implies that the stress–energy tensor must have a zero divergence: $T^{\alpha\beta}{}_{;\beta} = 0$. This is the tensorial expression for energy–momentum conservation. Equation (1.40) implies also that of the ten equations (1.41), only six are independent. The metric can therefore be determined up to four arbitrary functions, and this reflects our complete freedom in choosing the coordinate system. We note that the field equations can also be written in the form

$$R^{\alpha\beta} = 8\pi \left(T^{\alpha\beta} - \frac{1}{2} T g^{\alpha\beta} \right), \tag{1.42}$$

where $T \equiv T^\alpha_\alpha$ is the trace of the stress–energy tensor.

1.10 Geodesic deviation

The geometrical meaning of the Riemann tensor is best illustrated by examining the behaviour of neighbouring geodesics. Consider two such geodesics, γ_0 and γ_1, each described by relations $x^\alpha(t)$ in which t is an affine parameter; the geodesics can be either spacelike, timelike, or null. We want to develop the notion of a *deviation vector* between these two geodesics, and derive an evolution equation for this vector.

For this purpose we introduce, in the space between γ_0 and γ_1, an entire family of interpolating geodesics (Fig. 1.3). To each geodesic we assign a label $s \in [0, 1]$, such that γ_0 comes with the label $s = 0$ and γ_1 with $s = 1$. We collectively describe these geodesics with relations $x^\alpha(s, t)$, in which s serves to specify which geodesic

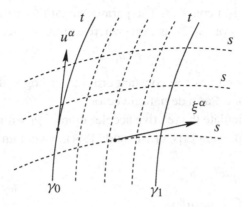

Figure 1.3 Deviation vector between two neighbouring geodesics.

and t is an affine parameter along the specified geodesic. The vector field $u^\alpha = \partial x^\alpha/\partial t$ is tangent to the geodesics, and it satisfies the equation $u^\alpha_{\ ;\beta}u^\beta = 0$.

If we keep t fixed in the relations $x^\alpha(s, t)$ and vary s instead, we obtain another family of curves, labelled by t and parameterized by s; in general these curves will not be geodesics. The family has $\xi^\alpha = \partial x^\alpha/\partial s$ as its tangent vector field, and the restriction of this vector to γ_0, $\xi^\alpha|_{s=0}$, gives a meaningful notion of a deviation vector between γ_0 and γ_1. We wish to derive an expression for its *acceleration*,

$$\frac{\mathrm{D}^2\xi^\alpha}{\mathrm{d}t^2} \equiv (\xi^\alpha_{\ ;\beta}u^\beta)_{;\gamma}u^\gamma, \tag{1.43}$$

in which it is understood that all quantities are to be evaluated on γ_0. In flat space-time the geodesics γ_0 and γ_1 are straight, and although their separation may change with t, this change is necessarily linear: $\mathrm{D}^2\xi^\alpha/\mathrm{d}t^2 = 0$ in flat spacetime. A nonzero result for $\mathrm{D}^2\xi^\alpha/\mathrm{d}t^2$ will therefore reveal the presence of curvature, and indeed, this vector will be found to be proportional to the Riemann tensor.

It follows at once from the relations $u^\alpha = \partial x^\alpha/\partial t$ and $\xi^\alpha = \partial x^\alpha/\partial s$ that $\partial u^\alpha/\partial s = \partial \xi^\alpha/\partial t$, which can be written in covariant form as

$$\pounds_u\xi^\alpha = \pounds_\xi u^\alpha = 0 \quad \Rightarrow \quad \xi^\alpha_{\ ;\beta}u^\beta = u^\alpha_{\ ;\beta}\xi^\beta. \tag{1.44}$$

We also have at our disposal the geodesic equation, $u^\alpha_{\ ;\beta}u^\beta = 0$. These equations can be combined to prove that $\xi^\alpha u_\alpha$ is constant along γ_0:

$$\begin{aligned}
\frac{\mathrm{d}}{\mathrm{d}t}\left(\xi^\alpha u_\alpha\right) &= (\xi^\alpha u_\alpha)_{;\beta}u^\beta \\
&= \xi^\alpha_{\ ;\beta}u^\beta u_\alpha + \xi^\alpha u_{\alpha;\beta}u^\beta \\
&= u^\alpha_{\ ;\beta}\xi^\beta u_\alpha \\
&= \frac{1}{2}(u^\alpha u_\alpha)_{;\beta}\xi^\beta \\
&= 0,
\end{aligned}$$

because $u^\alpha u_\alpha \equiv \varepsilon$ is a constant. The parameterization of the geodesics can therefore be tuned so that on γ_0, ξ^α is everywhere orthogonal to u^α:

$$\xi^\alpha u_\alpha = 0. \tag{1.45}$$

This means that the curves $t = $ constant cross γ_0 orthogonally, and adds weight to the interpretation of ξ^α as a deviation vector.

We may now calculate the relative acceleration of γ_1 with respect to γ_0. Starting from Eq. (1.43) and using Eqs. (1.31) and (1.44), we obtain

$$\begin{aligned}
\frac{D^2 \xi^\alpha}{dt^2} &= (\xi^\alpha_{\;;\beta} u^\beta)_{;\gamma} u^\gamma \\
&= (u^\alpha_{\;;\beta} \xi^\beta)_{;\gamma} u^\gamma \\
&= u^\alpha_{\;;\beta\gamma} \xi^\beta u^\gamma + u^\alpha_{\;;\beta} \xi^\beta_{\;;\gamma} u^\gamma \\
&= (u^\alpha_{\;;\gamma\beta} - R^\alpha_{\;\mu\beta\gamma} u^\mu) \xi^\beta u^\gamma + u^\alpha_{\;;\beta} u^\beta_{\;;\gamma} \xi^\gamma \\
&= (u^\alpha_{\;;\gamma} u^\gamma)_{;\beta} \xi^\beta - u^\alpha_{\;;\gamma} u^\gamma_{\;;\beta} \xi^\beta - R^\alpha_{\;\mu\beta\gamma} u^\mu \xi^\beta u^\gamma + u^\alpha_{\;;\beta} u^\beta_{\;;\gamma} \xi^\gamma.
\end{aligned}$$

The first term vanishes by virtue of the geodesic equation, while the second and fourth terms cancel out, leaving

$$\frac{D^2 \xi^\alpha}{dt^2} = -R^\alpha_{\;\beta\gamma\delta} u^\beta \xi^\gamma u^\delta. \tag{1.46}$$

This is the *geodesic deviation equation*. It shows that curvature produces a relative acceleration between two neighbouring geodesics; even if they start parallel, curvature prevents the geodesics from remaining parallel.

1.11 Fermi normal coordinates

The proof of the local-flatness theorem presented in Section 1.6 gives very little indication as to how one might construct a coordinate system that would enforce Eqs. (1.22). Our purpose in this section is to return to this issue, and provide a more geometric proof of the theorem. In fact, we will extend the theorem from a single point P to an entire geodesic γ. For concreteness we will take the geodesic to be timelike.

We will show that we can introduce coordinates $x^\alpha = (t, x^a)$ such that near γ, the metric can be expressed as

$$\begin{aligned}
g_{tt} &= -1 - R_{tatb}(t) x^a x^b + O(x^3), \\
g_{ta} &= -\frac{2}{3} R_{tbac}(t) x^b x^c + O(x^3), \\
g_{ab} &= \delta_{ab} - \frac{1}{3} R_{acbd}(t) x^c x^d + O(x^3).
\end{aligned} \tag{1.47}$$

These coordinates are known as *Fermi normal coordinates*, and t is proper time along the geodesic γ, on which the spatial coordinates x^a are all zero. In Eq. (1.47), the components of the Riemann tensor are evaluated on γ, and they depend on t only. It is obvious that Eq. (1.47) enforces $g_{\alpha\beta}|_\gamma = \eta_{\alpha\beta}$ and $\Gamma^\mu_{\alpha\beta}|_\gamma = 0$. The local-flatness theorem therefore holds everywhere on the geodesic.

The reader who is not interested in following the derivation of Eq. (1.47) can safely skip ahead to the end of this chapter. The material introduced in this section will not be encountered in any subsequent portion of this book.

1.11.1 Geometric construction

We will use $x^\alpha = (t, x^a)$ to denote the Fermi normal coordinates, and $x^{\alpha'}$ will refer to an arbitrary coordinate system. We imagine that we are given a spacetime with a metric $g_{\alpha'\beta'}$ expressed in these coordinates.

We consider a timelike geodesic γ in this spacetime. Its tangent vector is $u^{\alpha'}$, and we let t be proper time along γ. On this geodesic we select a point O at which we set $t = 0$. At this point we erect an orthonormal basis $\hat{e}^{\alpha'}_\mu$ (the subscript μ serves to label the four basis vectors), and we identify $\hat{e}^{\alpha'}_t$ with the tangent vector $u^{\alpha'}$ at O. From this we construct a basis everywhere on γ by parallel transporting $\hat{e}^{\alpha'}_\mu$ away from O. Our basis vectors therefore satisfy

$$\hat{e}^{\alpha'}_{\mu;\beta'}u^{\beta'} = 0, \qquad \hat{e}^{\alpha'}_t = u^{\alpha'}, \tag{1.48}$$

as well as

$$g_{\alpha'\beta'}\hat{e}^{\alpha'}_\mu \hat{e}^{\beta'}_\nu = \eta_{\mu\nu}, \tag{1.49}$$

everywhere on γ. Here, $\eta_{\mu\nu} = \text{diag}(-1, 1, 1, 1)$ is the Minkowski metric.

Consider now a spacelike geodesic β originating at a point P on γ, at which $t = t_P$. This geodesic has a tangent vector $v^{\alpha'}$, and we let s denote proper distance along β; we set $s = 0$ at P. We assume that at P, $v^{\alpha'}$ is orthogonal to $u^{\alpha'}$, so that it admits the decomposition

$$v^{\alpha'}\big|_\gamma = \Omega^a \hat{e}^{\alpha'}_a. \tag{1.50}$$

To ensure that $v^{\alpha'}$ is properly normalized, the expansion coefficients must satisfy $\delta_{ab}\Omega^a\Omega^b = 1$. By choosing different coefficients Ω^a we can construct new geodesics β that are also orthogonal to γ at P. We shall denote this entire family of spacelike geodesics by $\beta(t_P, \Omega^a)$.

The Fermi normal coordinates of a point Q located away from the geodesic γ are constructed as follows (Fig. 1.4). First we find the unique geodesic that passes through Q and intersects γ orthogonally. We label the intersection point P and

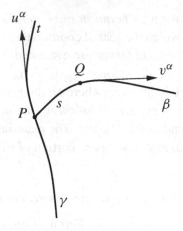

Figure 1.4   Geometric construction of the Fermi normal coordinates.

we call this geodesic $\beta(t_P, \Omega_Q^a)$, with t_P denoting proper time at the intersection point and Ω_Q^a the expansion coefficients of $v^{\alpha'}$ at that point. We then assign to Q the new coordinates

$$x^0 = t_P, \qquad x^a = \Omega_Q^a s_Q, \tag{1.51}$$

where s_Q is proper distance from P to Q. These are the Fermi normal coordinates of the point Q. Generically, therefore, $x^\alpha = (t, \Omega^a s)$, and we must now figure out how these coordinates are related to $x^{\alpha'}$, the original system.

1.11.2 Coordinate transformation

We note first that we can describe the family of geodesics $\beta(t, \Omega^a)$ by relations of the form $x^{\alpha'}(t, \Omega^a, s)$. In these, the parameters t and Ω^a serve to specify which geodesic, and s is proper distance along this geodesic. If we substitute $s = 0$ in these relations, we recover the description of the timelike geodesic γ in terms of its proper time t; the parameters Ω^a are then irrelevant. The tangent to the geodesics $\beta(t, \Omega^a)$ is

$$v^{\alpha'} = \left(\frac{\partial x^{\alpha'}}{\partial s}\right)_{t, \Omega^a}; \tag{1.52}$$

the notation indicates explicitly that the derivative with respect to s is taken while keeping t and Ω^a fixed. This vector is a solution to the geodesic equation subjected to the initial condition $v^{\alpha'}|_{s=0} = \Omega^a \hat{e}_a^{\alpha'}$. But the geodesic equation is invariant under a rescaling of the affine parameter, $s \to s/c$, in which c is a constant. Under this rescaling, $v^{\alpha'} \to c\, v^{\alpha'}$ and as a consequence we have that $\Omega^a \to c\, \Omega^a$. We have therefore established the identity $x^{\alpha'}(t, \Omega^a, s) = x^{\alpha'}(t, c\,\Omega^a, s/c)$, and as a special case we find

$$x^{\alpha'}(t, \Omega^a, s) = x^{\alpha'}(t, \Omega^a s, 1) \equiv x^{\alpha'}(x^\alpha). \qquad (1.53)$$

By virtue of Eqs. (1.51), this relation is the desired transformation between $x^{\alpha'}$ and the Fermi normal coordinates.

Now, as a consequence of Eqs. (1.50), (1.52), and (1.53) we have

$$\Omega^a \hat{e}_a^{\alpha'} = v^{\alpha'}\big|_\gamma = \frac{\partial x^{\alpha'}}{\partial s}\bigg|_{s=0} = \frac{\partial x^{\alpha'}}{\partial x^a}\bigg|_{s=0} \Omega^a,$$

which shows that

$$\frac{\partial x^{\alpha'}}{\partial x^a}\bigg|_\gamma = \hat{e}_a^{\alpha'}. \qquad (1.54)$$

From our previous observation that the relations $x^{\alpha'}(t, \Omega^a, 0)$ describe the geodesic γ, we also have

$$\frac{\partial x^{\alpha'}}{\partial t}\bigg|_\gamma = u^{\alpha'} \equiv \hat{e}_t^{\alpha'}. \qquad (1.55)$$

Equations (1.54) and (1.55) tell us that on γ, $\partial x^{\alpha'}/\partial x^\mu = \hat{e}_\mu^{\alpha'}$.

1.11.3 Deviation vectors

Suppose now that in the relations $x^{\alpha'}(t, \Omega^a, s)$, the parameters Ω^a are varied while keeping t and s fixed. This defines new curves that connect different geodesics β at the same proper distance s from their common intersection point P on γ. This is very similar to the construction described in Section 1.10, and the vectors

$$\xi_a^{\alpha'} = \left(\frac{\partial x^{\alpha'}}{\partial \Omega^a}\right)_{t,s} \qquad (1.56)$$

are deviation vectors relating geodesics $\beta(t, \Omega^a)$ with different coefficients Ω^a. Similarly,

$$\xi_t^{\alpha'} = \left(\frac{\partial x^{\alpha'}}{\partial t}\right)_{s, \Omega^a} \qquad (1.57)$$

is a deviation vector relating geodesics $\beta(t, \Omega^a)$ that start at different points on γ, but share the same coefficients Ω^a. The four vectors defined by Eqs. (1.56) and (1.57) satisfy the geodesic deviation equation, Eq. (1.46); it must be kept in mind that in this equation, the tangent vector is $v^{\alpha'}$, not $u^{\alpha'}$, and the affine parameter is s, not t.

1.11.4 Metric on γ

The components of the metric in the Fermi normal coordinates are related to the old components by the general relation

$$g_{\alpha\beta} = \frac{\partial x^{\alpha'}}{\partial x^{\alpha}} \frac{\partial x^{\beta'}}{\partial x^{\beta}} g_{\alpha'\beta'}.$$

Evaluating this on γ yields $g_{\alpha\beta}|_{\gamma} = \hat{e}^{\alpha'}_{\alpha} \hat{e}^{\beta'}_{\beta} g_{\alpha'\beta'}$, after using Eqs. (1.54) and (1.55). Substituting Eq. (1.49) we arrive at

$$g_{\alpha\beta}\big|_{\gamma} = \eta_{\alpha\beta}. \tag{1.58}$$

This states that in the Fermi normal coordinates, the metric is Minkowski everywhere on the geodesic γ.

1.11.5 First derivatives of the metric on γ

To evaluate the Christoffel symbols in the Fermi normal coordinates, we recall from Eq. (1.51) that the curves $x^0 = t$, $x^a = \Omega^a s$ are geodesics, so that these relations must be solutions to the geodesic equation,

$$\frac{d^2 x^{\alpha}}{ds^2} + \Gamma^{\alpha}{}_{\beta\gamma} \frac{dx^{\beta}}{ds} \frac{dx^{\gamma}}{ds} = 0.$$

This gives $\Gamma^{\alpha}{}_{bc}(x^{\alpha})\Omega^b\Omega^c = 0$. On γ the Christoffel symbols are functions of t only, and are therefore independent of Ω^a. Since these coefficients are arbitrary, we conclude that $\Gamma^{\alpha}{}_{bc}|_{\gamma} = 0$. To obtain the remaining components we recall that the basis vectors $\hat{e}^{\alpha}_{\mu}$ are parallel transported along γ, so that

$$\frac{d\hat{e}^{\alpha}_{\mu}}{dt} + \Gamma^{\alpha}{}_{\beta\gamma}\big|_{\gamma} \hat{e}^{\beta}_{\mu}\hat{e}^{\gamma}_{t} = 0,$$

since $\hat{e}^{\gamma}_{t} = u^{\alpha}$. By virtue of Eqs. (1.54) and (1.55) we have that $\hat{e}^{\alpha}_{\mu} = \delta^{\alpha}{}_{\mu}$ in the Fermi normal coordinates, and the parallel-transport equation implies $\Gamma^{\alpha}{}_{\beta t}|_{\gamma} = 0$. The Christoffel symbols are therefore all zero on γ. We shall write this as

$$g_{\alpha\beta,\gamma}\big|_{\gamma} = 0. \tag{1.59}$$

This proves that the Fermi normal coordinates enforce the local-flatness theorem everywhere on the geodesic γ.

1.11.6 Second derivatives of the metric on γ

We next turn to the second derivatives of the metric, or the first derivatives of the connection. From the fact that $\Gamma^{\alpha}{}_{\beta\gamma}$ is zero everywhere on γ, we obtain immediately

$$\Gamma^{\alpha}{}_{\beta\gamma,t}\big|_{\gamma} = 0. \tag{1.60}$$

From the definition of the Riemann tensor, Eq. (1.32), we also get

$$\Gamma^{\alpha}{}_{\beta t,\gamma}\big|_{\gamma} = R^{\alpha}{}_{\beta \gamma t}\big|_{\gamma}. \tag{1.61}$$

The other components are harder to come by. For these we must involve the deviation vectors ξ^{α}_{μ} introduced in Eqs. (1.56) and (1.57). These vectors satisfy the geodesic deviation equation, Eq. (1.46), which we write in full as

$$\frac{d^2 \xi^{\alpha}}{ds^2} + 2\Gamma^{\alpha}{}_{\beta\gamma} v^{\beta} \frac{d\xi^{\gamma}}{ds} + \left(R^{\alpha}{}_{\beta\gamma\delta} + \Gamma^{\alpha}{}_{\beta\gamma,\delta} - \Gamma^{\alpha}{}_{\gamma\mu}\Gamma^{\mu}{}_{\beta\delta} + \Gamma^{\alpha}{}_{\delta\mu}\Gamma^{\mu}{}_{\beta\gamma} \right) v^{\beta} \xi^{\gamma} v^{\delta} = 0.$$

According to Eqs. (1.51), (1.52), (1.56), and (1.57) we have that $v^{\alpha} = \Omega^a \delta^{\alpha}_a$, $\xi^{\alpha} = \delta^{\alpha}_t$, and $\xi^{\alpha}_a = s\delta^{\alpha}_a$ in the Fermi normal coordinates. If we substitute $\xi^{\alpha} = \xi^{\alpha}_t$ in the geodesic deviation equation and evaluate it at $s = 0$, we find $\Gamma^{\alpha}{}_{bt,c}\big|_{\gamma} = R^{\alpha}{}_{bct}\big|_{\gamma}$, which is just a special case of Eq. (1.61).

To learn something new, let us substitute $\xi^{\alpha} = \xi^{\alpha}_a$ instead. In this case we find

$$2\Gamma^{\alpha}{}_{ab}\Omega^b + s\left(R^{\alpha}{}_{bad} + \Gamma^{\alpha}{}_{ab,d} - \Gamma^{\alpha}{}_{a\mu}\Gamma^{\mu}{}_{bd} + \Gamma^{\alpha}{}_{d\mu}\Gamma^{\mu}{}_{ab} \right) \Omega^b \Omega^d = 0.$$

Before evaluating this on γ (which would give $0 = 0$), we expand the first term in powers of s:

$$\Gamma^{\alpha}{}_{ab} = \Gamma^{\alpha}{}_{ab}\big|_{\gamma} + s\Gamma^{\alpha}{}_{ab,\mu}\big|_{\gamma} v^{\mu} + O(s^2) = s\Gamma^{\alpha}{}_{ab,d}\big|_{\gamma} \Omega^d + O(s^2).$$

Dividing through by s and then evaluating on γ, we arrive at

$$\left(R^{\alpha}{}_{bad} + 3\Gamma^{\alpha}{}_{ab,d} \right)\big|_{\gamma} \Omega^b \Omega^d = 0.$$

Because the coefficients Ω^a are arbitrary, we conclude that the quantity within the brackets, properly symmetrized in the indices b and d, must vanish. A little algebra finally reveals that

$$\Gamma^{\alpha}{}_{ab,c}\big|_{\gamma} = -\frac{1}{3}\left(R^{\alpha}{}_{abc} + R^{\alpha}{}_{bac} \right)\big|_{\gamma}. \tag{1.62}$$

Equations (1.60), (1.61), and (1.62) give the complete set of derivatives of the Christoffel symbols on γ.

It is now a simple matter to turn these equations into statements regarding the second derivatives of the metric at γ. Because the metric is Minkowski everywhere on the geodesic, only the spatial derivatives are nonzero. These are given by

$$g_{tt,ab} = -2R_{tatb}\big|_{\gamma},$$

$$g_{ta,bc} = -\frac{2}{3}\left(R_{tbac} + R_{tcab} \right)\big|_{\gamma}, \tag{1.63}$$

$$g_{ab,cd} = -\frac{1}{3}\left(R_{acbd} + R_{adbc} \right)\big|_{\gamma}.$$

From Eqs. (1.58), (1.59), and (1.63) we recover Eqs. (1.47), the expansion of the metric about γ, to second order in the spatial displacements x^a.

1.11.7 Riemann tensor in Fermi normal coordinates

To express a given metric as an expansion in Fermi normal coordinates, it is necessary to evaluate the Riemann tensor on the reference geodesic, and write it as a function of t in this coordinate system. This is not as hard as it may seem. Because the Riemann tensor is evaluated on γ, we need to know the coordinate transformation only at γ; as was noted above, this is given by $\partial x^{\alpha'}/\partial x^\mu = \hat{e}^{\alpha'}_\mu$. We therefore have, for example,

$$R_{tabc}(t) = R_{\mu'\alpha'\beta'\gamma'}\,\hat{e}^{\mu'}_t\,\hat{e}^{\alpha'}_a\,\hat{e}^{\beta'}_b\,\hat{e}^{\gamma'}_c.$$

The difficult part of the calculation is therefore the determination of the orthonormal basis (which is parallel transported on the reference geodesic). Once this is known, the Fermi components of the Riemann tensor are obtained by projection, and these will naturally be expressed in terms of proper time t.

1.12 Bibliographical notes

Nothing in this text can be claimed to be entirely original, and the bibliographical notes at the end of each chapter intend to give credit where credit is due. During the preparation of this chapter I have relied on the following references: d'Inverno (1992); Manasse and Misner (1963); Misner, Thorne, and Wheeler (1973); Wald (1984); and Weinberg (1972).

More specifically:

Sections 1.2, 1.4, and 1.6 are based on Sections 6.3, 6.2, and 6.11 of d'Inverno, respectively. Sections 1.7 and 1.8 are based on Sections 4.7 and 4.4 of Weinberg, respectively. Section 1.10 is based on Section 3.3 of Wald. Finally, Section 1.11 and Problem 9 below are based on the paper by Manasse and Misner.

Suggestions for further reading:

In this Chapter I have presented a minimal account of differential geometry, just enough for the reader to get by in the remaining four chapters. For a more complete account, at a nice introductory level, I recommend the book *Geometrical methods of mathematical physics* by Bernard F. Schutz. At a more advanced level I recommend the book by Nakahara. For adanced topics that are of direct relevance to general relativity (some of which covered here), the book by Hawking and Ellis is a classic reference.

I have already listed some excellent textbooks on general relativity: d'Inverno; Misner, Thorne, and Wheeler; Schutz; Weinberg; and Wald. At an introductory level the book by Hartle is a superb alternative. At a more advanced level, Synge's book can be a very useful reference.

1.13 Problems

Warning: The results derived in Problem 9 are used in later portions of this book.

1. The surface of a two-dimensional cone is embedded in three-dimensional flat space. The cone has an opening angle of 2α. Points on the cone which all have the same distance r from the apex define a circle, and ϕ is the angle that runs along the circle.

    (a) Write down the metric of the cone, in terms of the coordinates r and ϕ.

    (b) Find the coordinate transformation $x(r, \phi)$, $y(r, \phi)$ that brings the metric into the form $ds^2 = dx^2 + dy^2$. Do these coordinates cover the entire two-dimensional plane?

    (c) Prove that any vector parallel transported along a circle of constant r on the surface of the cone ends up rotated by an angle β after a complete trip. Express β in terms of α.

2. Show that if $t^\alpha = dx^\alpha/d\lambda$ obeys the geodesic equation in the form $Dt^\alpha/d\lambda = \kappa t^\alpha$, then $u^\alpha = dx^\alpha/d\lambda^*$ satisfies $Du^\alpha/d\lambda^* = 0$ if λ^* and λ are related by $d\lambda^*/d\lambda = \exp \int \kappa(\lambda) \, d\lambda$.

3. (a) Let $x^\alpha(\lambda)$ describe a timelike geodesic parameterized by a nonaffine parameter λ, and let $t^\alpha = dx^\alpha/d\lambda$ be the geodesic's tangent vector. Calculate how $\varepsilon \equiv -t_\alpha t^\alpha$ changes as a function of λ.

    (b) Let ξ^α be a Killing vector. Calculate how $p \equiv \xi_\alpha t^\alpha$ changes as a function of λ on that same geodesic.

    (c) Let b^α be such that in a spacetime with metric $g_{\alpha\beta}$, $\pounds_b g_{\alpha\beta} = 2c \, g_{\alpha\beta}$, where c is a constant. (Such a vector is called *homothetic*.) Let $x^\alpha(\tau)$ describe a timelike geodesic parameterized by proper time τ, and let $u^\alpha = dx^\alpha/d\tau$ be the four-velocity. Calculate how $q \equiv b_\alpha u^\alpha$ changes with τ.

4. Prove that the Lie derivative of a type-(0, 2) tensor is given by $\pounds_u T_{\alpha\beta} = T_{\alpha\beta;\mu} u^\mu + u^\mu_{;\alpha} T_{\mu\beta} + u^\mu_{;\beta} T_{\alpha\mu}$.

5. Prove that $\xi^\alpha_{(1)}$ and $\xi^\alpha_{(2)}$, as given in Section 1.5, are indeed Killing vectors of spherically symmetric spacetimes.

6. A particle with electric charge e moves in a spacetime with metric $g_{\alpha\beta}$ in the presence of a vector potential A_α. The equations of motion are $u_{\alpha;\beta} u^\beta = e F_{\alpha\beta} u^\beta$, where u^α is the four-velocity and $F_{\alpha\beta} = A_{\beta;\alpha} - A_{\alpha;\beta}$. It is assumed that the spacetime possesses a Killing vector ξ^α, so that $\pounds_\xi g_{\alpha\beta} = \pounds_\xi A_\alpha = 0$.

Prove that

$$(u_\alpha + eA_\alpha)\xi^\alpha$$

is constant on the world line of the charged particle.

7. In flat spacetime, all Cartesian components of the Levi-Civita tensor can be obtained from $\varepsilon_{txyz} = 1$ by permutation of the indices. Using its tensorial property under coordinate transformations, calculate $\varepsilon_{\alpha\beta\gamma\delta}$ in the following coordinate systems:

   (a) spherical coordinates (t, r, θ, ϕ);

   (b) spherical-null coordinates (u, v, θ, ϕ), where $u = t - r$ and $v = t + r$.
   Show that your results are compatible with the general relation $\varepsilon_{\alpha\beta\gamma\delta} = \sqrt{-g}\,[\alpha\,\beta\,\gamma\,\delta]$ if $[t\,r\,\theta\,\phi] = 1$ in spherical coordinates, while $[u\,v\,\theta\,\phi] = 1$ in spherical-null coordinates.

8. In a manifold of dimension n, the *Weyl curvature tensor* is defined by

$$C_{\alpha\beta\gamma\delta} = R_{\alpha\beta\gamma\delta} - \frac{2}{n-2}\left(g_{\alpha[\gamma}R_{\delta]\beta} - g_{\beta[\gamma}R_{\delta]\alpha}\right)$$
$$+ \frac{2}{(n-1)(n-2)}\,R\,g_{\alpha[\gamma}g_{\delta]\beta}.$$

   Show that it possesses the same symmetries as the Riemann tensor. Also, prove that *any* contracted form of the Weyl tensor vanishes identically. This shows that the Riemann tensor can be decomposed into a tracefree part given by the Weyl tensor, and a trace part given by the Ricci tensor. The Einstein field equations imply that the trace part of the Riemann tensor is algebraically related to the distribution of matter in spacetime; the tracefree part, on the other hand, is algebraically independent of the matter. Thus, it can be said that the Weyl tensor represents the true gravitational degrees of freedom of the Riemann tensor.

9. Prove that the relations

$$\xi^\alpha{}_{;\mu\nu} = R^\alpha{}_{\mu\nu\beta}\xi^\beta, \qquad \Box\xi^\alpha = -R^\alpha{}_\beta\xi^\beta$$

   are satisfied by any Killing vector ξ^α. Here, $\Box \equiv \nabla^\alpha\nabla_\alpha$ is the curved-spacetime d'Alembertian operator. [Hint: Use the cyclic identity for the Riemann tensor, $R_{\mu\alpha\beta\gamma} + R_{\mu\gamma\alpha\beta} + R_{\mu\beta\gamma\alpha} = 0$.]

10. Express the Schwarzschild metric as an expansion in Fermi normal coordinates about a radially infalling, timelike geodesic.

11. Construct a coordinate system in a neighbourhood of a point P in spacetime, such that $g_{\alpha\beta}|_P = \eta_{\alpha\beta}$, $g_{\alpha\beta,\mu}|_P = 0$, and

$$g_{\alpha\beta,\mu\nu}\big|_P = -\frac{1}{3}\left(R_{\alpha\mu\beta\nu} + R_{\alpha\nu\beta\mu}\right)\big|_P.$$

Such coordinates are called *Riemann normal coordinates*.

12. A particle moving on a circular orbit in a stationary, axially symmetric space-time is subjected to a dissipative force which drives it to another, slightly smaller, circular orbit. During the transition, the particle loses an amount $\delta\tilde{E}$ of orbital energy (per unit rest-mass), and an amount $\delta\tilde{L}$ of orbital angular momentum (per unit rest-mass). You are asked to prove that these quantities are related by $\delta\tilde{E} = \Omega\,\delta\tilde{L}$, where Ω is the particle's original angular velocity. By 'circular orbit' we mean that the particle has a four-velocity given by

$$u^\alpha = \gamma\big(\xi_{(t)}^\alpha + \Omega\,\xi_{(\phi)}^\alpha\big),$$

where $\xi_{(t)}^\alpha$ and $\xi_{(\phi)}^\alpha$ are the spacetime's timelike and rotational Killing vectors, respectively; Ω and γ are constants.

You may proceed along the following lines: First, express γ in terms of $\tilde{E}$ and $\tilde{L}$. Second, find an expression for δu^α, the change in four-velocity as the particle goes from its original orbit to its final orbit. Third, prove the relation

$$u_\alpha \delta u^\alpha = \gamma(\delta\tilde{E} - \Omega\,\delta\tilde{L}),$$

from which the theorem follows.

2

Geodesic congruences

Our purpose in this chapter is to develop the mathematical techniques required in the description of congruences, the term designating an entire system of nonintersecting geodesics. We will consider separately the cases of timelike geodesics and null geodesics. (The case of spacelike geodesics does not require a separate treatment, as it is virtually identical to the timelike case; it is also less interesting from a physical point of view.) We will introduce the expansion scalar, as well as the shear and rotation tensors, as a means of describing the congruence's behaviour. We will derive a useful evolution equation for the expansion, known as Raychaudhuri's equation. On the basis of this equation we will show that gravity tends to focus geodesics, in the sense that an initially diverging congruence (geodesics flying apart) will be found to diverge less rapidly in the future, and that an initially converging congruence (geodesics coming together) will converge more rapidly in the future. And we will present Frobenius' theorem, which states that a congruence is hypersurface orthogonal – the geodesics are everywhere orthogonal to a family of hypersurfaces – if and only if its rotation tensor vanishes.

The chapter begins (in Section 2.1) with a review of the standard energy conditions of general relativity, because some of these are required in the proof of the focusing theorem. It continues (in Section 2.2) with a pedagogical introduction to the expansion scalar, shear tensor, and rotation tensor, based on the kinematics of a deformable medium. Congruences of timelike geodesics are then presented in Section 2.3, and the case of null geodesics is treated in Section 2.4.

The techniques introduced in this chapter are used in many different areas of gravitational physics. Most notably, they are part of the mathematical description of event horizons, a topic covered in Chapter 5. They also play a key role in the formulation of the singularity theorems of general relativity, a topic that (unfortunately) is not covered in this book.

Table 2.1 *Energy conditions.*

Name	Statement	Conditions
Weak	$T_{\alpha\beta} v^\alpha v^\beta \geq 0$	$\rho \geq 0, \quad \rho + p_i > 0$
Null	$T_{\alpha\beta} k^\alpha k^\beta \geq 0$	$\rho + p_i \geq 0$
Strong	$(T_{\alpha\beta} - \frac{1}{2} T g_{\alpha\beta}) v^\alpha v^\beta \geq 0$	$\rho + \sum_i p_i \geq 0, \quad \rho + p_i \geq 0$
Dominant	$-T^\alpha_{\ \beta} v^\beta$ future directed	$\rho \geq 0, \quad \rho \geq \lvert p_i \rvert$

2.1 Energy conditions

2.1.1 Introduction and summary

In the context of classical general relativity, it is reasonable to expect that the stress-energy tensor will satisfy certain conditions, such as positivity of the energy density and dominance of the energy density over the pressure. Such requirements are embodied in the *energy conditions*, which are summarized in Table 2.1.

To put the energy conditions in concrete form it is useful to assume that the stress-energy tensor admits the decomposition

$$T^{\alpha\beta} = \rho\, \hat{e}^\alpha_0 \hat{e}^\beta_0 + p_1\, \hat{e}^\alpha_1 \hat{e}^\beta_1 + p_2\, \hat{e}^\alpha_2 \hat{e}^\beta_2 + p_3\, \hat{e}^\alpha_3 \hat{e}^\beta_3, \tag{2.1}$$

in which the vectors $\hat{e}^\alpha_\mu$ form an orthonormal basis; they satisfy the relations

$$g_{\alpha\beta} \hat{e}^\alpha_\mu \hat{e}^\beta_\nu = \eta_{\mu\nu}, \tag{2.2}$$

where $\eta_{\mu\nu} = \text{diag}(-1, 1, 1, 1)$ is the Minkowski metric. (It goes without saying that the basis vectors are functions of the coordinates.) Equations (2.1) and (2.2) imply that the quantities ρ (energy density) and p_i (principal pressures) are eigenvalues of the stress-energy tensor, and $\hat{e}^\alpha_\mu$ are the normalized eigenvectors.

The inverse metric can neatly be expressed in terms of the basis vectors. It is easy to check that the relation

$$g^{\alpha\beta} = \eta^{\mu\nu} \hat{e}^\alpha_\mu \hat{e}^\beta_\nu, \tag{2.3}$$

where $\eta^{\mu\nu} = \text{diag}(-1, 1, 1, 1)$ is the inverse of $\eta_{\mu\nu}$, is compatible with Eq. (2.2). Equations such as (2.3) are called *completeness relations*.

If the stress-energy tensor is that of a perfect fluid, then $p_1 = p_2 = p_3 \equiv p$. Substituting this into Eq. (2.1) and using Eq. (2.3) yields

$$
\begin{aligned}
T^{\alpha\beta} &= \rho\,\hat{e}_0^\alpha\hat{e}_0^\beta + p\big(\hat{e}_1^\alpha\hat{e}_1^\beta + \hat{e}_2^\alpha\hat{e}_2^\beta + \hat{e}_3^\alpha\hat{e}_3^\beta\big) \\
&= \rho\,\hat{e}_0^\alpha\hat{e}_0^\beta + p\big(g^{\alpha\beta} + \hat{e}_0^\alpha\hat{e}_0^\beta\big) \\
&= (\rho + p)\,\hat{e}_0^\alpha\hat{e}_0^\beta + p\,g^{\alpha\beta}.
\end{aligned}
$$

The vector $\hat{e}_0^\alpha$ is identified with the four-velocity of the perfect fluid.

Some of the energy conditions are formulated in terms of a normalized, future-directed, but otherwise arbitrary timelike vector v^α; this represents the four-velocity of an arbitrary observer in spacetime. Such a vector can be decomposed as

$$
v^\alpha = \gamma\big(\hat{e}_0^\alpha + a\,\hat{e}_1^\alpha + b\,\hat{e}_2^\alpha + c\,\hat{e}_3^\alpha\big), \qquad \gamma = (1 - a^2 - b^2 - c^2)^{-1/2}, \qquad (2.4)
$$

where a, b, and c are arbitrary functions of the coordinates, restricted by $a^2 + b^2 + c^2 < 1$. We will also need an arbitrary, future-directed null vector k^α. This we shall express as

$$
k^\alpha = \hat{e}_0^\alpha + a'\,\hat{e}_1^\alpha + b'\,\hat{e}_2^\alpha + c'\,\hat{e}_3^\alpha, \qquad (2.5)
$$

where a', b', and c' are arbitrary functions of the coordinates, restricted by $a'^2 + b'^2 + c'^2 = 1$. Recall that the normalization of a null vector is always arbitrary.

2.1.2 Weak energy condition

The weak energy condition states that the energy density of any matter distribution, as measured by any observer in spacetime, must be nonnegative. Because an observer with four-velocity v^α measures the energy density to be $T_{\alpha\beta}v^\alpha v^\beta$, we must have

$$
T_{\alpha\beta}v^\alpha v^\beta \geq 0 \qquad (2.6)
$$

for any future-directed timelike vector v^α. To put this in concrete form we substitute Eqs. (2.1) and (2.4), which gives

$$
\rho + a^2 p_1 + b^2 p_2 + c^2 p_3 \geq 0.
$$

Because a, b, c, are arbitrary, we may choose $a = b = c = 0$, and this gives $\rho \geq 0$. Alternatively, we may choose $b = c = 0$, which gives $\rho + a^2 p_1 \geq 0$. Recalling that a^2 must be smaller than unity, we obtain $0 \leq \rho + a^2 p_1 < \rho + p_1$. So $\rho + p_1 > 0$, and similar expressions hold for p_2 and p_3. The weak energy condition therefore implies

$$\rho \geq 0, \qquad \rho + p_i > 0. \tag{2.7}$$

2.1.3 Null energy condition

The null energy condition makes the same statement as the weak form, except that v^α is replaced by an arbitrary, future-directed null vector k^α. Thus,

$$T_{\alpha\beta} k^\alpha k^\beta \geq 0 \tag{2.8}$$

is the statement of the null energy condition. Substituting Eqs. (2.1) and (2.5) gives

$$\rho + a'^2 p_1 + b'^2 p_2 + c'^2 p_3 \geq 0.$$

Choosing $b' = c' = 0$ enforces $a' = 1$, and we obtain $\rho + p_1 \geq 0$, with similar expressions holding for p_2 and p_3. The null energy condition therefore implies

$$\rho + p_i \geq 0. \tag{2.9}$$

Notice that the weak energy condition implies the null form.

2.1.4 Strong energy condition

The statement of the strong energy condition is

$$\left(T_{\alpha\beta} - \frac{1}{2} T g_{\alpha\beta}\right) v^\alpha v^\beta \geq 0, \tag{2.10}$$

or $T_{\alpha\beta} v^\alpha v^\beta > -\frac{1}{2} T$, where v^α is any future-directed, normalized, timelike vector. Because $T_{\alpha\beta} - \frac{1}{2} T g_{\alpha\beta} = R_{\alpha\beta}/8\pi$ by virtue of the Einstein field equations, the strong energy condition is really a statement about the Ricci tensor. Substituting Eqs. (2.1) and (2.4) gives

$$\gamma^2(\rho + a^2 p_1 + b^2 p_2 + c^2 p_3) \geq \frac{1}{2}(\rho - p_1 - p_2 - p_3).$$

Choosing $a = b = c = 0$ enforces $\gamma = 1$, and we obtain $\rho + p_1 + p_2 + p_3 \geq 0$. Alternatively, choosing $b = c = 0$ implies $\gamma^2 = 1/(1 - a^2)$, and after some simple algebra we obtain $\rho + p_1 + p_2 + p_3 \geq a^2(p_2 + p_3 - \rho - p_1)$. Because this must hold for any $a^2 < 1$, we have $\rho + p_1 \geq 0$, with similar relations holding for p_2 and p_3. The strong energy condition therefore implies

$$\rho + p_1 + p_2 + p_3 \geq 0, \qquad \rho + p_i \geq 0. \tag{2.11}$$

It should be noted that the strong energy condition does *not* imply the weak form.

2.1.5 Dominant energy condition

The dominant energy condition embodies the notion that matter should flow along timelike or null world lines. Its precise statement is that if v^α is an arbitrary, future-directed, timelike vector field, then

$$-T^\alpha_{\ \beta} v^\beta \text{ is a future-directed, timelike or null, vector field.} \qquad (2.12)$$

The quantity $-T^\alpha_{\ \beta} v^\beta$ is the matter's momentum density as measured by an observer with four-velocity v^α, and this is required to be timelike or null. Substituting Eqs. (2.1) and (2.4) and demanding that $-T^\alpha_{\ \beta} v^\beta$ not be spacelike gives

$$\rho^2 - a^2 p_1^{\ 2} - b^2 p_2^{\ 2} - c^2 p_3^{\ 2} \geq 0.$$

Choosing $a = b = c = 0$ gives $\rho^2 \geq 0$, and demanding that $-T^\alpha_{\ \beta} v^\beta$ be future directed selects the positive branch: $\rho \geq 0$. Alternatively, choosing $b = c = 0$ gives $\rho^2 \geq a^2 p_1^{\ 2}$. Because this must hold for any $a^2 < 1$, we have $\rho \geq |p_1|$, having taken the future direction for $-T^\alpha_{\ \beta} v^\beta$. Similar relations hold for p_2 and p_3. The dominant energy condition therefore implies

$$\rho \geq 0, \qquad \rho \geq |p_i|. \qquad (2.13)$$

2.1.6 Violations of the energy conditions

While the energy conditions typically hold for classical matter, they can be violated by quantized matter fields. A well-known example is the Casimir vacuum energy between two conducting plates separated by a distance d:

$$\rho = -\frac{\pi^2}{720} \frac{\hbar}{d^4}.$$

Although quantum effects allow for a localized violation of the energy conditions, recent work suggests that there is a limit to the extent by which the energy conditions can be violated *globally*. In this context it is useful to formulate averaged versions of the energy conditions. For example, the *averaged null energy condition* states that the integral of $T_{\alpha\beta} k^\alpha k^\beta$ along a null geodesic γ must be nonnegative:

$$\int_\gamma T_{\alpha\beta} k^\alpha k^\beta \, d\lambda \geq 0.$$

Such averaged energy conditions play a central role in the theory of traversable wormholes (see Section 2.6, Problem 1). The averaged null energy condition is known to always hold in flat spacetime, for noninteracting scalar and electromagnetic fields in arbitrary quantum states; this is true in spite of the fact that $T_{\alpha\beta} k^\alpha k^\beta$ can be negative somewhere along the geodesic. Its status in curved spacetimes is not yet fully settled. A complete discussion, as of 1994, can be found in Matt Visser's book.

2.2 Kinematics of a deformable medium

2.2.1 Two-dimensional medium

As a warm-up for what is to follow, consider, in a purely Newtonian context, the internal motion of a two-dimensional deformable medium. (Picture this as a thin sheet of rubber; see Fig. 2.1.) How the medium actually moves depends on its internal dynamics, which will remain unspecified for the purpose of this discussion. From a purely kinematical point of view, however, we may always write that for a sufficiently small displacement ξ^a about a reference point O,

$$\frac{\mathrm{d}\xi^a}{\mathrm{d}t} = B^a{}_b(t)\xi^b + O(\xi^2),$$

for some tensor $B^a{}_b$. The time dependence of this tensor is determined by the medium's dynamics. For short time intervals,

$$\xi^a(t_1) = \xi^a(t_0) + \Delta\xi^a(t_0),$$

where

$$\Delta\xi^a = B^a{}_b(t_0)\xi^b(t_0)\,\Delta t + O(\Delta t^2),$$

and $\Delta t = t_1 - t_0$. To describe the action of $B^a{}_b$ we will consider the simple figure described by $\xi^a(t_0) = r_0(\cos\phi, \sin\phi)$; this is a circle of radius r_0 drawn in the two-dimensional medium.

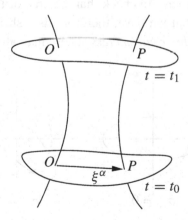

Figure 2.1   Two-dimensional deformable medium.

2.2.2 Expansion

Suppose first that B^a_b is proportional to the identity matrix, so that

$$B^a_b = \begin{pmatrix} \frac{1}{2}\theta & 0 \\ 0 & \frac{1}{2}\theta \end{pmatrix},$$

where $\theta \equiv B^a_a$. Then $\Delta\xi^a = \frac{1}{2}\theta r_0 \Delta t (\cos\phi, \sin\phi)$, which corresponds to a change in the circle's radius: $r_1 = r_0 + \frac{1}{2}\theta r_0 \Delta t$. The corresponding change in area is then $\Delta A \equiv A_1 - A_0 = \pi r_0^2 \theta \Delta t$, so that

$$\theta = \frac{1}{A_0}\frac{\Delta A}{\Delta t}.$$

The quantity θ is therefore the fractional change of area per unit time; we shall call it the *expansion parameter*. This is actually a function, because θ may depend on time and on the choice of reference point O.

2.2.3 Shear

Suppose next that B^a_b is symmetric and tracefree, so that

$$B^a_b = \begin{pmatrix} \sigma_+ & \sigma_\times \\ \sigma_\times & -\sigma_+ \end{pmatrix}.$$

Then $\Delta\xi^a = r_0 \Delta t (\sigma_+ \cos\phi + \sigma_\times \sin\phi, -\sigma_+ \sin\phi + \sigma_\times \cos\phi)$. The parametric equation describing the new figure is $r_1(\phi) = r_0(1 + \sigma_+ \Delta t \cos 2\phi + \sigma_\times \Delta t \sin 2\phi)$. If $\sigma_\times = 0$, this represents an ellipse with major axis oriented along the $\phi = 0$ direction (Fig. 2.2). If, on the other hand, $\sigma_+ = 0$, then the ellipse's major axis is oriented along $\phi = \pi/4$. The general situation is an ellipse oriented at an arbitrary angle. It is easy to check that the area of the figure is not affected by the transformation. What we have, therefore, is a shearing of the figure, and σ_+ and $\sigma_\times$ are called the *shear parameters*. These may also vary over the medium.

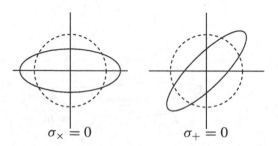

$$\sigma_\times = 0 \qquad\qquad\qquad \sigma_+ = 0$$

Figure 2.2   Effect of the shear tensor.

2.2.4 Rotation

Finally, we suppose that B^a_b is antisymmetric, so that

$$B^a_b = \begin{pmatrix} 0 & \omega \\ -\omega & 0 \end{pmatrix}.$$

Then $\Delta\xi^a = r_0\omega\Delta t(\sin\phi, -\cos\phi)$, and the new displacement vector is $\xi^a(t_1) = r_0(\cos\phi', \sin\phi')$, where $\phi' = \phi - \omega\Delta t$. This clearly represents an overall rotation of the original figure, and this operation also leaves the area unchanged; ω is called the *rotation parameter*.

2.2.5 General case

The most general matrix B^a_b has $2 \times 2 = 4$ components, and it may be expressed as

$$B^a_b = \begin{pmatrix} \frac{1}{2}\theta & 0 \\ 0 & \frac{1}{2}\theta \end{pmatrix} + \begin{pmatrix} \sigma_+ & \sigma_\times \\ \sigma_\times & -\sigma_+ \end{pmatrix} + \begin{pmatrix} 0 & \omega \\ -\omega & 0 \end{pmatrix}.$$

The action of this most general tensor is a linear combination of expansion, shear, and rotation. The tensor can also be expressed as

$$B_{ab} = \frac{1}{2}\theta\,\delta_{ab} + \sigma_{ab} + \omega_{ab},$$

where $\theta = B^a_a$ (the expansion scalar) is the trace part of B_{ab}, $\sigma_{ab} = B_{(ab)} - \frac{1}{2}\theta\delta_{ab}$ (the shear tensor) is the symmetric-tracefree part of B_{ab}, and $\omega_{ab} = B_{[ab]}$ (the rotation tensor) is the antisymmetric part of B_{ab}.

2.2.6 Three-dimensional medium

In three dimensions the tensor B_{ab} would be decomposed as

$$B_{ab} = \frac{1}{3}\theta\,\delta_{ab} + \sigma_{ab} + \omega_{ab},$$

where $\theta = B^a_a$ is the expansion scalar, $\sigma_{ab} = B_{(ab)} - \frac{1}{3}\theta\delta_{ab}$ the shear tensor, and $\omega_{ab} = B_{[ab]}$ the rotation tensor. In the three-dimensional case, the expansion is the fractional change of *volume* per unit time:

$$\theta = \frac{1}{V}\frac{\Delta V}{\Delta t}.$$

To see this, treat the three-dimensional relation

$$\xi^a(t_1) = (\delta^a_b + B^a_b\Delta t)\xi^b(t_0)$$

as a coordinate transformation from $\xi^a(t_0)$ to $\xi^a(t_1)$. The Jacobian of this transformation is

$$
\begin{aligned}
J &= \det[\delta^a{}_b + B^a{}_b \Delta t] \\
&= 1 + \mathrm{Tr}[B^a{}_b \Delta t] \\
&= 1 + \theta \Delta t.
\end{aligned}
$$

This implies that volumes at t_0 and t_1 are related by $V_1 = (1 + \theta \Delta t) V_0$, so that $V_0 \theta = (V_1 - V_0)/\Delta t$. This argument shows also that the volume is not affected by the shear and rotation tensors.

2.3 Congruence of timelike geodesics

Let $\mathscr{O}$ be an open region in spacetime. A *congruence* in $\mathscr{O}$ is a family of curves such that through each point in $\mathscr{O}$ there passes one and only one curve from this family. (The curves do not intersect; picture this as a tight bundle of copper wires.) In this section we will be interested in congruences of timelike geodesics, which means that each curve in the family is a timelike geodesic; congruences of null geodesics will be considered in the following section. We wish to determine how such a congruence evolves with time. More precisely stated, we want to determine the behaviour of the *deviation vector* ξ^α between two neighbouring geodesics in the congruence (Fig. 2.3), as a function of proper time τ along the reference geodesic. The geometric setup is the same as in Section 1.10, and the relations

$$
u^\alpha u_\alpha = -1, \quad u^\alpha{}_{;\beta} u^\beta = 0, \quad u^\alpha{}_{;\beta} \xi^\beta = \xi^\alpha{}_{;\beta} u^\beta, \quad u^\alpha \xi_\alpha = 0,
$$

where u^α is tangent to the geodesics, will be assumed to hold. Notice in particular that ξ^α is orthogonal to u^α: the deviation vector points in the directions *transverse* to the flow of the congruence.

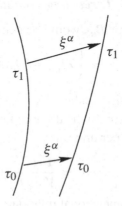

Figure 2.3 Deviation vector between two neighbouring members of a congruence.

2.3.1 Transverse metric

Given the congruence and the associated timelike vector field u^α, the spacetime metric $g_{\alpha\beta}$ can be decomposed into a longitudinal part $-u_\alpha u_\beta$ and a transverse part $h_{\alpha\beta}$ given by

$$h_{\alpha\beta} = g_{\alpha\beta} + u_\alpha u_\beta. \tag{2.14}$$

The transverse metric is purely 'spatial', in the sense that it is orthogonal to u^α: $u^\alpha h_{\alpha\beta} = 0 = h_{\alpha\beta}u^\beta$. It is effectively three-dimensional: in a comoving Lorentz frame at some point P within the congruence, $u_\alpha \overset{*}{=} (-1, 0, 0, 0)$, $g_{\alpha\beta} \overset{*}{=}$ diag$(-1, 1, 1, 1)$, and $h_{\alpha\beta} \overset{*}{=}$ diag$(0, 1, 1, 1)$. We may also note the relations $h^\alpha_{\ \alpha} = 3$ and $h^\alpha_{\ \mu}h^\mu_{\ \beta} = h^\alpha_{\ \beta}$.

2.3.2 Kinematics

We now introduce the tensor field

$$B_{\alpha\beta} = u_{\alpha;\beta}. \tag{2.15}$$

Like $h_{\alpha\beta}$, this tensor is purely transverse, as $u^\alpha B_{\alpha\beta} = u^\alpha u_{\alpha;\beta} = \frac{1}{2}(u_\alpha u^\alpha)_{;\beta} = 0$ and $B_{\alpha\beta}u^\beta = u_{\alpha;\beta}u^\beta = 0$. It determines the evolution of the deviation vector: from $\xi^\alpha_{\ ;\beta}u^\beta = u^\alpha_{\ ;\beta}\xi^\beta$ we immediately obtain

$$\xi^\alpha_{\ ;\beta}u^\beta = B^\alpha_{\ \beta}\xi^\beta, \tag{2.16}$$

and we see that $B^\alpha_{\ \beta}$ measures the failure of ξ^α to be parallel transported along the congruence.

Equation (2.16) is directly analogous to the first equation of Section 2.2. We may decompose $B_{\alpha\beta}$ into trace, symmetric-tracefree, and antisymmetric parts. This gives

$$B_{\alpha\beta} = \frac{1}{3}\theta\, h_{\alpha\beta} + \sigma_{\alpha\beta} + \omega_{\alpha\beta}, \tag{2.17}$$

where $\theta = B^\alpha_{\ \alpha} = u^\alpha_{\ ;\alpha}$ is the *expansion scalar*, $\sigma_{\alpha\beta} = B_{(\alpha\beta)} - \frac{1}{3}\theta h_{\alpha\beta}$ the *shear tensor*, and $\omega_{\alpha\beta} = B_{[\alpha\beta]}$ the *rotation tensor*. These quantities come with the same interpretation as in Section 2.2. In particular, the congruence will be diverging (geodesics flying apart) if $\theta > 0$, and it will be converging (geodesics coming together) if $\theta < 0$.

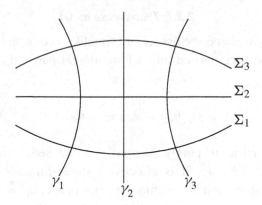

Figure 2.4   Family of hypersurfaces orthogonal to a congruence of timelike geodesics.

2.3.3 Frobenius' theorem

Some congruences have a vanishing rotation tensor, $\omega_{\alpha\beta} = 0$. These are said to be *hypersurface orthogonal*, meaning that the congruence is everywhere orthogonal to a family of spacelike hypersurfaces foliating $\mathcal{O}$ (Fig. 2.4). We now provide a partial proof of this statement.

The congruence will be hypersurface orthogonal if u^α is everywhere proportional to n^α, the normal to the hypersurfaces. Supposing that these are described by equations of the form $\Phi(x^\alpha) = c$, where c is a constant specific to each hypersurface, then $n_\alpha \propto \Phi_{,\alpha}$ and

$$u_\alpha = -\mu\Phi_{,\alpha},$$

for some proportionality factor μ. (We suppose that Φ increases toward the future, and the positive quantity μ can be determined from the normalization condition $u^\alpha u_\alpha = -1$.) Differentiating this equation gives $u_{\alpha;\beta} = -\mu\Phi_{;\alpha\beta} - \Phi_{,\alpha}\mu_{,\beta}$. Consider now the completely antisymmetric tensor

$$u_{[\alpha;\beta}u_{\gamma]} \equiv \frac{1}{3!}\left(u_{\alpha;\beta}u_\gamma + u_{\gamma;\alpha}u_\beta + u_{\beta;\gamma}u_\alpha - u_{\beta;\alpha}u_\gamma - u_{\alpha;\gamma}u_\beta - u_{\gamma;\beta}u_\alpha\right).$$

Direct evaluation of the right-hand side, using $\Phi_{;\beta\alpha} = \Phi_{;\alpha\beta}$, returns zero. We therefore have

$$\text{hypersurface orthogonal} \quad \Rightarrow \quad u_{[\alpha;\beta}u_{\gamma]} = 0. \tag{2.18}$$

The converse of this statement, that $u_{[\alpha;\beta}u_{\gamma]} = 0$ implies the existence of a scalar field Φ such that $u_\alpha \propto \Phi_{,\alpha}$, is also true (but harder to prove).

Equation (2.18) is a useful result, because whether or not u^α is hypersurface orthogonal can be decided on the basis of the vector field alone, without having to find Φ explicitly. We note that the geodesic equation $u^\alpha{}_{;\beta}u^\beta = 0$ was never used in the derivation of Eq. (2.18). We also never used the fact that u^α was normalized. Equation (2.18) is therefore quite general: A congruence of curves (timelike, spacelike, or null) is hypersurface orthogonal if and only if $u_{[\alpha;\beta}u_{\gamma]} = 0$, where u^α is tangent to the curves. This statement is known as *Frobenius' theorem*.

We now return to our geodesic congruence, and use Eqs. (2.15) and (2.17) to calculate

$$3! \, u_{[\alpha;\beta}u_{\gamma]} = 2(u_{[\alpha;\beta]}u_\gamma + u_{[\gamma;\alpha]}u_\beta + u_{[\beta;\gamma]}u_\alpha)$$

$$= 2(B_{[\alpha\beta]}u_\gamma + B_{[\gamma\alpha]}u_\beta + B_{[\beta\gamma]}u_\alpha)$$

$$= 2(\omega_{\alpha\beta}u_\gamma + \omega_{\gamma\alpha}u_\beta + \omega_{\beta\gamma}u_\alpha).$$

If we put the left-hand side to zero and multiply the right-hand side by u^γ, we obtain $\omega_{\alpha\beta} = 0$, because $\omega_{\gamma\alpha}u^\gamma = 0 = \omega_{\beta\gamma}u^\gamma$. (Recall the purely transverse property of $B_{\alpha\beta}$.) Therefore,

$$\text{hypersurface orthogonal} \quad \Rightarrow \quad \omega_{\alpha\beta} = 0. \qquad (2.19)$$

This concludes the proof of our initial statement.

Notice that Eq. (2.19) holds for timelike geodesics only, whereas Eq. (2.18) is general. In fact, Eq. (2.19) could have been derived much more directly, but in doing so we would have bypassed the more general formulation of Frobenius' theorem. The direct proof goes as follows.

If u^α is hypersurface orthogonal, then $u_\alpha = -\mu\Phi_{,\alpha}$ for some scalars μ and Φ. It follows from $\omega_{\alpha\beta} = u_{[\alpha;\beta]}$ and the symmetry of $\Phi_{;\alpha\beta}$ that

$$\omega_{\alpha\beta} = -\Phi_{[,\alpha}\mu_{,\beta]} = \frac{1}{\mu}u_{[\alpha}\mu_{,\beta]}.$$

But we know that $\omega_{\alpha\beta}$ must be orthogonal to u^α, and the relation $\omega_{\alpha\beta}u^\beta = 0$ implies $\mu_{,\alpha} = -(\mu_{,\beta}u^\beta)u_\alpha$. This, in turn, establishes that the rotation tensor vanishes identically: $\omega_{\alpha\beta} = 0$.

We have learned that μ must be constant on each hypersurface, because it varies only in the direction normal to the hypersurfaces. Thus, μ can be expressed as a function of Φ, and defining a new scalar $\Psi = \int \mu(\Phi)\,d\Phi$ we find that u_α is not only proportional to a gradient, it is *equal* to one: $u_\alpha = -\Psi_{,\alpha}$. Notice that if u_α can be expressed in this form, then it automatically satisfies the geodesic equation: $u_{\alpha;\beta}u^\beta = \Psi_{;\alpha\beta}\Psi^{,\beta} = \Psi_{;\beta\alpha}\Psi^{,\beta} = \frac{1}{2}(\Psi^{,\beta}\Psi_{,\beta})_{;\alpha} = \frac{1}{2}(u^\beta u_\beta)_{;\alpha} = 0$.

In summary:

A vector field u^α (timelike, spacelike, or null, and not necessarily geodesic) is hypersurface orthogonal if there exists a scalar field Φ such that $u_\alpha \propto \Phi_{,\alpha}$, which implies $u_{[\alpha;\beta}u_{\gamma]} = 0$. If the vector field is timelike and geodesic, then it is hypersurface orthogonal if there exists a scalar field Ψ such that $u_\alpha = -\Psi_{,\alpha}$, which implies $\omega_{\alpha\beta} = u_{[\alpha;\beta]} = 0$.

2.3.4 Raychaudhuri's equation

We now want to derive an evolution equation for θ, the expansion scalar. We begin by developing an equation for $B_{\alpha\beta}$ itself:

$$
\begin{aligned}
B_{\alpha\beta;\mu}u^\mu &= u_{\alpha;\beta\mu}u^\mu \\
&= (u_{\alpha;\mu\beta} - R_{\alpha\nu\beta\mu}u^\nu)u^\mu \\
&= (u_{\alpha;\mu}u^\mu)_{;\beta} - u_{\alpha;\mu}u^\mu{}_{;\beta} - R_{\alpha\nu\beta\mu}u^\nu u^\mu \\
&= -B_{\alpha\mu}B^\mu{}_\beta - R_{\alpha\mu\beta\nu}u^\mu u^\nu .
\end{aligned}
$$

The equation for θ is obtained by taking the trace:

$$
\frac{d\theta}{d\tau} = -B^{\alpha\beta}B_{\beta\alpha} - R_{\alpha\beta}u^\alpha u^\beta .
$$

It is then easy to check that $B^{\alpha\beta}B_{\beta\alpha} = \frac{1}{3}\theta^2 + \sigma^{\alpha\beta}\sigma_{\alpha\beta} - \omega^{\alpha\beta}\omega_{\alpha\beta}$. Making the substitution, we arrive at

$$
\frac{d\theta}{d\tau} = -\frac{1}{3}\theta^2 - \sigma^{\alpha\beta}\sigma_{\alpha\beta} + \omega^{\alpha\beta}\omega_{\alpha\beta} - R_{\alpha\beta}u^\alpha u^\beta . \tag{2.20}
$$

This is *Raychaudhuri's equation* for a congruence of timelike geodesics. We note that since the shear and rotation tensors are purely spatial, $\sigma^{\alpha\beta}\sigma_{\alpha\beta} \geq 0$ and $\omega^{\alpha\beta}\omega_{\alpha\beta} \geq 0$, with the equality sign holding if and only if the tensor is identically zero.

2.3.5 Focusing theorem

The importance of Eq. (2.20) for general relativity is revealed by the following theorem: Let a congruence of timelike geodesics be *hypersurface orthogonal*, so that $\omega_{\alpha\beta} = 0$, and let the *strong energy condition* hold, so that (by virtue of the Einstein field equations) $R_{\alpha\beta}u^\alpha u^\beta \geq 0$. Then the Raychaudhuri equation implies

$$
\frac{d\theta}{d\tau} = -\frac{1}{3}\theta^2 - \sigma^{\alpha\beta}\sigma_{\alpha\beta} - R_{\alpha\beta}u^\alpha u^\beta \leq 0.
$$

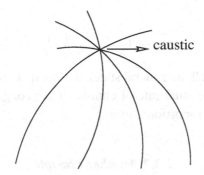

Figure 2.5   Geodesics converge into a caustic of the congruence.

The expansion must therefore *decrease* during the congruence's evolution. Thus, an initially diverging ($\theta > 0$) congruence will diverge less rapidly in the future, while an initially converging ($\theta < 0$) congruence will converge more rapidly in the future. This is the statement of the focusing theorem. Its physical interpretation is that gravitation is an attractive force when the strong energy condition holds, and the geodesics get focused as a result of this attraction.

It also follows from Raychaudhuri's equation that under the conditions of the focusing theorem, $d\theta/d\tau \leq -\frac{1}{3}\theta^2$. This can be integrated at once, giving

$$\theta^{-1}(\tau) \geq \theta_0^{-1} + \frac{\tau}{3},$$

where $\theta_0 \equiv \theta(0)$. This shows that if the congruence is initially converging ($\theta_0 < 0$), then $\theta(\tau) \to -\infty$ within a proper time $\tau \leq 3/|\theta_0|$. The interpretation of this result is that the congruence will develop a *caustic*, a point at which some of the geodesics come together (Fig. 2.5). Obviously, a caustic is a singularity of the congruence, and equations such as (2.20) lose their meaning at such points.

2.3.6 Example

As an illustrative example, let us consider the congruence of comoving world lines in an expanding universe with metric

$$ds^2 = -dt^2 + a^2(t)(dx^2 + dy^2 + dz^2),$$

where $a(t)$ is the scale factor. The tangent vector field is $u_\alpha = -\partial_\alpha t$, and a quick calculation reveals that

$$B_{\alpha\beta} = u_{\alpha;\beta} = \frac{\dot{a}}{a} h_{\alpha\beta},$$

where an overdot indicates differentiation with respect to t. This shows that the shear and rotation tensors are both zero for this congruence. The expansion, on the other hand, is given by

$$\theta = 3\frac{\dot{a}}{a} = \frac{1}{a^3}\frac{d}{dt}a^3.$$

This illustrates rather well the general statement (made in Section 2.3.8 below) that the expansion is the fractional rate of change of the congruence's cross-sectional volume (which is here proportional to a^3).

2.3.7 Another example

As a second example we consider a congruence of radial, marginally bound, time-like geodesics of the Schwarzschild spacetime. The metric is

$$ds^2 = -f\,dt^2 + f^{-1}\,dr^2 + r^2\,d\Omega^2,$$

where $f = 1 - 2M/r$ and $d\Omega^2 = d\theta^2 + \sin^2\theta\,d\phi^2$. For radial geodesics, $u^\theta = u^\phi = 0$, and the geodesics are marginally bound if $1 = \tilde{E} \equiv -u_\alpha \xi^\alpha_{(t)} = -u_t$. This means that the conserved energy is precisely equal to the rest-mass energy, and this gives us the equation $u^t = 1/f$. From the normalization condition $g_{\alpha\beta}u^\alpha u^\beta = -1$ we also get $u^r = \pm\sqrt{2M/r}$; the upper sign applies to outgoing geodesics, and the lower sign applies to ingoing geodesics.

The four-velocity is therefore given by

$$u^\alpha \partial_\alpha = f^{-1}\partial_t \pm \sqrt{2M/r}\,\partial_r, \qquad u_\alpha\,dx^\alpha = -dt \pm f^{-1}\sqrt{2M/r}\,dr.$$

It follows that u_α is equal to a gradient: $u_\alpha = -\Phi_{,\alpha}$, where

$$\Phi = t \mp 4M\left[\sqrt{r/2M} + \frac{1}{2}\ln\left(\frac{\sqrt{r/2M}-1}{\sqrt{r/2M}+1}\right)\right].$$

This means that the congruence is everywhere orthogonal to the spacelike hyper-surfaces $\Phi = \text{constant}$.

The expansion is calculated as

$$\theta = u^\alpha_{\ ;\alpha} = \frac{1}{\sqrt{-g}}\left(\sqrt{-g}\,u^\alpha\right)_{,\alpha} = \frac{1}{r^2}(r^2 u^r)',$$

where a prime indicates differentiation with respect to r. Completing the calculation gives

$$\theta = \pm\frac{3}{2}\sqrt{\frac{2M}{r^3}}.$$

Not surprisingly, the congruence is diverging ($\theta > 0$) if the geodesics are outgoing, and converging ($\theta < 0$) if the geodesics are ingoing. The rate of change of the expansion is calculated as $d\theta/d\tau = (d\theta/dr)(dr/d\tau) = \theta' u^r$, and the result is

$$\frac{d\theta}{d\tau} = -\frac{9M}{2r^3}.$$

As dictated by the focusing theorem, $d\theta/d\tau$ is negative in both cases.

2.3.8 Interpretation of θ

We now prove that θ is equal to the fractional rate of change of δV, the congruence's cross-sectional volume:

$$\theta = \frac{1}{\delta V}\frac{d}{d\tau}\delta V. \tag{2.21}$$

Although this may already be obvious from Eqs. (2.16) and (2.17), it is still instructive to go through a formal proof. The first step is to introduce the notions of cross-section, and cross-sectional volume.

Select a particular geodesic γ from the congruence, and on this geodesic, pick a point P at which $\tau = \tau_P$. Construct, in a small neighbourhood around P, a small set $\delta\Sigma(\tau_P)$ of points P' such that (i) through each of these points there passes another geodesic from the congruence, and (ii) at each point P', τ is also equal to τ_P. This set forms a three-dimensional region, a small segment of the hypersurface $\tau = \tau_P$ (Fig. 2.6). We assume that the parameterization has been adjusted so that γ intersects $\delta\Sigma(\tau_P)$ orthogonally. (There is no requirement that other geodesics do, as the congruence may not be hypersurface orthogonal.) We shall call $\delta\Sigma(\tau_P)$ the congruence's *cross section* around the geodesic γ, at proper time $\tau = \tau_P$. We want to calculate the volume of this hypersurface segment, and compare it with the volume of $\delta\Sigma(\tau_Q)$, where Q is a neighbouring point on γ.

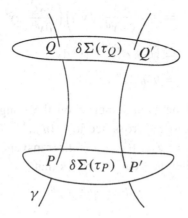

Figure 2.6 Congruence's cross section about a reference geodesic.

We introduce coordinates on $\delta\Sigma(\tau_P)$ by assigning a label y^a ($a = 1, 2, 3$) to each point P' in the set. Recalling that through each of these points there passes a geodesic from the congruence, we see that we may use y^a to label the geodesics themselves. By demanding that each geodesic keep its label as it moves away from $\delta\Sigma(\tau_P)$, we simultaneously obtain a coordinate system y^a in $\delta\Sigma(\tau_Q)$ or any other cross section. This construction therefore defines a coordinate system (τ, y^a) in a neighbourhood of the geodesic γ, and there exists a transformation between this system and the one originally in use: $x^\alpha = x^\alpha(\tau, y^a)$. Because y^a is constant along the geodesics, we have

$$u^\alpha = \left(\frac{\partial x^\alpha}{\partial \tau}\right)_{y^a}.$$

(2.22)

On the other hand, the vectors

$$e_a^\alpha = \left(\frac{\partial x^\alpha}{\partial y^a}\right)_\tau$$

(2.23)

are tangent to the cross sections. These relations imply $\pounds_u e_a^\alpha = 0$, and we also have $u_\alpha\, e_a^\alpha = 0$ holding on γ (and only γ).

We now introduce a three-tensor h_{ab} defined by

$$h_{ab} = g_{\alpha\beta}\, e_a^\alpha e_b^\beta.$$

(2.24)

(A three-tensor is a tensor with respect to coordinate transformations $y^a \to y^{a'}$, but a scalar with respect to transformations $x^\alpha \to x^{\alpha'}$.) This acts as a metric tensor on $\delta\Sigma(\tau)$: For displacements confined to the cross section (so that $d\tau = 0$), $x^\alpha = x^\alpha(y^a)$ and

$$\begin{aligned}
ds^2 &= g_{\alpha\beta}\, dx^\alpha\, dx^\beta \\
&= g_{\alpha\beta}\left(\frac{\partial x^\alpha}{\partial y^a}\, dy^a\right)\left(\frac{\partial x^\beta}{\partial y^b}\, dy^b\right) \\
&= \left(g_{\alpha\beta}\, e_a^\alpha e_b^\beta\right) dy^a dy^b \\
&= h_{ab}\, dy^a dy^b.
\end{aligned}$$

Thus, h_{ab} is the three-dimensional metric on the congruence's cross sections. Because γ is orthogonal to its cross sections ($u_\alpha\, e_a^\alpha = 0$), we have that $h_{ab} = h_{\alpha\beta}\, e_a^\alpha e_b^\beta$ on γ, where $h_{\alpha\beta} = g_{\alpha\beta} + u_\alpha u_\beta$ is the transverse metric. If we define h^{ab} to be the inverse of h_{ab}, then it is easy to check that

$$h^{\alpha\beta} = h^{ab}\, e_a^\alpha e_b^\beta$$

(2.25)

on γ.

The three-dimensional volume element on the cross sections, or *cross-sectional volume*, is $\delta V = \sqrt{h}\, d^3 y$, where $h \equiv \det[h_{ab}]$. Because the coordinates y^a are co-moving (since each geodesic moves with a constant value of its coordinates), $d^3 y$ does not change as the cross section $\delta\Sigma(\tau)$ evolves from $\tau = \tau_P$ to $\tau = \tau_Q$. A change in δV therefore comes entirely from a change in $\sqrt{h}$:

$$\frac{1}{\delta V}\frac{d}{d\tau}\delta V = \frac{1}{\sqrt{h}}\frac{d}{d\tau}\sqrt{h} = \frac{1}{2}h^{ab}\frac{dh_{ab}}{d\tau}.$$

We must now calculate the rate of change of the three-metric:

$$\begin{aligned}
\frac{dh_{ab}}{d\tau} &\equiv \left(g_{\alpha\beta}\, e_a^\alpha e_b^\beta\right)_{;\mu} u^\mu \\
&= g_{\alpha\beta}\left(e_{a;\mu}^\alpha u^\mu\right)e_b^\beta + g_{\alpha\beta}\, e_a^\alpha\left(e_{b;\mu}^\beta u^\mu\right) \\
&= g_{\alpha\beta}\left(u_{;\mu}^\alpha e_a^\mu\right)e_b^\beta + g_{\alpha\beta}\, e_a^\alpha\left(u_{;\mu}^\beta e_b^\mu\right) \\
&= u_{\beta;\alpha}\, e_a^\alpha e_b^\beta + u_{\alpha;\beta}\, e_a^\alpha e_b^\beta \\
&= (B_{\alpha\beta} + B_{\beta\alpha})e_a^\alpha e_b^\beta.
\end{aligned} \tag{2.26}$$

Multiplying by h^{ab} and evaluating on γ, so that Eq. (2.25) may be used, we obtain

$$\begin{aligned}
h^{ab}\frac{dh_{ab}}{d\tau} &= (B_{\alpha\beta} + B_{\beta\alpha})\left(h^{ab}\, e_a^\alpha e_b^\beta\right) \\
&= 2B_{\alpha\beta}h^{\alpha\beta} \\
&= 2B_{\alpha\beta}g^{\alpha\beta} \\
&= 2\theta.
\end{aligned}$$

This establishes that

$$\theta = \frac{1}{\sqrt{h}}\frac{d}{d\tau}\sqrt{h}, \tag{2.27}$$

which is the same statement as in Eq. (2.21).

2.4 Congruence of null geodesics

We now turn to the case of null geodesics. The geometric setup is the same as in the preceding section, except that the tangent vector field, denoted k^α, is null. We assume that the geodesics are affinely parameterized by λ, so that a displacement along a member of the congruence is described by $dx^\alpha = k^\alpha\, d\lambda$. The deviation vector will again be denoted ξ^α, and we again take it to be orthogonal to, and Lie transported along, the geodesics. The following equations therefore hold:

$$k^\alpha k_\alpha = 0, \quad k^\alpha_{;\beta}k^\beta = 0, \quad k^\alpha_{;\beta}\xi^\beta = \xi^\alpha_{;\beta}k^\beta, \quad k^\alpha\xi_\alpha = 0.$$

As we were in the preceding section, we will be interested in the transverse properties of the congruence, which are determined by the deviation vector ξ^α. We can,
however, anticipate some difficulties, because here the condition $k^\alpha \xi_\alpha = 0$ fails to
remove an eventual component of ξ^α in the direction of k^α. One of our first tasks,
therefore, will be to isolate the purely transverse part of the deviation vector. This
we will do with the help of $h_{\alpha\beta}$, the transverse metric.

2.4.1 Transverse metric

To isolate the part of the metric that is transverse to k^α is not entirely straightfor-
ward when k^α is null. The expression $h'_{\alpha\beta} = g_{\alpha\beta} + k_\alpha k_\beta$ does not work, because
$h'_{\alpha\beta} k^\beta = k_\alpha \neq 0$. To see what must be done, let us go to a local Lorentz frame at
some point P, and let us introduce the null coordinates $u = t - x$ and $v = t + x$.
The line element can then be expressed as $ds^2 \overset{*}{=} -du\, dv + dy^2 + dz^2$. Supposing
that k^α is tangent to the curves $u = $ constant, we see that the transverse line element is $d\tilde{s}^2 \overset{*}{=} dy^2 + dz^2$: the transverse metric is *two-dimensional*. This clearly
has to do with the fact that $ds^2 = 0$ for displacements along the v direction.

   To isolate the transverse part of the metric we need to introduce *another* null
vector field N_α, such that $N_\alpha k^\alpha \neq 0$. Because the normalization of a null vector is arbitrary, we may always impose $k^\alpha N_\alpha = -1$. If $k_\alpha \overset{*}{=} -\partial_\alpha u$ in the local Lorentz frame, then we might choose $N_\alpha \overset{*}{=} -\frac{1}{2}\partial_\alpha v$. Now consider the object $h_{\alpha\beta} = g_{\alpha\beta} + k_\alpha N_\beta + N_\alpha k_\beta$. This is clearly orthogonal to both k^α and N^α:
$h_{\alpha\beta} k^\beta = h_{\alpha\beta} N^\beta = 0$. Furthermore, $h_{\alpha\beta} \overset{*}{=} \text{diag}(0, 0, 1, 1)$ in the local Lorentz
frame, and $h_{\alpha\beta}$ is properly transverse and two-dimensional. This, then, is the object
we seek.

   The transverse metric is therefore obtained as follows: Given the null vector
field k^α, *select* an auxiliary null vector field N_α and choose its normalization to be
such that $k^\alpha N_\alpha = -1$. Then the transverse metric is given by

$$h_{\alpha\beta} = g_{\alpha\beta} + k_\alpha N_\beta + N_\alpha k_\beta. \tag{2.28}$$

It satisfies the relations

$$h_{\alpha\beta} k^\beta = h_{\alpha\beta} N^\beta = 0, \quad h^\alpha{}_\alpha = 2, \quad h^\alpha{}_\mu h^\mu{}_\beta = h^\alpha{}_\beta, \tag{2.29}$$

which confirm that $h_{\alpha\beta}$ is purely transverse (orthogonal to both k^α and N^α) and
effectively two-dimensional.

   Evidently, the conditions $N^\alpha N_\alpha = 0$ and $k^\alpha N_\alpha = -1$ do not determine N_α
uniquely. This implies that the transverse metric is *not unique*. As we shall see,
however, quantities such as the expansion of the congruence will turn out to be the

same for *all* choices of auxiliary null vector. Further aspects of this non-uniqueness are explored in Section 2.6, Problem 6.

2.4.2 Kinematics

As before, we introduce the tensor field

$$B_{\alpha\beta} = k_{\alpha;\beta} \tag{2.30}$$

as a measure of the failure of ξ^α to be parallel transported along the congruence:

$$\xi^\alpha_{;\beta} k^\beta = B^\alpha_{\ \beta} \xi^\beta. \tag{2.31}$$

As before, $B_{\alpha\beta}$ is orthogonal to the tangent vector field: $k^\alpha B_{\alpha\beta} = 0 = B_{\alpha\beta} k^\beta$. However, $B_{\alpha\beta}$ is *not* orthogonal to N^α, and Eq. (2.31) has a non-transverse component that should be removed.

We begin by isolating the purely transverse part of the deviation vector, which we denote $\tilde{\xi}^\alpha$. Because $h_{\alpha\beta}$ is itself purely transverse, it is easy to see that

$$\tilde{\xi}^\alpha \equiv h^\alpha_{\ \mu} \xi^\mu = \xi^\alpha + \left(N_\mu \xi^\mu \right) k^\alpha \tag{2.32}$$

is the desired object. Its covariant derivative in the direction of k^α represents the relative velocity of two neighbouring geodesics. This is given by

$$\tilde{\xi}^\mu_{\ ;\beta} k^\beta = h^\mu_{\ \nu} B^\nu_{\ \beta} \xi^\beta + h^\mu_{\ \nu;\beta} \xi^\nu k^\beta,$$

where we have inserted Eq. (2.31) in the first term of the right-hand side. Calculating the second term gives

$$\tilde{\xi}^\mu_{\ ;\beta} k^\beta = h^\mu_{\ \nu} B^\nu_{\ \beta} \xi^\beta + \left(N_{\nu;\beta} \xi^\nu k^\beta \right) k^\mu,$$

and we see that the vector $\tilde{\xi}^\mu_{\ ;\beta} k^\beta$ has a component along k^μ. Once again we remove this by projecting with $h^\alpha_{\ \mu}$. Using the last of Eqs. (2.29) we obtain

$$\begin{aligned} \left(\tilde{\xi}^\alpha_{\ ;\beta} k^\beta \right)^\sim &\equiv h^\alpha_{\ \mu} \left(\tilde{\xi}^\mu_{\ ;\beta} k^\beta \right) = h^\alpha_{\ \mu} B^\mu_{\ \nu} \xi^\nu \\ &= h^\alpha_{\ \mu} B^\mu_{\ \nu} \tilde{\xi}^\nu \\ &= h^\alpha_{\ \mu} h^\nu_{\ \beta} B^\mu_{\ \nu} \tilde{\xi}^\beta \end{aligned}$$

for the *transverse components* of the relative velocity. In the first line we have replaced ξ^ν with $\tilde{\xi}^\nu$ because $B^\mu_{\ \nu} k^\nu = 0$. In the third line we have inserted the relation $\tilde{\xi}^\nu = h^\nu_{\ \beta} \tilde{\xi}^\beta$; this holds because $\tilde{\xi}^\nu$ is already purely transverse.

We have obtained

$$\left(\tilde{\xi}^\alpha_{\ ;\beta} k^\beta \right)^\sim = \tilde{B}^\alpha_{\ \beta} \tilde{\xi}^\beta, \tag{2.33}$$

where

$$\tilde{B}_{\alpha\beta} = h^\mu{}_\alpha h^\nu{}_\beta B_{\mu\nu} \tag{2.34}$$

is the purely transverse part of $B_{\mu\nu} = k_{\mu;\nu}$. This can be expressed in a more explicit form by using Eq. (2.28):

$$
\begin{aligned}
\tilde{B}_{\alpha\beta} &= (g_\alpha{}^\mu + k_\alpha N^\mu + N_\alpha k^\mu)(g_\beta{}^\nu + k_\beta N^\nu + N_\beta k^\nu)\, B_{\mu\nu} \\
&= (g_\alpha{}^\mu + k_\alpha N^\mu + N_\alpha k^\mu)(B_{\mu\beta} + k_\beta B_{\mu\nu} N^\nu) \\
&= B_{\alpha\beta} + k_\alpha N^\mu B_{\mu\beta} + k_\beta B_{\alpha\mu} N^\mu + k_\alpha k_\beta B_{\mu\nu} N^\mu N^\nu.
\end{aligned} \tag{2.35}
$$

Equation (2.33) governs the purely transverse behaviour of the null congruence, and the vector $\tilde{B}^\alpha{}_\beta \tilde{\xi}^\beta$ can be interpreted as the transverse relative velocity between two neighbouring geodesics.

As we did before, we decompose the evolution tensor $\tilde{B}_{\alpha\beta}$ into its irreducible parts:

$$\tilde{B}_{\alpha\beta} = \frac{1}{2}\theta\, h_{\alpha\beta} + \sigma_{\alpha\beta} + \omega_{\alpha\beta}, \tag{2.36}$$

where $\theta = \tilde{B}^\alpha{}_\alpha$ is the expansion scalar, $\sigma_{\alpha\beta} = \tilde{B}_{(\alpha\beta)} - \frac{1}{2}\theta\, h_{\alpha\beta}$ the shear tensor, and $\omega_{\alpha\beta} = \tilde{B}_{[\alpha\beta]}$ the rotation tensor. The expansion is given more explicitly by

$$
\begin{aligned}
\theta &= g^{\alpha\beta} \tilde{B}_{\alpha\beta} \\
&= g^{\alpha\beta} B_{\alpha\beta},
\end{aligned}
$$

which follows from Eq. (2.35) and the fact that $B_{\alpha\beta}$ is orthogonal to k^α. From this we obtain

$$\theta = k^\alpha{}_{;\alpha}. \tag{2.37}$$

We see explicitly that θ does not depend on the choice of auxiliary null vector N^α: the expansion is unique. The geometric meaning of the expansion will be considered in detail below; we will show that θ is the fractional rate of change – per unit affine-parameter distance – of the congruence's cross-sectional area. (Recall that here, the transverse space is two-dimensional.)

2.4.3 Frobenius' theorem

We now show that if the vector field k^α is such that $\omega_{\alpha\beta} = 0$, then the congruence is hypersurface orthogonal, in the sense that k_α must be proportional to the normal $\Phi_{,\alpha}$ of a family of hypersurfaces described by $\Phi(x^\alpha) = c$. These hypersurfaces must clearly be null: $g^{\alpha\beta}\Phi_{,\alpha}\Phi_{,\beta} \propto g^{\alpha\beta}k_\alpha k_\beta = 0$. Furthermore, because

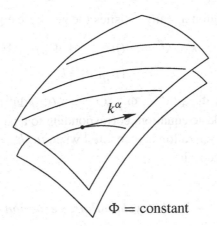

$$k^\alpha$$

$$\Phi = \text{constant}$$

Figure 2.7 Family of hypersurfaces orthogonal to a congruence of null geodesics.

k^α is at once parallel and orthogonal to $\Phi_{,\alpha}$ ($k^\alpha \Phi_{,\alpha} = 0$), the vector k^α is also *tangent* to the hypersurfaces. The null geodesics therefore lie *within* the hypersurfaces (Fig. 2.7); they are called the *null generators* of the hypersurfaces $\Phi(x^\alpha) = c$.

We begin with the general statement of Frobenius' theorem derived in Section 2.3.3: The congruence is hypersurface orthogonal if and only if $k_{[\alpha;\beta}k_{\gamma]} = 0$. This condition implies $B_{[\alpha\beta]}k_\gamma + B_{[\gamma\alpha]}k_\beta + B_{[\beta\gamma]}k_\alpha = 0$, and transvecting with N^γ gives

$$
\begin{aligned}
B_{[\alpha\beta]} &= B_{[\gamma\alpha]}k_\beta N^\gamma + B_{[\beta\gamma]}k_\alpha N^\gamma \\
&= \tfrac{1}{2}\left(B_{\gamma\alpha}k_\beta - B_{\alpha\gamma}k_\beta + B_{\beta\gamma}k_\alpha - B_{\gamma\beta}k_\alpha\right)N^\gamma \\
&= B_{\gamma[\alpha}k_{\beta]}N^\gamma + k_{[\alpha}B_{\beta]\gamma}N^\gamma.
\end{aligned}
$$

But from Eq. (2.35) we also have

$$
\tilde{B}_{[\alpha\beta]} = B_{[\alpha\beta]} - B_{\mu[\alpha}k_{\beta]}N^\mu - k_{[\alpha}B_{\beta]\mu}N^\mu,
$$

and it follows immediately that $\tilde{B}_{[\alpha\beta]} = 0$. We therefore can say

$$\text{hypersurface orthogonal} \quad \Rightarrow \quad \omega_{\alpha\beta} = 0, \tag{2.38}$$

and this concludes the proof. (In Section 2.6, Problem 6 you will show that if $\omega_{\alpha\beta} = 0$ for a specific choice of auxiliary null vector N^α, then $\omega_{\alpha\beta} = 0$ for all possible choices.)

The congruence is hypersurface orthogonal if there exists a scalar field $\Phi(x^\alpha)$ which is constant on the hypersurfaces and $k_\alpha = -\mu\Phi_{,\alpha}$ for some scalar μ. A vector of this form automatically satisfies the geodesic equation:

$$k_{\alpha;\beta}k^\beta = -(\mu\Phi_{;\alpha\beta} + \Phi_{,\alpha}\mu_{,\beta})k^\beta$$
$$= -\left(\mu_{,\beta}\Phi^{,\beta}\right)k_\alpha,$$

where we have used $\Phi_{;\alpha\beta}\Phi^{,\beta} = \Phi_{;\beta\alpha}\Phi^{,\beta} = \frac{1}{2}(\Phi_{,\beta}\Phi^{,\beta})_{,\alpha} = 0$. This is the general form of the geodesic equation, corresponding to a parameterization that is not affine. Affine parameterization is recovered when $\mu_{,\alpha}k^\alpha = 0$, that is, when μ does not vary along the geodesics.

2.4.4 Raychaudhuri's equation

The derivation of the null version of Raychaudhuri's equation proceeds much as in Section 2.3.4. In particular, the equation

$$\frac{d\theta}{d\lambda} = -B^{\alpha\beta}B_{\beta\alpha} - R_{\alpha\beta}k^\alpha k^\beta$$

follows from the same series of steps. It is then easy to check that $B^{\alpha\beta}B_{\beta\alpha} = \tilde{B}^{\alpha\beta}\tilde{B}_{\beta\alpha} = \frac{1}{2}\theta^2 + \sigma^{\alpha\beta}\sigma_{\alpha\beta} - \omega^{\alpha\beta}\omega_{\alpha\beta}$, which gives

$$\frac{d\theta}{d\lambda} = -\frac{1}{2}\theta^2 - \sigma^{\alpha\beta}\sigma_{\alpha\beta} + \omega^{\alpha\beta}\omega_{\alpha\beta} - R_{\alpha\beta}k^\alpha k^\beta. \tag{2.39}$$

This is Raychaudhuri's equation for a congruence of null geodesics. It should be noted that this equation is invariant under a change of auxiliary null vector N^α; this is established in Section 2.6, Problem 6. We also note that because the shear and rotation tensors are purely transverse, $\sigma^{\alpha\beta}\sigma_{\alpha\beta} \geq 0$ and $\omega^{\alpha\beta}\omega_{\alpha\beta} \geq 0$, with the equality sign holding if and only if the tensor vanishes.

2.4.5 Focusing theorem

The null version of the focusing theorem goes as follows: Let a congruence of null geodesics be *hypersurface orthogonal*, so that $\omega_{\alpha\beta} = 0$, and let the *null energy condition* hold, so that (by virtue of the Einstein field equations) $R_{\alpha\beta}k^\alpha k^\beta \geq 0$. Then the Raychaudhuri equation implies

$$\frac{d\theta}{d\lambda} = -\frac{1}{2}\theta^2 - \sigma^{\alpha\beta}\sigma_{\alpha\beta} - R_{\alpha\beta}k^\alpha k^\beta \leq 0,$$

which means that the geodesics are focused during the evolution of the congruence. Integrating $d\theta/d\lambda \leq -\frac{1}{2}\theta^2$ yields

$$\theta^{-1}(\lambda) \geq \theta_0^{-1} + \frac{\lambda}{2},$$

where $\theta_0 \equiv \theta(0)$. This shows that if the congruence is initially converging ($\theta_0 <$ 0), then $\theta(\lambda) \rightarrow -\infty$ within an affine parameter $\lambda \leq 2/|\theta_0|$. As in the case of a timelike congruence, this generally signals the occurrence of a caustic.

2.4.6 Example

As an illustrative example, let us consider the congruence formed by the generators of a null cone in flat spacetime. The geodesics emanate from a single point P (which we place at the origin of the coordinate system) and they radiate in all directions; note that P is a caustic of the congruence. In spherical coordinates, the geodesics are described by the relations $t = \lambda$, $r = \lambda$, $\theta =$ constant, and $\phi =$ constant, in which λ is the affine parameter. The tangent vector field is

$$k_\alpha = -\partial_\alpha(t - r).$$

We must find an auxiliary null vector field N^α that satisfies $k_\alpha N^\alpha = -1$. If we choose N^α to lie in the (t, r) plane, the unique solution is $N_\alpha = -\frac{1}{2}\partial_\alpha(t + r)$. With this choice we find that the transverse metric is given by $h_{\alpha\beta} = \text{diag}(0, 0, r^2, r^2 \sin^2 \theta)$. A straightforward calculation gives $B_{\alpha\beta} = k_{\alpha;\beta} = \text{diag}(0, 0, r, r \sin^2 \theta)$, and we see that $B_{\alpha\beta}$ is already transverse for this choice of N^α. We have found

$$\tilde{B}_{\alpha\beta} = \frac{1}{r} h_{\alpha\beta},$$

and this shows that the shear and rotation tensors are both zero for this congruence. The expansion, on the other hand, is given by

$$\theta = \frac{2}{r} = \frac{1}{4\pi r^2} \frac{\mathrm{d}}{\mathrm{d}\lambda}(4\pi r^2).$$

This verifies the general statement (made in Section 2.4.8 below) that the expansion is the fractional rate of change of the congruence's cross-sectional area.

We might ask how making a different choice for N^α would affect our results. It is easy to check that the vector $N_\alpha \, \mathrm{d}x^\alpha = -\mathrm{d}t + r \sin\theta \, \mathrm{d}\phi$ satisfies both $N_\alpha N^\alpha = 0$ and $N_\alpha k^\alpha = -1$. It is therefore an acceptable choice of auxiliary null vector field. This choice leads to a complicated expression for the transverse metric, which now has components along t and r. And while the expression for $B_{\alpha\beta}$ does not change, we find that $\tilde{B}_{\alpha\beta}$ is no longer equal to $B_{\alpha\beta}$, and is much more complicated than the expression given previously. You may check, however, that the relation $\tilde{B}_{\alpha\beta} = h_{\alpha\beta}/r$ is not affected by the change of auxiliary null vector. Our results for θ, $\sigma_{\alpha\beta}$, and $\omega_{\alpha\beta}$ are therefore preserved.

2.4.7 Another example

As a second example we consider the radial null geodesics of Schwarzschild space-time. For $d\theta = d\phi = 0$ the Schwarzschild line element reduces to

$$ds^2 = -f\,dt^2 + f^{-1}\,dr^2 = -f(dt - f^{-1}\,dr)(dt + f^{-1}\,dr),$$

where $f = 1 - 2M/r$. The displacements will be null if $ds^2 = 0$. If we define

$$u = t - r^*, \qquad v = t + r^*,$$

where $r^* = \int f^{-1}\,dr = r + 2M \ln(r/2M - 1)$, we find that $u = $ constant on out-going null geodesics, while $v = $ constant on ingoing null geodesics. The vector fields

$$k_\alpha^{\text{out}} = -\partial_\alpha u, \qquad k_\alpha^{\text{in}} = -\partial_\alpha v$$

are null, and they both satisfy the geodesic equation, with $+r$ as an affine parameter for k^α_{out} and $-r$ as an affine parameter for k^α_{in}. (Check this.) As their labels indicate, k_α^{out} is tangent to the outgoing geodesics, while k_α^{in} is tangent to the ingoing geodesics. The congruences are clearly hypersurface orthogonal. Their expansions are easily calculated:

$$\theta = \pm\frac{2}{r},$$

where the positive (negative) sign refers to the outgoing (ingoing) congruence. We also have

$$\frac{d\theta}{d\lambda} = -\frac{2}{r^2},$$

which is properly negative.

2.4.8 Interpretation of θ

We shall now give a formal proof of the statement that θ is the fractional rate of change of the congruence's cross-sectional area:

$$\theta = \frac{1}{\delta A}\frac{d}{d\lambda}\,\delta A, \tag{2.40}$$

where δA is measured in the purely transverse directions. The proof is very similar to what was presented in Section 2.3.8; the only crucial difference concerns the dimensionality of the transverse space.

We pick a particular geodesic γ from the congruence, and on this geodesic we select a point P at which $\lambda = \lambda_P$. We then consider the null curves to which N^α is tangent, and we let μ be the parameter on these *auxiliary curves*; we adjust the parameterization so that μ is constant on the null geodesics. The auxiliary curve that passes through P is called β, and we have that $\mu = \mu_\gamma$ at P. The *cross section* $\delta S(\lambda_P)$ is defined to be a small set of points P' in a neighbourhood of P such that (i) through each of these points there passes another geodesic from the congruence *and* another auxiliary curve, and (ii) at each point P', λ is also equal to λ_P *and* μ is equal to μ_γ. This set forms a two-dimensional region, the intersection of small segments of the hypersurfaces $\lambda = \lambda_P$ and $\mu = \mu_\gamma$. We assume that the parameterization has been adjusted so that both γ and β intersect $\delta S(\lambda_P)$ orthogonally. (There is no requirement that other curves do.)

We introduce coordinates in $\delta S(\lambda_P)$ by assigning a label θ^A ($A = 2, 3$) to each point in the set. Recalling that through each of these points there passes a geodesic from the congruence, we see that we may use θ^A to label the geodesics themselves. By demanding that each geodesic keep its label as it moves away from $\delta S(\lambda_P)$, we simultaneously obtain a coordinate system θ^A in any other cross-section $\delta S(\lambda)$. This construction therefore produces a coordinate system (λ, μ, θ^A) in a neighbourhood of the geodesic γ, and there exists a transformation between this system and the one originally in use: $x^\alpha = x^\alpha(\lambda, \mu, \theta^A)$. Because μ and θ^A are constant along the geodesics, we have

$$k^\alpha = \left(\frac{\partial x^\alpha}{\partial \lambda} \right)_{\mu, \theta^A}.$$

On the other hand, the vectors

$$e_A^\alpha = \left(\frac{\partial x^\alpha}{\partial \theta^A} \right)_{\lambda, \mu}$$

are tangent to the cross sections. These relations imply $\pounds_k e_A^\alpha = 0$ and we have also that on γ (and only γ), $k_\alpha e_A^\alpha = N_\alpha e_A^\alpha = 0$.

The remaining steps are very similar to those carried out in Section 2.3.8, and it will suffice to present a brief outline. The two-tensor

$$\sigma_{AB} = g_{\alpha\beta}\, e_A^\alpha e_B^\beta$$

acts as a metric on $\delta S(\lambda)$. The *cross-sectional area* is therefore defined by $\delta A = \sqrt{\sigma}\, \mathrm{d}^2\theta$, where $\sigma = \det[\sigma_{AB}]$. The inverse σ^{AB} of the two-metric is such that $h^{\alpha\beta} = \sigma^{AB} e_A^\alpha e_B^\beta$ on γ, where $h_{\alpha\beta} = g_{\alpha\beta} + k_\alpha N_\beta + N_\alpha k_\beta$ is the transverse metric. The relation

$$\frac{\mathrm{d}\sigma_{AB}}{\mathrm{d}\lambda} = (B_{\alpha\beta} + B_{\beta\alpha})\, e_A^\alpha e_B^\beta$$

follows, and taking its trace yields

$$\theta = \frac{1}{\sqrt{\sigma}} \frac{\mathrm{d}}{\mathrm{d}\lambda} \sqrt{\sigma}.$$

This statement is equivalent to Eq. (2.40).

2.5 Bibliographical notes

During the preparation of this chapter I have relied on the following references: Carter (1979); Visser (1995); and Wald (1984).

More specifically:

Section 2.1 is based on Section 9.2 of Wald and Chapter 12 of Visser. Sections 2.3 and 2.4 are based partially on Section 9.2 and Appendix B of Wald, as well as Section 6.2.1 of Carter.

Suggestions for further reading:

The book by Matt Visser offers a complete account of the theory of traversible wormholes and a review of the known violations of the standard energy conditions. The 1988 article that started this whole field, by Morris and Thorne, is very accessible and well worth reading.

Congruence of timelike curves play a central role in the field of mathematical cosmology, the study of exact solutions to the Einstein field equations that describe expanding universes. This active area of research is reviewed in the book by Wainwright and Ellis.

2.6 Problems

Warning: The results derived in Problem 8 are used in later portions of this book.

1. Consider a curved spacetime with metric

    $$ds^2 = -\mathrm{d}t^2 + \mathrm{d}\ell^2 + r^2(\ell)\,\mathrm{d}\Omega^2,$$

    where the function $r(\ell)$ is such that (i) it is minimum at $\ell = 0$, with a value r_0, and (ii) it asymptotically becomes equal to $|\ell|$ as $\ell \to \pm\infty$.
    (a) Argue that this spacetime contains a traversable wormhole between two asymptotically-flat regions, with a throat of radius r_0.
    (b) Find which energy conditions are violated at $\ell = 0$.
2. We examine the congruence of comoving world lines of a Friedmann–Robertson–Walker spacetime. The metric is

    $$ds^2 = -\mathrm{d}t^2 + a^2(t)\left(\frac{\mathrm{d}r^2}{1 - kr^2} + r^2\,\mathrm{d}\Omega^2\right),$$

where $a(t)$ is the scale factor and k a constant normalized to either ± 1 or zero. The vector tangent to the congruence is $u^\alpha = \partial x^\alpha / \partial t$.

(a) Show that the congruence is geodesic.

(b) Calculate the expansion, shear, and rotation of this congruence.

(c) Use the Raychaudhuri equation to deduce

$$\frac{\ddot{a}}{a} = -\frac{4\pi}{3}(\rho + 3p),$$

where ρ is the energy density of a perfect fluid with four-velocity u^α, and p is the pressure.

3. In this problem we consider the vector field

$$u^\alpha \partial_\alpha = \frac{1}{\sqrt{1 - 3M/r}}\left(\partial_t + \sqrt{M/r^3}\,\partial_\theta\right)$$

in Schwarzschild spacetime; the vector is expressed in terms of the usual Schwarzschild coordinates, and M is the mass of the black hole.

(a) Show that the vector field is timelike and geodesic. Describe the geodesics to which u^α is tangent.

(b) Calculate the expansion of the congruence. Explain why the expansion is positive in the northern hemisphere and negative in the southern hemisphere. Explain also why the expansion is singular at the north and south poles.

(c) Compute the rotation tensor for this congruence. Check that its square is given by

$$\omega^{\alpha\beta}\omega_{\alpha\beta} = \frac{M}{8r^3}\left(\frac{1 - 6M/r}{1 - 3M/r}\right)^2.$$

(d) Calculate $d\theta/d\tau$ and check that Raychaudhuri's equation is satisfied.

4. Derive the following evolution equations for the shear and rotation tensors of a congruence of timelike geodesics:

$$\sigma_{\alpha\beta;\mu}u^\mu = -\frac{2}{3}\theta\,\sigma_{\alpha\beta} - \sigma_{\alpha\mu}\sigma^\mu{}_\beta - \omega_{\alpha\mu}\omega^\mu{}_\beta + \frac{1}{3}\left(\sigma^{\mu\nu}\sigma_{\mu\nu} - \omega^{\mu\nu}\omega_{\mu\nu}\right)h_{\alpha\beta}$$

$$- C_{\alpha\mu\beta\nu}u^\mu u^\nu + \frac{1}{2}R^{TT}_{\alpha\beta},$$

$$\omega_{\alpha\beta;\mu}u^\mu = -\frac{2}{3}\theta\,\omega_{\alpha\beta} - \sigma_{\alpha\mu}\omega^\mu{}_\beta - \omega_{\alpha\mu}\sigma^\mu{}_\beta.$$

Here, $C_{\alpha\mu\beta\nu}$ is the Weyl tensor (Section 1.13, Problem 8), and $R^{TT}_{\alpha\beta} \equiv R^{T}_{\alpha\beta} - \frac{1}{3}(h^{\mu\nu}R^{T}_{\mu\nu})h_{\alpha\beta}$ is the 'transverse-tracefree' part of the Ricci tensor; its transverse part is $R^{T}_{\alpha\beta} \equiv h_\alpha{}^\mu h_\beta{}^\nu R_{\mu\nu}$.

5. In this problem we consider a spacetime with metric

$$ds^2 = -dt^2 + \frac{r^2 + a^2 \cos^2\theta}{r^2 + a^2}\, dr^2 + \left(r^2 + a^2 \cos^2\theta\right) d\theta^2$$
$$+ \left(r^2 + a^2\right)\sin^2\theta\, d\phi^2,$$

where a is a constant, together with a congruence of null geodesics with tangent vector field

$$k^\alpha \partial_\alpha = \partial_t + \partial_r + \frac{a}{r^2 + a^2}\, \partial_\phi.$$

(a) Check that k^α is null, that it satisfies the geodesic equation, and that r is an affine parameter.

(b) Find a suitable auxiliary null vector N^α and calculate the congruence's expansion, shear, and rotation. In particular, verify the following results:

$$\theta = \frac{2r}{r^2 + a^2 \cos^2\theta}, \quad \sigma_{\alpha\beta} = 0, \quad \omega^{\alpha\beta}\omega_{\alpha\beta} = \frac{2a^2 \cos^2\theta}{(r^2 + a^2 \cos^2\theta)^2}.$$

These reveal that the congruence is diverging, shear-free, and not hypersurface orthogonal.

(c) Show that the coordinate transformation

$$x = \sqrt{r^2 + a^2}\,\sin\theta\,\cos\phi, \quad y = \sqrt{r^2 + a^2}\,\sin\theta\,\sin\phi, \quad z = r\cos\theta$$

brings the metric to the standard Minkowski form for flat spacetime. Express k^α in this coordinate system.

6. The auxiliary null vector N^α introduced in Section 2.4 is not unique, and in this problem we examine various consequences of this fact. For the purpose of this discussion we introduce vectors $\hat{e}_A^\alpha$ ($A = 2, 3$) that point in the two directions orthogonal to both k^α and N^α, and we choose them to be orthonormal, so that they satisfy $g_{\alpha\beta}\hat{e}_A^\alpha \hat{e}_B^\beta = \delta_{AB}$. We also introduce the 2×2 matrix

$$B_{AB} = B_{\alpha\beta}\,\hat{e}_A^\alpha \hat{e}_B^\beta,$$

the projection of the tensor $B_{\alpha\beta} = k_{\alpha;\beta}$ in the transverse space spanned by the vectors $\hat{e}_A^\alpha$. In the following we shall use δ_{AB} and δ^{AB} to lower and raise uppercase Latin indices; for example, $B^{AB} = \delta^{AM}\delta^{BN} B_{MN}$.

(a) Derive the following relations:

$$h^{\alpha\beta} = \delta^{AB}\,\hat{e}_A^\alpha \hat{e}_B^\beta, \qquad \tilde{B}^{\alpha\beta} = B^{AB}\,\hat{e}_A^\alpha \hat{e}_B^\beta,$$
$$\theta = \delta_{AB} B^{AB}, \qquad \sigma^{\alpha\beta} = \sigma^{AB}\,\hat{e}_A^\alpha \hat{e}_B^\beta, \qquad \omega^{\alpha\beta} = \omega^{AB}\,\hat{e}_A^\alpha \hat{e}_B^\beta,$$

where $\sigma^{AB} = \frac{1}{2}(B^{AB} + B^{BA} - \theta\delta^{AB})$ and $\omega^{AB} = \frac{1}{2}(B^{AB} - B^{BA})$.
These confirm that the tensors $h_{\alpha\beta}$, $\tilde{B}_{\alpha\beta}$, $\sigma_{\alpha\beta}$, and $\omega_{\alpha\beta}$ are all orthogonal to both k^α and N^α. We now must determine how a change of auxiliary null vector affects these results.

(b) The vector N^α must satisfy the relations $N^\alpha N_\alpha = 0$ and $k^\alpha N_\alpha = -1$. Prove that the transformation

$$N^\alpha \to N'^\alpha = N^\alpha + c\,k^\alpha + c^A\,\hat{e}_A^\alpha,$$

where $c = \frac{1}{2}c_A c^A$, is the only one that preserves the defining relations for the auxiliary null vector. (The coefficients c^A are arbitrary.)

(c) Calculate how $h^{\alpha\beta}$ changes under this transformation.

(d) Calculate how $\tilde{B}^{\alpha\beta}$ changes.

(e) Show that θ is invariant under the transformation.

(f) Prove that $\sigma^{\alpha\beta}$ changes according to

$$\sigma'^{\alpha\beta} = \left(c^A c^B \sigma_{AB}\right) k^\alpha k^\beta + \left(c^A \sigma_A{}^B\right) k^\alpha \hat{e}_B^\beta$$
$$+ \left(c^B \sigma_B{}^A\right) \hat{e}_A^\alpha k^\beta + \sigma^{AB}\,\hat{e}_A^\alpha \hat{e}_B^\beta.$$

This shows that if $\sigma_{\alpha\beta} = 0$ for one choice of N^α, then $\sigma_{\alpha\beta} = 0$ for *any* other choice. Prove that $\sigma^{\alpha\beta}\sigma_{\alpha\beta}$ is invariant under the transformation.

(g) Prove that $\omega^{\alpha\beta}$ changes according to

$$\omega'^{\alpha\beta} = \left(c^A \omega_A{}^B\right) k^\alpha \hat{e}_B^\beta - \left(c^B \omega_B{}^A\right) \hat{e}_A^\alpha k^\beta + \omega^{AB}\hat{e}_A^\alpha \hat{e}_B^\beta.$$

This shows that if $\omega_{\alpha\beta} = 0$ for one choice of N^α, then $\omega_{\alpha\beta} = 0$ for *any* other choice. Prove that $\omega^{\alpha\beta}\omega_{\alpha\beta}$ is invariant under the transformation.

These results imply that the Raychaudhuri equation is invariant under a change of auxiliary null vector field. They also show that $\omega_{\alpha\beta} = 0$ implies hypersurface orthogonality for *any* choice of N^α.

7. We want to derive evolution equations for the shear and rotation tensors of a congruence of null geodesics. For this purpose it is useful to refer back to the basis k^α, N^α, $\hat{e}_A^\alpha$, and the 2×2 matrix $B_{AB} = B_{\alpha\beta}\,\hat{e}_A^\alpha \hat{e}_B^\beta$, introduced in Problem 6. We shall also need

$$R_{AB} = R_{\alpha\mu\beta\nu}\,\hat{e}_A^\alpha k^\mu \hat{e}_B^\beta k^\nu, \qquad \Gamma_{AB} = \hat{e}_{B\mu}\hat{e}_{A;\nu}^\mu k^\nu.$$

Notice that R_{AB} is a symmetric matrix, while Γ_{AB} is antisymmetric. Notice also that it is possible to set $\Gamma_{AB} = 0$ by choosing $\hat{e}_A^\alpha$ to be parallel transported along the congruence.

(a) First, derive the main evolution equation,

$$\frac{dB_{AB}}{d\lambda} = -B_{AC}B_B^C - R_{AB} + \Gamma_A^C B_{CB} + \Gamma_B^C B_{AC}.$$

(b)   Second, decompose the various matrices into their irreducible parts, as

$$B_{AB} = \frac{1}{2}\theta\,\delta_{AB} + \sigma_{AB} + \omega_{AB}, \qquad R_{AB} = \frac{1}{2}\mathscr{R}\,\delta_{AB} + C_{AB},$$

where σ_{AB} and C_{AB} are both symmetric and tracefree, while ω_{AB} is antisymmetric. Prove that $\mathscr{R} = R_{\alpha\beta}k^{\alpha}k^{\beta}$ and $C_{AB} = C_{\alpha\mu\beta\nu}\,\hat{e}_{A}^{\alpha}\,k^{\mu}\hat{e}_{B}^{\beta}k^{\nu}$, where $C_{\alpha\mu\beta\nu}$ is the Weyl tensor (Section 1.13, Problem 8). Then introduce the parameterization

$$\sigma_{AB} = \begin{pmatrix} \sigma_{+} & \sigma_{\times} \\ \sigma_{\times} & -\sigma_{+} \end{pmatrix}, \qquad C_{AB} = \begin{pmatrix} C_{+} & C_{\times} \\ C_{\times} & -C_{+} \end{pmatrix}$$

for the symmetric-tracefree matrices, and

$$\omega_{AB} = \begin{pmatrix} 0 & \omega \\ -\omega & 0 \end{pmatrix}, \qquad \Gamma_{AB} = \begin{pmatrix} 0 & \Gamma \\ -\Gamma & 0 \end{pmatrix}$$

for the antisymmetric matrices.

(c)   Third, and finally, derive the following explicit forms for the evolution equations,

$$\frac{d\theta}{d\lambda} = -\frac{1}{2}\theta^{2} - 2\left(\sigma_{+}^{2} + \sigma_{\times}^{2}\right) + 2\omega^{2} - \mathscr{R},$$

$$\frac{d\sigma_{+}}{d\lambda} = -\theta\,\sigma_{+} - C_{+} + 2\Gamma\,\sigma_{\times},$$

$$\frac{d\sigma_{\times}}{d\lambda} = -\theta\,\sigma_{\times} - C_{\times} - 2\Gamma\,\sigma_{+},$$

$$\frac{d\omega}{d\lambda} = -\theta\,\omega.$$

Check that the equation for θ agrees with the form of Raychaudhuri's equation given in the text. Recall that we can always set $\Gamma = 0$ by taking $\hat{e}_{A}^{\alpha}$ to be parallel transported along the congruence; this eliminates the coupling between the shear parameters.

8.   Retrace the steps of Section 2.4, but without the assumption that the null geodesics are affinely parameterized. Show that:

(a)   equation (2.35) stays unchanged;

(b)   the expansion is now given by $\theta = k^{\alpha}_{\;;\alpha} - \kappa$, where κ is defined by the relation $k^{\alpha}_{\;;\beta}k^{\beta} = \kappa\,k^{\alpha}$.

(c)   Raychaudhuri's equation now takes the form

$$\frac{d\theta}{d\lambda} = \kappa\,\theta - \frac{1}{2}\theta^{2} - \sigma^{\alpha\beta}\sigma_{\alpha\beta} + \omega^{\alpha\beta}\omega_{\alpha\beta} - R_{\alpha\beta}k^{\alpha}k^{\beta}.$$

3

Hypersurfaces

This chapter covers three main topics that can all be grouped under the rubric of hypersurfaces, the term designating a three-dimensional submanifold in a four-dimensional spacetime.

The first part of the chapter (Sections 3.1 to 3.3) is concerned with the *intrinsic geometry* of a hypersurface, and it examines the following questions: Given that the spacetime is endowed with a metric tensor $g_{\alpha\beta}$, how does one define an induced, three-dimensional metric h_{ab} on a specified hypersurface? And once this three-metric has been introduced, how does one define a vectorial surface element that allows vector fields to be integrated over the hypersurface? While these questions admit straightforward answers when the hypersurface is either timelike or spacelike, we will see that the null case requires special care.

The second part of the chapter (Sections 3.4 to 3.6) is concerned with the *extrinsic geometry* of a hypersurface, or how the hypersurface is embedded in the enveloping spacetime manifold. We will see how the spacetime curvature tensor can be decomposed into a purely intrinsic part – the curvature tensor of the hypersurface – and an extrinsic part that measures the bending of the hypersurface in spacetime; this bending is described by a three-dimensional tensor K_{ab} known as the extrinsic curvature. We will see what constraints the Einstein field equations place on the induced metric and extrinsic curvature of a hypersurface.

The third part of the chapter (Sections 3.7 to 3.11) is concerned with possible discontinuities of the metric and its derivatives across a hypersurface. We will consider the following question: Suppose that a hypersurface partitions spacetime into two regions, and that we are given a distinct metric tensor in each region; does the union of the two metrics form a valid solution to the Einstein field equations? We will see that the conditions for an affirmative answer are that the induced metric and the extrinsic curvature must be the same on both sides of the hypersurface. Failing this, we will see that a discontinuity in the extrinsic curvature can be explained by the presence of a thin distribution of matter – a surface layer – at the hypersurface. (The induced metric can never be discontinuous: The hypersurface would not have a well-defined intrinsic geometry.) We will first develop the mathematical formalism of junction conditions and surface layers, and then consider some applications.

3.1 Description of hypersurfaces

3.1.1 Defining equations

In a four-dimensional spacetime manifold, a *hypersurface* is a three-dimensional submanifold that can be either timelike, spacelike, or null. A particular hypersurface Σ is selected either by putting a restriction on the coordinates,

$$\Phi(x^\alpha) = 0, \tag{3.1}$$

or by giving parametric equations of the form

$$x^\alpha = x^\alpha(y^a), \tag{3.2}$$

where y^a ($a = 1, 2, 3$) are coordinates intrinsic to the hypersurface. For example, a two-sphere in a three-dimensional flat space can be described either by $\Phi(x, y, z) = x^2 + y^2 + z^2 - R^2 = 0$, where R is the sphere's radius, or by $x = R \sin\theta \cos\phi$, $y = R \sin\theta \sin\phi$, and $z = R \cos\theta$, where θ and ϕ are the intrinsic coordinates. Notice that the relations $x^\alpha(y^a)$ describe curves contained entirely in Σ (Fig. 3.1).

3.1.2 Normal vector

The vector $\Phi_{,\alpha}$ is normal to the hypersurface, because the value of Φ changes only in the direction orthogonal to Σ. A *unit normal* n_α can be introduced if the

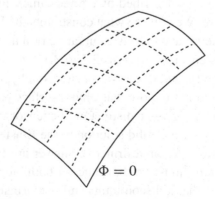

Figure 3.1   A three-dimensional hypersurface in spacetime.

hypersurface is not null. This is defined so that

$$n^\alpha n_\alpha = \varepsilon \equiv \begin{cases} -1 & \text{if } \Sigma \text{ is spacelike} \\ +1 & \text{if } \Sigma \text{ is timelike} \end{cases}, \tag{3.3}$$

and we demand that n^α point in the direction of increasing Φ: $n^\alpha \Phi_{,\alpha} > 0$. It is easy to check that n_α is given by

$$n_\alpha = \frac{\varepsilon \Phi_{,\alpha}}{\left| g^{\mu\nu} \Phi_{,\mu} \Phi_{,\nu} \right|^{1/2}} \tag{3.4}$$

if the hypersurface is either spacelike or timelike.

The unit normal is not defined when Σ is null, because $g^{\mu\nu} \Phi_{,\mu} \Phi_{,\nu}$ is then equal to zero. In this case we let

$$k_\alpha = -\Phi_{,\alpha} \tag{3.5}$$

be the normal vector; the sign is chosen so that k^α is future-directed when Φ increases toward the future. Because k^α is orthogonal to itself ($k^\alpha k_\alpha = 0$), this vector is also *tangent* to the null hypersurface Σ (Fig. 3.2). In fact, by computing $k^\alpha_{;\beta} k^\beta$ and showing that it is proportional to k^α, we can prove that k^α is tangent to null geodesics contained in Σ. We have $k_{\alpha;\beta} k^\beta = \Phi_{;\alpha\beta} \Phi^{,\beta} = \Phi_{;\beta\alpha} \Phi^{,\beta} = \frac{1}{2}(\Phi_{,\beta} \Phi^{,\beta})_{;\alpha}$; because $\Phi_{,\beta} \Phi^{,\beta}$ is zero everywhere on Σ, its gradient must be directed along k_α, and we have that $(\Phi_{,\beta} \Phi^{,\beta})_{;\alpha} = 2\kappa k_\alpha$ for some scalar κ. We have found that the normal vector satisfies

$$k^\alpha_{;\beta} k^\beta = \kappa k^\alpha,$$

the general form of the geodesic equation. The hypersurface is therefore generated by null geodesics, and k^α is tangent to the generators. The geodesics are parameterized by λ, so that a displacement along each generator is described by $dx^\alpha = k^\alpha \, d\lambda$. In general λ is not an affine parameter, but in special situations in

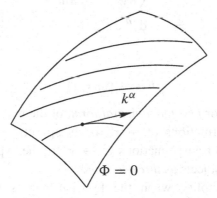

Figure 3.2   A null hypersurface and its generators.

which the relations $\Phi(x^\alpha) = $ constant describe a whole family of null hypersurfaces (so that $\Phi_{,\beta}\Phi^{,\beta}$ is zero not only on Σ but also in a neighbourhood around Σ), $\kappa = 0$ and λ is an affine parameter.

When the hypersurface is null, it is advantageous to install on Σ a coordinate system that is well adapted to the behaviour of the generators. We therefore let the parameter λ be one of the coordinates, and we introduce two additional coordinates θ^A ($A = 2, 3$) to label the generators; these are constant on each generator, and they span the two-dimensional space transverse to the generators. Thus, we shall set

$$y^a = \left(\lambda, \theta^A\right) \tag{3.6}$$

when Σ is null; varying λ while keeping θ^A constant produces a displacement along a single generator, and varying θ^A produces a displacement across generators.

3.1.3 Induced metric

The metric intrinsic to the hypersurface Σ is obtained by restricting the line element to displacements confined to the hypersurface. Recalling the parametric equations $x^\alpha = x^\alpha(y^a)$, we have that the vectors

$$e_a^\alpha = \frac{\partial x^\alpha}{\partial y^a} \tag{3.7}$$

are tangent to curves contained in Σ. (This implies that $e_a^\alpha n_\alpha = 0$ in the non-null case, and $e_a^\alpha k_\alpha = 0$ in the null case.) Now, for displacements within Σ we have

$$\begin{aligned}
\mathrm{ds}_\Sigma^2 &= g_{\alpha\beta}\,\mathrm{d}x^\alpha\,\mathrm{d}x^\beta \\
&= g_{\alpha\beta}\left(\frac{\partial x^\alpha}{\partial y^a}\,\mathrm{d}y^a\right)\left(\frac{\partial x^\beta}{\partial y^b}\,\mathrm{d}y^b\right) \\
&= h_{ab}\,\mathrm{d}y^a\,\mathrm{d}y^b,
\end{aligned} \tag{3.8}$$

where

$$h_{ab} = g_{\alpha\beta}\,e_a^\alpha e_b^\beta \tag{3.9}$$

is the *induced metric*, or *first fundamental form*, of the hypersurface. It is a *scalar* with respect to transformations $x^\alpha \to x^{\alpha'}$ of the spacetime coordinates, but it behaves as a *tensor* under transformations $y^a \to y^{a'}$ of the hypersurface coordinates. We will refer to such objects as *three-tensors*.

These relations simplify when the hypersurface is null and we use the coordinates of Eq. (3.6). Then $e_1^\alpha = (\partial x^\alpha/\partial \lambda)_{\theta^A} \equiv k^\alpha$ and it follows that

$h_{11} = g_{\alpha\beta} k^\alpha k^\beta = 0$ and $h_{1A} = g_{\alpha\beta} k^\alpha e_A^\beta = 0$, because by construction $e_A^\alpha \equiv (\partial x^\alpha / \partial \theta^A)_\lambda$ is orthogonal to k^α. In the null case, therefore,

$$ds_\Sigma^2 = \sigma_{AB}\, d\theta^A\, d\theta^B, \tag{3.10}$$

where

$$\sigma_{AB} = g_{\alpha\beta}\, e_A^\alpha e_B^\beta, \qquad e_A^\alpha = \left(\frac{\partial x^\alpha}{\partial \theta^A}\right)_\lambda. \tag{3.11}$$

Here the induced metric is a two-tensor.

We conclude by writing down completeness relations for the inverse metric. In the non-null case,

$$g^{\alpha\beta} = \varepsilon n^\alpha n^\beta + h^{ab} e_a^\alpha e_b^\beta, \tag{3.12}$$

where h^{ab} is the inverse of the induced metric. Equation (3.12) is verified by computing all inner products between n^α and e_a^α and recovering the expected results. In the null case we must introduce, everywhere on Σ, an auxiliary null vector field N^α satisfying $N_\alpha k^\alpha = -1$ and $N_\alpha e_A^\alpha = 0$ (see Section 2.4). Then the inverse metric can be expressed as

$$g^{\alpha\beta} = -k^\alpha N^\beta - N^\alpha k^\beta + \sigma^{AB} e_A^\alpha e_B^\beta, \tag{3.13}$$

where σ^{AB} is the inverse of σ_{AB}. Equation (3.13) is verified by computing all inner products between k^α, N^α, and e_A^α.

3.1.4 Light cone in flat spacetime

An example of a null hypersurface in flat spacetime is the future light cone of an event P, which we place at the origin of a Cartesian coordinate system x^α. The defining relation for this hypersurface is $\Phi \equiv t - r = 0$, where $r^2 = x^2 + y^2 + z^2$. The normal vector is $k_\alpha = -\partial_\alpha(t - r) = (-1, x/r, y/r, z/r)$. A suitable set of parametric equations is $t = \lambda, x = \lambda \sin\theta \cos\phi, y = \lambda \sin\theta \sin\phi$, and $z = \lambda \cos\theta$, in which $y^a = (\lambda, \theta, \phi)$ are the intrinsic coordinates; λ is an affine parameter on the light cone's null generators, which move with constant values of $\theta^A = (\theta, \phi)$.

From the parametric equations we compute the hypersurface's tangent vectors,

$$e_\lambda^\alpha = \frac{\partial x^\alpha}{\partial \lambda} = (1, \sin\theta \cos\phi, \sin\theta \sin\phi, \cos\theta) = k^\alpha,$$

$$e_\theta^\alpha = \frac{\partial x^\alpha}{\partial \theta} = (0, \lambda \cos\theta \cos\phi, \lambda \cos\theta \sin\phi, -\lambda \sin\theta),$$

$$e_\phi^\alpha = \frac{\partial x^\alpha}{\partial \phi} = (0, -\lambda \sin\theta \sin\phi, \lambda \sin\theta \cos\phi, 0).$$

You may check that these vectors are all orthogonal to k^α. Inner products between e_θ^α and e_ϕ^α define the two-metric σ_{AB}, and we find

$$\sigma_{AB}\, d\theta^A\, d\theta^B = \lambda^2(d\theta^2 + \sin^2\theta\, d\phi^2).$$

Not surprisingly, the hypersurface has a spherical geometry, and λ is the areal radius of the two-spheres.

It is easy to check that the unique null vector N^α that satisfies the relations $N_\alpha k^\alpha = -1$ and $N_\alpha e_A^\alpha = 0$ is $N^\alpha = \frac{1}{2}(1, -\sin\theta\cos\phi, -\sin\theta\sin\phi, -\cos\theta)$. You may also verify that the vectors k^α, N^α, and e_A^α combine as in Eq. (3.13) to form the inverse Minkowski metric.

3.2 Integration on hypersurfaces

3.2.1 Surface element (non-null case)

If Σ is not null, then

$$d\Sigma \equiv |h|^{1/2}\, d^3y, \tag{3.14}$$

where $h \equiv \det[h_{ab}]$, is an invariant three-dimensional volume element on the hypersurface. To avoid confusing this with the four-dimensional volume element $\sqrt{-g}\, d^4x$, we shall refer to $d\Sigma$ as a *surface element*. The combination $n_\alpha d\Sigma$ is a *directed* surface element that points in the direction of increasing Φ. In the null case these quantities are not defined, because $h = 0$ and n_α does not exist.

To see how Eq. (3.14) must be generalized so as to incorporate also the null case, we consider the infinitesimal vector field

$$d\Sigma_\mu = \varepsilon_{\mu\alpha\beta\gamma}\, e_1^\alpha e_2^\beta e_3^\gamma\, d^3y, \tag{3.15}$$

where $\varepsilon_{\mu\alpha\beta\gamma} = \sqrt{-g}[\mu\,\alpha\,\beta\,\gamma]$ is the Levi-Civita tensor of Section 1.8. We will show below that

$$d\Sigma_\alpha = \varepsilon n_\alpha d\Sigma \tag{3.16}$$

when the hypersurface is not null. Thus, apart from a factor $\varepsilon = \pm 1$, $d\Sigma_\alpha$ is a directed surface element on Σ. Notice that when Σ is spacelike, the factor $\varepsilon = -1$ makes $d\Sigma_\alpha$ a past-directed vector; this is unfortunately a potential source of confusion. Notice also that Eq. (3.15) remains meaningful even when the hypersurface is null. By continuity, therefore, $d\Sigma_\alpha$ is also a directed surface element on a null hypersurface.

Because $d\Sigma_\alpha$ is proportional to the completely antisymmetric Levi-Civita tensor, its sign depends on the ordering of the coordinates y^1, y^2, and y^3. But this ordering is a priori arbitrary, and we need a convention to remove the sign ambiguity.

We shall choose an ordering that makes the scalar $f \equiv \varepsilon_{\mu\alpha\beta\gamma} n^{\mu} e_1^{\alpha} e_2^{\beta} e_3^{\gamma}$ a positive quantity. Notice that this convention was already in force when we went from Eq. (3.15) to Eq. (3.16): $n^{\alpha} \, d\Sigma_{\alpha} = d\Sigma > 0$.

As a first example of how this works, consider a hypersurface of constant t in Minkowski spacetime. If $\Phi = t$, then $n_{\alpha} = -\partial_{\alpha} t$ is the future-directed normal vector. If we choose the ordering $y^a = (x, y, z)$ we find that $f = \varepsilon_{txyz} = 1$ has the correct sign. Equation (3.15) implies $d\Sigma_{\mu} = \delta^t_{\mu} \, dx \, dy \, dz = -n_{\mu} \, dx \, dy \, dz$, which is compatible with Eq. (3.16).

For our second example we choose a surface of constant x in Minkowski spacetime. We take $\Phi = x$, and $n_{\alpha} = \partial_{\alpha} x$ points in the direction of increasing Φ. We impose the ordering $y^a = (y, t, z)$ because $f = \varepsilon_{xytz} = -\varepsilon_{xtyz} = \varepsilon_{txyz} = 1$ has then the correct sign. [Notice that the more tempting ordering $y^a = (t, y, z)$ would produce the wrong sign.] With this choice, Eq. (3.15) implies $d\Sigma_{\mu} = \delta^x_{\mu} \, dt \, dy \, dz = n_{\mu} \, dt \, dy \, dz$, which is compatible with Eq. (3.16).

We now turn to a derivation of Eq. (3.16). It is clear that $d\Sigma_{\mu}$ must be proportional to n_{μ}, because $e_a^{\mu} \varepsilon_{\mu\alpha\beta\gamma} e_1^{\alpha} e_2^{\beta} e_3^{\gamma} = 0$ by virtue of the antisymmetric property of the Levi-Civita tensor. So we may write

$$\varepsilon_{\mu\alpha\beta\gamma} e_1^{\alpha} e_2^{\beta} e_3^{\gamma} = \varepsilon f n_{\mu},$$

where $f = \varepsilon_{\mu\alpha\beta\gamma} n^{\mu} e_1^{\alpha} e_2^{\beta} e_3^{\gamma}$. Because f is a scalar, we can evaluate it in any convenient coordinate system x^{α}. We choose our coordinates so that $x^0 \equiv \Phi$, and on Σ we identify x^a with the intrinsic coordinates y^a. Then $f \stackrel{*}{=} \sqrt{-g}\, n^{\Phi}$. In these coordinates $g^{\Phi\Phi} \stackrel{*}{=} g^{\alpha\beta} \Phi_{,\alpha} \Phi_{,\beta}$, and $n_{\Phi} \stackrel{*}{=} \varepsilon \, |g^{\Phi\Phi}|^{-1/2}$ is the only nonvanishing component of the normal. It follows that $n^{\Phi} \stackrel{*}{=} g^{\Phi\alpha} n_{\alpha} \stackrel{*}{=} g^{\Phi\Phi} n_{\Phi} \stackrel{*}{=} |g^{\Phi\Phi}|^{1/2}$, and we have that $f \stackrel{*}{=} |g g^{\Phi\Phi}|^{1/2}$. We now use the definition of the matrix inverse to write $g^{\Phi\Phi} \stackrel{*}{=} \text{cofactor}(g_{\Phi\Phi})/g$, where the cofactor of a matrix element is the determinant obtained after eliminating the row and column to which the element belongs. This determinant is h and we conclude that

$$f = |h|^{1/2}.$$

While this result was obtained in the special coordinates x^{α}, it is valid in all coordinate systems because h, like h_{ab}, is a *scalar* with respect to four-dimensional coordinate transformations. This result shows that when Σ is not null, Eq. (3.16) is indeed equivalent to Eq. (3.15).

3.2.2 Surface element (null case)

As we have seen in Section 3.1.2, when Σ is null we identify y^1 with λ, the parameter on the hypersurface's null generators, and the remaining coordinates, denoted

θ^A, are constant on the generators. Then $e_1^\alpha = k^\alpha$, $\mathrm{d}^3 y = \mathrm{d}\lambda\,\mathrm{d}^2\theta$, and we may write the directed surface element as

$$\mathrm{d}\Sigma_\mu = k^\nu\,\mathrm{d}S_{\mu\nu}\,\mathrm{d}\lambda, \tag{3.17}$$

where

$$\mathrm{d}S_{\mu\nu} = \varepsilon_{\mu\nu\beta\gamma}\,e_2^\beta e_3^\gamma\,\mathrm{d}^2\theta \tag{3.18}$$

is interpreted as an element of two-dimensional surface area. We will show below that this can also be expressed as

$$\mathrm{d}S_{\alpha\beta} = 2k_{[\alpha}N_{\beta]}\sqrt{\sigma}\,\mathrm{d}^2\theta, \tag{3.19}$$

where N_α is the auxiliary null vector field introduced in Eq. (3.13), and $\sigma = \det[\sigma_{AB}]$, with σ_{AB} the two-metric defined by Eq. (3.11). Combining Eq. (3.19) with Eq. (3.17) yields

$$\mathrm{d}\Sigma_\alpha = -k_\alpha\sqrt{\sigma}\,\mathrm{d}^2\theta\,\mathrm{d}\lambda. \tag{3.20}$$

The interpretation of this result is clear: Apart from a minus sign, the surface element is directed along k_α, the normal to the null hypersurface; the factor $\mathrm{d}\lambda$ represents an element of parameter-distance along the null generators, and $\sqrt{\sigma}\,\mathrm{d}^2\theta$ is an element of *cross-sectional area* – an element of two-dimensional surface area in the directions transverse to the generators.

There is also an ordering issue with the coordinates θ^A, and our convention shall be that the scalar $f \equiv \varepsilon_{\mu\nu\beta\gamma}N^\mu k^\nu e_2^\beta e_3^\gamma$ must be a positive quantity. Notice that this convention was already in force when we went from Eq. (3.18) to Eq. (3.19): $N^\alpha k^\beta\,\mathrm{d}S_{\alpha\beta} = \sqrt{\sigma}\,\mathrm{d}^2\theta > 0$.

As an example, consider a surface $u = $ constant in Minkowski spacetime, where $u = t - x$. The normal vector is $k_\alpha = -\partial_\alpha(t - x)$ and we may choose the ordering $\theta^A = (y, z)$. Then $N_\alpha = -\frac{1}{2}\partial_\alpha(t + x)$ satisfies all the requirements for an auxiliary null vector field. It is easy to check that with these choices, $f = 1$ (which is properly positive). We obtain $\mathrm{d}S_{tx} = \mathrm{d}y\,\mathrm{d}z = -\mathrm{d}S_{xt}$, and since t can be identified with the affine parameter λ, Eq. (3.17) implies $\mathrm{d}\Sigma_t = \mathrm{d}t\,\mathrm{d}y\,\mathrm{d}z = -\mathrm{d}\Sigma_x$. These results are compatible with Eq. (3.20).

Let us consider a more complicated example: the light cone of Section 3.1.4. The vectors k^α, N^α, and e_A^α are displayed in that section, and the cone's intrinsic coordinates are $y^a = (\lambda, \theta, \phi)$. We want to compute $\mathrm{d}\Sigma_\mu$ for this hypersurface, starting with the definition of Eq. (3.15). We know that $\mathrm{d}\Sigma_\mu$ must point in the direction of the normal, so that $\mathrm{d}\Sigma_\mu = -fk_\mu\,\mathrm{d}\theta\,\mathrm{d}\phi\,\mathrm{d}\lambda$, where $f = \varepsilon_{\mu\nu\beta\gamma}N^\mu k^\nu e_2^\beta e_3^\gamma$. If we let $N^\mu \equiv e_0^\mu$ and $k^\nu \equiv e_1^\nu$ we can write this as $f = [\mu\,\nu\,\beta\,\gamma]e_0^\mu e_1^\nu e_2^\beta e_3^\gamma \equiv \det E$, where E is the matrix constructed by lining up the four basis vectors. Its determinant is easy to compute and we obtain $f = \lambda^2 \sin\theta = \sqrt{\sigma}$. We therefore have $d\Sigma_\mu = -k_\mu \sqrt{\sigma}\, d^2\theta\, d\lambda$, which is just the same statement as in Eq. (3.20).

We must now give a proper derivation of Eq. (3.19). The steps are somewhat similar to those leading to Eq. (3.16). We begin by noting that the tensor $\varepsilon_{\mu\nu\beta\gamma}\, e_2^\beta e_3^\gamma$ is orthogonal to e_A^α and antisymmetric in the indices μ and ν. It may be expressed as

$$\varepsilon_{\mu\nu\beta\gamma}\, e_2^\beta e_3^\gamma = 2f k_{[\mu} N_{\nu]} = f(k_\mu N_\nu - N_\mu k_\nu),$$

where $f = \varepsilon_{\mu\nu\beta\gamma} N^\mu k^\nu e_2^\beta e_3^\gamma > 0$. To evaluate f we choose our coordinates such that $x^0 \equiv \Phi$ and $x^a \equiv y^a = (\lambda, \theta^A)$ on Σ. In these coordinates, $k_\Phi \stackrel{*}{=} -1$ and $k^\lambda \stackrel{*}{=} 1$ are the only nonvanishing components of the normal vector, $N^\Phi \stackrel{*}{=} 1$ comes as a consequence of the normalization condition $N^\alpha k_\alpha = -1$, and $g^{\Phi\Phi} \stackrel{*}{=} 0$ follows from the fact that k_α is null. Using this information we deduce that $f \stackrel{*}{=} \sqrt{-g}$, and we must now compute the metric determinant in the specified coordinates. For this purpose we note that the completeness relations of Eq. (3.13) imply the following structure for the inverse metric:

$$g^{-1} = \begin{pmatrix} 0 & 1 & 0 \\ 1 & -2N^\lambda & -N^A \\ 0 & -N^A & \sigma^{AB} \end{pmatrix};$$

this immediately implies $\det g^{-1} = -\det[\sigma^{AB}]$, or $\sqrt{-g} \stackrel{*}{=} \sqrt{\sigma}$. We therefore have

$$f = \sqrt{\sigma},$$

which holds in *any* coordinate system x^α. This establishes that Eq. (3.19) is indeed equivalent to Eq. (3.18). This, in turn, implies that Eq. (3.20) is equivalent to Eq. (3.15) when Σ is null and coordinates $y^a = (\lambda, \theta^A)$ are placed on the hypersurface.

3.2.3 Element of two-surface

The interpretation of

$$dS_{\mu\nu} = \varepsilon_{\mu\nu\beta\gamma}\, e_2^\beta e_3^\gamma\, d^2\theta$$

as a directed element of two-dimensional surface area is not limited to the consideration of null hypersurfaces. Here we consider a typical situation in which a two-dimensional surface S is imagined to be embedded in a three-dimensional, spacelike hypersurface Σ.

The hypersurface Σ is described by an equation of the form $\Phi(x^\alpha) = 0$, and by parametric relations $x^\alpha(y^a)$; $n_\alpha \propto \partial_\alpha \Phi$ is the future-directed unit normal, and the vectors $e_a^\alpha = \partial x^\alpha / \partial y^a$ are tangent to the hypersurface. The metric on Σ, induced from $g_{\alpha\beta}$, is $h_{ab} = g_{\alpha\beta} e_a^\alpha e_b^\beta$, and we have the completeness relations $g^{\alpha\beta} = -n^\alpha n^\beta + h^{ab} e_a^\alpha e_b^\beta$.

The two-surface S is introduced as a submanifold of Σ. It is described by an equation of the form $\psi(y^a) = 0$, and by parametric relations $y^a(\theta^A)$ in which θ^A are coordinates intrinsic to S; $r_a \propto \partial_a \psi$ is the outward unit normal, and the three-vectors $e_A^a = \partial y^a / \partial \theta^A$ are tangent to the two-surface. The metric on S, induced from h_{ab}, is $\sigma_{AB} = h_{ab} e_A^a e_B^b$, and we have the completeness relations $h^{ab} = r^a r^b + \sigma^{AB} e_A^a e_B^b$.

The parametric relations $y^a(\theta^A)$ and $x^\alpha(y^a)$ can be combined to give the relations $x^\alpha(\theta^A)$, which describe how S is embedded in the four-dimensional spacetime. The vectors

$$e_A^\alpha = \frac{\partial x^\alpha}{\partial \theta^A} = \frac{\partial x^\alpha}{\partial y^a} \frac{\partial y^a}{\partial \theta^A} = e_a^\alpha e_A^a$$

are tangent to S, and

$$r^\alpha \equiv r^a e_a^\alpha, \qquad r_\alpha n^\alpha = 0$$

is normal to S. The vector n^α is also normal to S, and we have that the two-surface admits *two* normal vectors: a timelike normal n^α and a spacelike normal r^α. We note that the spacelike normal can be related to a gradient, $r_\alpha \propto \partial_\alpha \Psi$, if we introduce, in a neighbourhood of Σ, a function $\Psi(x^\alpha)$ such that $\Psi|_\Sigma \equiv \psi$. In this description the induced metric on S is still

$$\begin{aligned}
\sigma_{AB} &= h_{ab} e_A^a e_B^b \\
&= \left(g_{\alpha\beta} e_a^\alpha e_b^\beta\right) e_A^a e_B^b \\
&= g_{\alpha\beta} \left(e_a^\alpha e_A^a\right)\left(e_b^\beta e_B^b\right) \\
&= g_{\alpha\beta} e_A^\alpha e_B^\beta,
\end{aligned}$$

and the completeness relations

$$g^{\alpha\beta} = -n^\alpha n^\beta + r^\alpha r^\beta + \sigma^{AB} e_A^\alpha e_B^\beta$$

are easily established from our preceding results.

We want to show that $\mathrm{d}S_{\alpha\beta}$ can be expressed neatly in terms of the timelike normal n^α, the spacelike normal r^α, and $\sqrt{\sigma}\, \mathrm{d}^2\theta$, the induced surface element on S. The expression is

$$\mathrm{d}S_{\alpha\beta} = -2n_{[\alpha} r_{\beta]} \sqrt{\sigma}\, \mathrm{d}^2\theta, \qquad (3.21)$$

where $\sigma = \det[\sigma_{AB}]$. The derivation of this result involves familiar steps. We first note that because $\varepsilon_{\mu\nu\beta\gamma} e_2^\beta e_3^\gamma$ is orthogonal to e_A^α and antisymmetric in μ and ν, it may be expressed as

$$\varepsilon_{\mu\nu\beta\gamma}\, e_2^\beta e_3^\gamma = -2f n_{[\mu}r_{\nu]} = -f(n_\mu r_\nu - r_\mu n_\nu),$$

where $f = \varepsilon_{\mu\nu\beta\gamma} n^\mu r^\nu e_2^\beta e_3^\gamma > 0$. To evaluate f we adopt coordinates $x^0 \equiv \Phi$, $x^1 \equiv \Psi$, and on S we identify x^A with θ^A. In these coordinates $n_\Phi \stackrel{*}{=} -(-g^{\Phi\Phi})^{-1/2}$ is the only nonvanishing component of the timelike normal, $r_\Psi \stackrel{*}{=} (g^{\Psi\Psi})^{-1/2}$ is the only nonvanishing component of the spacelike normal, and from the fact that these vectors are orthogonal we infer $g^{\Phi\Psi} \stackrel{*}{=} 0$. From all this we find that $f^2 \stackrel{*}{=} g g^{\Phi\Phi} g^{\Psi\Psi}$, which we rewrite as $f^2 \stackrel{*}{=} \text{cofactor}(g_{\Phi\Phi})\,\text{cofactor}(g_{\Psi\Psi})/g$. We also have $\text{cofactor}(g_{\Phi\Psi}) \stackrel{*}{=} 0$, and these two equations give us enough information to deduce

$$f = \sqrt{\sigma}.$$

This result is true in any coordinate system x^α.

As a final remark we note that the vectors n^α and r^α can be combined to form null vectors k^α and N^α. The appropriate relations are

$$k^\alpha = \frac{1}{\sqrt{2}}\left(n^\alpha + r^\alpha\right), \qquad N^\alpha = \frac{1}{\sqrt{2}}\left(n^\alpha - r^\alpha\right),$$

and these vectors are the *null normals* of the two-surface S. It is easy to check that after these substitutions, Eq. (3.21) takes the form of Eq. (3.19).

3.3 Gauss–Stokes theorem

3.3.1 First version

We consider a finite region $\mathcal{V}$ of the spacetime manifold, bounded by a closed hypersurface $\partial\mathcal{V}$ (Fig. 3.3). The signature of the hypersurface is not restricted; it may have segments that are timelike, spacelike, or null. We will show that for any vector field A^α defined within $\mathcal{V}$,

$$\int_{\mathcal{V}} A^\alpha{}_{;\alpha} \sqrt{-g}\, \mathrm{d}^4 x = \oint_{\partial\mathcal{V}} A^\alpha\, \mathrm{d}\Sigma_\alpha, \tag{3.22}$$

where $\mathrm{d}\Sigma_\alpha$ is the surface element defined by Eq. (3.15).

To derive this result, known as *Gauss' theorem*, we construct the following coordinate system in $\mathcal{V}$. We imagine a nest of closed hypersurfaces foliating $\mathcal{V}$, with the boundary $\partial\mathcal{V}$ forming the outer layer of the nest. (Picture this as the many layers of an onion.) We let x^0 be a constant on each one of these hypersurfaces, with $x^0 = 1$ designating $\partial\mathcal{V}$ and $x^0 = 0$ the zero-volume hypersurface at the 'centre' of $\mathcal{V}$. While x^0 grows 'radially outward' from this 'centre,' we take the remaining coordinates x^a to be angular coordinates on the closed hypersurfaces

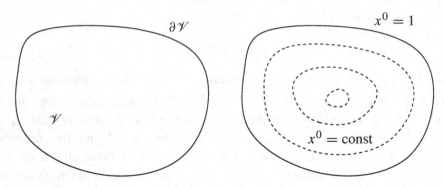

Figure 3.3   Proof of the Gauss–Stokes theorem.

$x^0 = $ constant. The coordinates y^a on $\partial \mathcal{V}$ are then identified with these angular coordinates.

Using such coordinates, the left-hand side of Eq. (3.22) becomes

$$\int_{\mathcal{V}} A^\alpha{}_{;\alpha} \sqrt{-g}\, d^4x = \int_{\mathcal{V}} (\sqrt{-g}\, A^\alpha)_{,\alpha}\, d^4x$$

$$\overset{*}{=} \int dx^0 \oint (\sqrt{-g}\, A^0)_{,0}\, d^3x + \int dx^0 \oint (\sqrt{-g} A^a)_{,a}\, d^3x$$

$$\overset{*}{=} \int dx^0 \frac{d}{dx^0} \oint \sqrt{-g}\, A^0\, d^3x$$

$$\overset{*}{=} \oint \sqrt{-g}\, A^0\, d^3x \Big|_{x^0=0}^{x^0=1}$$

$$\overset{*}{=} \oint_{\partial \mathcal{V}} \sqrt{-g}\, A^0\, d^3y.$$

In the first line we have used the divergence formula for the vector field A^α. The second integral of the second line vanishes because x^a are angular coordinates and the integration is over a closed three-dimensional surface. (Understanding this statement requires some thought. Try working through a three-dimensional version of the proof, using spherical coordinates in flat space.) In the fourth line, the contribution at $x^0 = 0$ vanishes because the 'hypersurface' $x^0 = 0$ has zero volume.

It is easy to check that $d\Sigma_\alpha \overset{*}{=} \delta^0_\alpha \sqrt{-g}\, d^3y$ in the specified coordinates, giving

$$\oint_{\partial \mathcal{V}} A^\alpha\, d\Sigma_\alpha \overset{*}{=} \oint_{\partial \mathcal{V}} A^0 \sqrt{-g}\, d^3y$$

for the right-hand side of Eq. (3.22). The two sides are therefore equal in the specified coordinate system; because Eq. (3.22) is a tensorial equation, this suffices to establish the validity of the theorem.

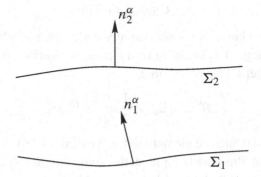

Figure 3.4   Two spacelike surfaces and their normal vectors.

3.3.2 Conservation

Gauss' theorem has many useful applications. An example is the following conservation statement.

Suppose that a vector field j^α has a vanishing divergence,

$$j^\alpha_{;\alpha} = 0.$$

Then $\oint_\Sigma j^\alpha \, \mathrm{d}\Sigma_\alpha = 0$ for any closed hypersurface Σ. Supposing now that j^α vanishes at spatial infinity, we can choose Σ to be composed of two spacelike hypersurfaces, Σ_1 and Σ_2, extending all the way to infinity (Fig. 3.4), and of a three-cylinder at infinity, on which $j^\alpha = 0$. Then

$$\int_{\Sigma_1} j^\alpha \, \mathrm{d}\Sigma_\alpha + \int_{\Sigma_2} j^\alpha \, \mathrm{d}\Sigma_\alpha = 0.$$

On each of the spacelike hypersurfaces, $\mathrm{d}\Sigma_\alpha = -n_\alpha \sqrt{h} \, \mathrm{d}^3 y$, where n_α is the *outward* normal to the closed surface Σ and h is the determinant of the induced metric on the spacelike hypersurfaces. Letting $n_\alpha \equiv n_{2\alpha}$ on Σ_2 and $n_\alpha \equiv -n_{1\alpha}$ on Σ_1, where $n_{1\alpha}$ and $n_{2\alpha}$ are both future directed, we finally obtain

$$j^\alpha_{;\alpha} = 0 \quad \Rightarrow \quad \int_{\Sigma_1} j^\alpha n_{1\alpha} \sqrt{h} \, \mathrm{d}^3 y = \int_{\Sigma_2} j^\alpha n_{2\alpha} \sqrt{h} \, \mathrm{d}^3 y. \tag{3.23}$$

The interpretation of this result is clear: If j^α is a divergence-free vector, then the 'total charge' $\int j^\alpha n_\alpha \, \mathrm{d}\Sigma$ is independent of the hypersurface on which it is evaluated. This is obviously a statement of 'charge' conservation.

3.3.3 Second version

Another version of Gauss' theorem (usually called Stokes' theorem) involves a three-dimensional region Σ bounded by a closed two-surface $\partial\Sigma$. It states that for any *antisymmetric* tensor field $B^{\alpha\beta}$ in Σ,

$$\int_\Sigma B^{\alpha\beta}{}_{;\beta}\, d\Sigma_\alpha = \frac{1}{2}\oint_{\partial\Sigma} B^{\alpha\beta}\, dS_{\alpha\beta}, \tag{3.24}$$

where $dS_{\alpha\beta}$ is the two-surface element defined by Eq. (3.18).

The derivation of this identity proceeds along familiar lines. We construct a coordinate system such that (i) x^0 is constant on the hypersurface Σ, (ii) $x^1 =$ constant describes a nest of closed two-surfaces in Σ (with $x^1 = 1$ representing $\partial\Sigma$ and $x^1 = 0$ the zero-area surface at the 'centre' of Σ), and (iii) x^A are angular coordinates on the closed surfaces (with $\theta^A = x^A$ on $\partial\Sigma$).

It is easy to check that with such coordinates, $d\Sigma_\alpha \overset{*}{=} \delta^0_\alpha \sqrt{-g}\, dx^1\, dx^2\, dx^3$. The left-hand side of Eq. (3.24) becomes

$$\int_\Sigma B^{\alpha\beta}{}_{;\beta}\, d\Sigma_\alpha = \int_\Sigma \frac{1}{\sqrt{-g}}\,(\sqrt{-g}\, B^{\alpha\beta})_{,\beta}\, d\Sigma_\alpha$$

$$\overset{*}{=} \int_\Sigma (\sqrt{-g}\, B^{0\beta})_{,\beta}\, dx^1\, dx^2\, dx^3$$

$$\overset{*}{=} \int dx^1 \oint (\sqrt{-g}\, B^{01})_{,1}\, dx^2\, dx^3 + \int dx^1 \oint (\sqrt{-g}\, B^{0A})_{,A}\, dx^2\, dx^3$$

$$\overset{*}{=} \oint \sqrt{-g}\, B^{01}\, dx^2\, dx^3 \Big|_{x^1=0}^{x^1=1}$$

$$\overset{*}{=} \oint_{\partial\Sigma} \sqrt{-g}\, B^{01}\, d^2\theta.$$

In the first line we have used the divergence formula for an antisymmetric tensor field. The explicit expression for $d\Sigma_\alpha$ was substituted in the second line. The second integral of the third line vanishes because x^A are angular coordinates and the domain of integration is a closed two-surface. In the fourth line, the lower limit of integration does not contribute because the 'surface' $x^1 = 0$ has zero area.

It is easy to check that in the specified coordinate system, $B^{\alpha\beta}\, dS_{\alpha\beta} \overset{*}{=} (B^{01} - B^{10})\sqrt{-g}\, d^2\theta = 2B^{01}\sqrt{-g}\, d^2\theta$. The right-hand side of Eq. (3.24) therefore reads

$$\frac{1}{2}\oint_{\partial\Sigma} B^{\alpha\beta}\, dS_{\alpha\beta} \overset{*}{=} \oint_{\partial\Sigma} B^{01}\sqrt{-g}\, d^2\theta,$$

and Eq. (3.24) follows from the equality of both sides in the specified coordinate system.

3.4 Differentiation of tangent vector fields

(For the remainder of Chapter 3, except for Section 3.11, we shall assume that the hypersurface Σ is either spacelike or timelike. In Section 3.11 we shall return to the case of a null hypersurface.)

3.4.1 Tangent tensor fields

Once we are presented with a hypersurface Σ, it is a common situation to have tensor fields $A^{\alpha\beta\cdots}$ that are defined only on Σ and which are purely *tangent* to the hypersurface. Such tensors admit the decomposition

$$A^{\alpha\beta\cdots} = A^{ab\cdots} e_a^\alpha e_b^\beta \cdots, \tag{3.25}$$

where $e_a^\alpha = \partial x^\alpha / \partial y^a$ are basis vectors on Σ. Equation (3.25) implies that $A^{\alpha\beta\cdots} n_\alpha = A^{\alpha\beta\cdots} n_\beta = \cdots = 0$, which confirms that $A^{\alpha\beta\cdots}$ is tangent to the hypersurface. We note that an arbitrary tensor $T^{\alpha\beta\cdots}$ can always be projected down to the hypersurface, so that only its tangential components survive. The quantity that effects the projection is $h^{\alpha\beta} \equiv h^{ab} e_a^\alpha e_b^\beta = g^{\alpha\beta} - \varepsilon n^\alpha n^\beta$, and $h^\alpha_{\ \mu} h^\beta_{\ \nu} \cdots T^{\mu\nu\cdots}$ is evidently tangent to the hypersurface.

The projections

$$A_{\alpha\beta\cdots} e_a^\alpha e_b^\beta \cdots = A_{ab\cdots} \equiv h_{am} h_{bn} \cdots A^{mn\cdots} \tag{3.26}$$

give the three-tensor $A^{ab\cdots}$ associated with $A^{\alpha\beta\cdots}$; Latin indices are lowered and raised with h_{ab} and h^{ab}, respectively. Equations (3.25) and (3.26) show that one can easily go back and forth between a tangent tensor field $A^{\alpha\beta\cdots}$ and its equivalent three-tensor $A^{ab\cdots}$. We emphasize that while $A^{ab\cdots}$ behaves as a *tensor* under a transformation $y^a \to y^{a'}$ of the coordinates intrinsic to Σ, it is a *scalar* under a transformation $x^\alpha \to x^{\alpha'}$ of the spacetime coordinates.

3.4.2 Intrinsic covariant derivative

We wish to examine how tangent tensor fields are *differentiated*. We want to relate the covariant derivative of $A^{\alpha\beta\cdots}$ (with respect to a connection that is compatible with the spacetime metric $g_{\alpha\beta}$) to the covariant derivative of $A^{ab\cdots}$, defined in terms of a connection that is compatible with the induced metric h_{ab}. For simplicity we shall restrict our attention to the case of a tangent *vector* field A^α, such that

$$A^\alpha = A^a e_a^\alpha, \quad A^\alpha n_\alpha = 0, \quad A_a = A_\alpha e_a^\alpha.$$

Generalization to three-tensors of higher ranks will be obvious.

We define the *intrinsic covariant derivative* of a three-vector A_a to be the projection of $A_{\alpha;\beta}$ onto the hypersurface:

$$A_{a|b} \equiv A_{\alpha;\beta}\, e_a^\alpha e_b^\beta. \tag{3.27}$$

We will show that $A_{a|b}$, as defined here, is nothing but the covariant derivative of A_a defined in the usual way in terms of a connection $\Gamma^a_{\ bc}$ that is compatible with h_{ab}.

To get started, let us express the right-hand side of Eq. (3.27) as

$$\begin{aligned}
A_{\alpha;\beta}\, e_a^\alpha e_b^\beta &= (A_\alpha e_a^\alpha)_{;\beta}\, e_b^\beta - A_\alpha e_{a;\beta}^\alpha e_b^\beta \\
&= A_{a,\beta}\, e_b^\beta - e_{a\gamma;\beta}\, e_b^\beta A^c e_c^\gamma \\
&= \frac{\partial A_a}{\partial x^\beta}\frac{\partial x^\beta}{\partial y^b} - e_c^\gamma\, e_{a\gamma;\beta}\, e_b^\beta A^c \\
&= A_{a,b} - \Gamma_{cab} A^c,
\end{aligned}$$

where we have defined

$$\Gamma_{cab} = e_c^\gamma\, e_{a\gamma;\beta}\, e_b^\beta. \tag{3.28}$$

Equation (3.27) then reads

$$A_{a|b} = A_{a,b} - \Gamma^c_{\ ab} A_c, \tag{3.29}$$

and this is the familiar expression for the covariant derivative.

The connection used here is the one defined by Eq. (3.28), and we would like to show that it is compatible with the induced metric. In other words, we would like to prove that Γ_{cab}, as defined by Eq. (3.28), can also be expressed as

$$\Gamma_{cab} = \frac{1}{2}\left(h_{ca,b} + h_{cb,a} - h_{ab,c}\right). \tag{3.30}$$

This could be done directly by working out the right-hand side of Eq. (3.28). It is easier, however, to show that the connection is such that $h_{ab|c} \equiv h_{\alpha\beta;\gamma}\, e_a^\alpha e_b^\beta e_c^\gamma = 0$. Indeed,

$$\begin{aligned}
h_{\alpha\beta;\gamma}\, e_a^\alpha e_b^\beta e_c^\gamma &= (g_{\alpha\beta} - \varepsilon n_\alpha n_\beta)_{;\gamma}\, e_a^\alpha e_b^\beta e_c^\gamma \\
&= -\varepsilon(n_{\alpha;\gamma} n_\beta + n_\alpha n_{\beta;\gamma})\, e_a^\alpha e_b^\beta e_c^\gamma \\
&= 0,
\end{aligned}$$

because $n_\alpha e_a^\alpha = 0$. Intrinsic covariant differentiation is therefore the same operation as straightforward covariant differentiation of a three-tensor.

3.4.3 Extrinsic curvature

The quantities $A_{a|b} = A_{\alpha;\beta} e_a^\alpha e_b^\beta$ are the tangential components of the vector $A^\alpha_{;\beta} e_b^\beta$. The question we would like to investigate now is whether this vector possesses also a normal component.

To answer this we re-express $A^\alpha_{;\beta} e_b^\beta$ as $g^\alpha_{\mu} A^\mu_{;\beta} e_b^\beta$ and decompose the metric into its normal and tangential parts, as in Eq. (3.12). This gives

$$A^\alpha_{;\beta} e_b^\beta = \left(\varepsilon n^\alpha n_\mu + h^{am} e_a^\alpha e_{m\mu}\right) A^\mu_{;\beta} e_b^\beta$$
$$= \varepsilon\left(n_\mu A^\mu_{;\beta} e_b^\beta\right) n^\alpha + h^{am}\left(A_{\mu;\beta} e_m^\mu e_b^\beta\right) e_a^\alpha,$$

and we see that while the second term is tangent to the hypersurface, the first term is normal to it. We now use Eq. (3.27) and the fact that A^μ is orthogonal to n^μ:

$$A^\alpha_{;\beta} e_b^\beta = -\varepsilon\left(n_{\mu;\beta} A^\mu e_b^\beta\right) n^\alpha + h^{am} A_{m|b} e_a^\alpha$$
$$= A^a_{|b}\, e_a^\alpha - \varepsilon A^a\left(n_{\mu;\beta} e_a^\mu e_b^\beta\right) n^\alpha.$$

At this point we introduce the three-tensor

$$K_{ab} \equiv n_{\alpha;\beta}\, e_a^\alpha e_b^\beta, \tag{3.31}$$

called the *extrinsic curvature*, or *second fundamental form*, of the hypersurface Σ. In terms of this we have

$$A^\alpha_{;\beta} e_b^\beta = A^a_{|b}\, e_a^\alpha - \varepsilon A^a K_{ab} n^\alpha, \tag{3.32}$$

and we see that $A^a_{|b}$ gives the purely tangential part of the vector field, while $-\varepsilon A^a K_{ab}$ represents the normal component. This answers our question: The normal component vanishes if and only if the extrinsic curvature vanishes.

We note that if e_a^α is substituted in place of A^α, then $A^c = \delta^c_a$ and Eqs. (3.29), (3.32) imply

$$e^\alpha_{a;\beta} e_b^\beta = \Gamma^c_{ab}\, e_c^\alpha - \varepsilon K_{ab} n^\alpha. \tag{3.33}$$

This is known as the Gauss–Weingarten equation.

The extrinsic curvature is a very important quantity; we will encounter it often in the remaining sections of this book. We may prove that it is a *symmetric* tensor:

$$K_{ba} = K_{ab}. \tag{3.34}$$

The proof is based on the properties that (i) the vectors e_a^α and n^α are orthogonal, and (ii) the basis vectors are Lie transported along one another, so that

$e^{\alpha}_{a;\beta}e^{\beta}_b = e^{\alpha}_{b;\beta}e^{\beta}_a$. We have

$$n_{\alpha;\beta}e^{\alpha}_a e^{\beta}_b = -n_{\alpha}e^{\alpha}_{a;\beta}e^{\beta}_b$$
$$= -n_{\alpha}e^{\alpha}_{b;\beta}e^{\beta}_a$$
$$= n_{\alpha;\beta}e^{\alpha}_b e^{\beta}_a,$$

and Eq. (3.34) follows. The symmetric property of the extrinsic curvature gives rise to the relations

$$K_{ab} = n_{(\alpha;\beta)}e^{\alpha}_a e^{\beta}_b = \frac{1}{2}\left(\pounds_n g_{\alpha\beta}\right)e^{\alpha}_a e^{\beta}_b, \tag{3.35}$$

and K_{ab} is therefore intimately related to the normal derivative of the metric tensor.
We also note the relation

$$K \equiv h^{ab}K_{ab} = n^{\alpha}_{\;;\alpha}, \tag{3.36}$$

which shows that K is equal to the expansion of a congruence of geodesics that intersect the hypersurface orthogonally (so that their tangent vector is equal to n^{α} on the hypersurface). From this result we conclude that the hypersurface is *convex* if $K > 0$ (the congruence is diverging), or *concave* if $K < 0$ (the congruence is converging).

We see that while h_{ab} is concerned with the purely *intrinsic* aspects of a hypersurface's geometry, K_{ab} is concerned with the *extrinsic* aspects – the way in which the hypersurface is embedded in the enveloping spacetime manifold. Taken together, these tensors provide a virtually complete characterization of the hypersurface.

3.5 Gauss–Codazzi equations

3.5.1 General form

We have introduced the induced metric h_{ab} and its associated intrinsic covariant derivative. A purely intrinsic curvature tensor can now be defined by the relation

$$A^c_{\;|ab} - A^c_{\;|ba} = -R^c_{\;dab}A^d, \tag{3.37}$$

which of course implies

$$R^c_{\;dab} = \Gamma^c_{\;db,a} - \Gamma^c_{\;da,b} + \Gamma^c_{\;ma}\Gamma^m_{\;db} - \Gamma^c_{\;mb}\Gamma^m_{\;da}. \tag{3.38}$$

The question we now examine is whether this three-dimensional Riemann tensor can be expressed in terms of $R^{\gamma}_{\;\delta\alpha\beta}$ – the four-dimensional version – evaluated on Σ.

To answer this we start with the identity

$$\left(e^\alpha_{a;\beta} e^\beta_b\right)_{;\gamma} e^\gamma_c = \left(\Gamma^d{}_{ab} e^\alpha_d - \varepsilon K_{ab} n^\alpha\right)_{;\gamma} e^\gamma_c$$

which follows immediately from Eq. (3.33). We first develop the left-hand side:

$$\begin{aligned}
\text{LHS} &= \left(e^\alpha_{a;\beta} e^\beta_b\right)_{;\gamma} e^\gamma_c \\
&= e^\alpha_{a;\beta\gamma} e^\beta_b e^\gamma_c + e^\alpha_{a;\beta} e^\beta_{b;\gamma} e^\gamma_c \\
&= e^\alpha_{a;\beta\gamma} e^\beta_b e^\gamma_c + e^\alpha_{a;\beta} \left(\Gamma^d{}_{bc} e^\beta_d - \varepsilon K_{bc} n^\beta\right) \\
&= e^\alpha_{a;\beta\gamma} e^\beta_b e^\gamma_c + \Gamma^d{}_{bc} \left(\Gamma^e{}_{ad} e^\alpha_e - \varepsilon K_{ad} n^\alpha\right) - \varepsilon K_{bc} e^\alpha_{a;\beta} n^\beta.
\end{aligned}$$

Next we turn to the right-hand side:

$$\begin{aligned}
\text{RHS} &= \left(\Gamma^d{}_{ab} e^\alpha_d - \varepsilon K_{ab} n^\alpha\right)_{;\gamma} e^\gamma_c \\
&= \Gamma^d{}_{ab,c} e^\alpha_d + \Gamma^d{}_{ab} e^\alpha_{d;\gamma} e^\gamma_c - \varepsilon K_{ab,c} n^\alpha - \varepsilon K_{ab} n^\alpha_{;\gamma} e^\gamma_c \\
&= \Gamma^d{}_{ab,c} e^\alpha_d + \Gamma^d{}_{ab} \left(\Gamma^e{}_{dc} e^\alpha_e - \varepsilon K_{dc} n^\alpha\right) - \varepsilon K_{ab,c} n^\alpha - \varepsilon K_{ab} n^\alpha_{;\gamma} e^\gamma_c.
\end{aligned}$$

We now equate the two sides and solve for $e^\alpha_{a;\beta} e^\beta_b e^\gamma_c$. Subtracting a similar expression for $e^\alpha_{a;\gamma\beta} e^\gamma_c e^\beta_b$ gives $-R^\alpha{}_{\mu\beta\gamma} e^\mu_a e^\beta_b e^\gamma_c$, the quantity in which we are interested. After some algebra we find

$$R^\mu{}_{\alpha\beta\gamma} e^\alpha_a e^\beta_b e^\gamma_c = R^m{}_{abc} e^\mu_m + \varepsilon\left(K_{ab|c} - K_{ac|b}\right) n^\mu + \varepsilon K_{ab} n^\mu_{;\gamma} e^\gamma_c - \varepsilon K_{ac} n^\mu_{;\beta} e^\beta_b.$$

Projecting along $e_{d\mu}$ gives

$$R_{\alpha\beta\gamma\delta} e^\alpha_a e^\beta_b e^\gamma_c e^\delta_d = R_{abcd} + \varepsilon(K_{ad} K_{bc} - K_{ac} K_{bd}), \qquad (3.39)$$

and this is the desired relation between R_{abcd} and the full Riemann tensor. Projecting instead along n_μ gives

$$R_{\mu\alpha\beta\gamma} n^\mu e^\alpha_a e^\beta_b e^\gamma_c = K_{ab|c} - K_{ac|b}. \qquad (3.40)$$

Equations (3.39) and (3.40) are known as the *Gauss–Codazzi* equations. They reveal that some components of the spacetime curvature tensor can be expressed in terms of the intrinsic and extrinsic curvatures of a hypersurface. The missing components are $R_{\mu\alpha\nu\beta} n^\mu e^\alpha_a n^\nu e^\beta_b$, and these cannot be expressed solely in terms of h_{ab}, K_{ab}, and related quantities.

3.5.2 Contracted form

The Gauss–Codazzi equations can also be written in contracted form, in terms of the Einstein tensor $G_{\alpha\beta} = R_{\alpha\beta} - \frac{1}{2}Rg_{\alpha\beta}$. The spacetime Ricci tensor is given by

$$
\begin{aligned}
R_{\alpha\beta} &= g^{\mu\nu} R_{\mu\alpha\nu\beta} \\
&= \left(\varepsilon n^{\mu} n^{\nu} + h^{mn} e_m^{\mu} e_n^{\nu}\right) R_{\mu\alpha\nu\beta} \\
&= \varepsilon R_{\mu\alpha\nu\beta} n^{\mu} n^{\nu} + h^{mn} R_{\mu\alpha\nu\beta} e_m^{\mu} e_n^{\nu},
\end{aligned}
$$

and the Ricci scalar is

$$
\begin{aligned}
R &= g^{\alpha\beta} R_{\alpha\beta} \\
&= \left(\varepsilon n^{\alpha} n^{\beta} + h^{ab} e_a^{\alpha} e_b^{\beta}\right)\left(\varepsilon R_{\mu\alpha\nu\beta} n^{\mu} n^{\nu} + h^{mn} R_{\mu\alpha\nu\beta} e_m^{\mu} e_n^{\nu}\right) \\
&= 2\varepsilon h^{ab} R_{\mu\alpha\nu\beta} n^{\mu} e_a^{\alpha} n^{\nu} e_b^{\beta} + h^{ab} h^{mn} R_{\mu\alpha\nu\beta} e_m^{\mu} e_a^{\alpha} e_n^{\nu} e_b^{\beta}.
\end{aligned}
$$

A little algebra then reveals the relations

$$
-2\varepsilon G_{\alpha\beta} n^{\alpha} n^{\beta} = {}^3R + \varepsilon\left(K^{ab} K_{ab} - K^2\right) \tag{3.41}
$$

and

$$
G_{\alpha\beta} e_a^{\alpha} n^{\beta} = K^b{}_{a|b} - K_{,a}. \tag{3.42}
$$

Here, ${}^3R = h^{ab} R^m{}_{amb}$ is the three-dimensional Ricci scalar. The importance of Eqs. (3.41) and (3.42) lies with the fact that they form part of the Einstein field equations on a hypersurface Σ; this observation will be elaborated in the next section. We note that $G_{\alpha\beta} e_a^{\alpha} e_b^{\beta}$, the remaining components of the Einstein tensor, cannot be expressed solely in terms of h_{ab}, K_{ab}, and related quantities.

3.5.3 Ricci scalar

We now complete the computation of the four-dimensional Ricci scalar. Our starting point is the relation

$$
R = 2\varepsilon h^{ab} R_{\mu\alpha\nu\beta} n^{\mu} e_a^{\alpha} n^{\nu} e_b^{\beta} + h^{ab} h^{mn} R_{\mu\alpha\nu\beta} e_m^{\mu} e_a^{\alpha} e_n^{\nu} e_b^{\beta},
$$

which was derived previously. The first term is simplified by using the completeness relations (3.12) and the fact that $R_{\mu\alpha\nu\beta} n^{\mu} n^{\alpha} n^{\nu} n^{\beta} = 0$; it becomes $2\varepsilon R_{\alpha\beta} n^{\alpha} n^{\beta}$. Using the definition of the Riemann tensor, we rewrite this as

$$
\begin{aligned}
R_{\alpha\beta} n^{\alpha} n^{\beta} &= -n^{\alpha}{}_{;\alpha\beta} n^{\beta} + n^{\alpha}{}_{;\beta\alpha} n^{\beta} \\
&= -(n^{\alpha}{}_{;\alpha} n^{\beta})_{;\beta} + n^{\alpha}{}_{;\alpha} n^{\beta}{}_{;\beta} + (n^{\alpha}{}_{;\beta} n^{\beta})_{;\alpha} - n^{\alpha}{}_{;\beta} n^{\beta}{}_{;\alpha}.
\end{aligned}
$$

In the second term of this last expression we recognize K^2, where $K = n^\alpha_{;\alpha}$ is the trace of the extrinsic curvature. The fourth term, on the other hand, can be expressed as

$$
\begin{aligned}
n^\alpha_{;\beta} n^\beta_{;\alpha} &= g^{\beta\mu} g^{\alpha\nu} n_{\alpha;\beta} n_{\mu;\nu} \\
&= (\varepsilon n^\beta n^\mu + h^{\beta\mu})(\varepsilon n^\alpha n^\nu + h^{\alpha\nu}) n_{\alpha;\beta} n_{\mu;\nu} \\
&= (\varepsilon n^\beta n^\mu + h^{\beta\mu}) h^{\alpha\nu} n_{\alpha;\beta} n_{\mu;\nu} \\
&= h^{\beta\mu} h^{\alpha\nu} n_{\alpha;\beta} n_{\mu;\nu} \\
&= h^{bm} h^{an} n_{\alpha;\beta} e^\alpha_a e^\beta_b n_{\mu;\nu} e^\mu_m e^\nu_n \\
&= h^{bm} h^{an} K_{ab} K_{mn} \\
&= K_{ab} K^{ba} \\
&= K^{ab} K_{ab}.
\end{aligned}
$$

In the second line we have inserted the completeness relations (3.12) and recalled the notation $h^{\alpha\beta} = h^{ab} e^\alpha_a e^\beta_b$. In the third and fourth lines we have used the fact that $n^\alpha n_{\alpha;\beta} = \frac{1}{2}(n^\alpha n_\alpha)_{;\beta} = 0$. In the sixth line we have substituted the definition (3.31) for the extrinsic curvature. Finally, in the last line we have used the fact that K_{ab} is a symmetric three-tensor.

The previous manipulations take care of the first term in our starting expression for the Ricci scalar. The second term is simplified by substituting the Gauss–Codazzi equations (3.39),

$$
\begin{aligned}
h^{ab} h^{mn} R_{\mu\alpha\nu\beta} e^\mu_m e^\alpha_a e^\nu_n e^\beta_b &= h^{ab} h^{mn} \left[R_{manb} + \varepsilon(K_{mb} K_{an} - K_{mn} K_{ab}) \right] \\
&= {}^3R + \varepsilon(K^{ab} K_{ab} - K^2).
\end{aligned}
$$

Putting all this together, we arrive at

$$
R = {}^3R + \varepsilon\left(K^2 - K^{ab} K_{ab}\right) + 2\varepsilon\left(n^\alpha_{;\beta} n^\beta - n^\alpha n^\beta_{;\beta}\right)_{;\alpha}. \tag{3.43}
$$

This is the four-dimensional Ricci scalar evaluated on the hypersurface Σ. This result will be put to good use in Chapter 4.

3.6 Initial-value problem

3.6.1 Constraints

In Newtonian mechanics, a complete solution to the equations of motion requires the specification of initial values for the position and velocity of each moving body. In field theories, a complete solution to the field equations requires the specification of the field and its time derivative at one instant of time.

A similar statement can be made for general relativity. Because the Einstein field equations are second-order partial differential equations, we would expect that a complete solution requires the specification of $g_{\alpha\beta}$ and $g_{\alpha\beta,t}$ at one instant of time. While this is essentially correct, it is desirable to convert this decidedly noncovariant statement into something more geometrical.

The *initial-value problem* of general relativity starts with the selection of a spacelike hypersurface Σ which represents an 'instant of time.' This hypersurface can be chosen freely. On this hypersurface we place arbitrary coordinates y^a.

The spacetime metric $g_{\alpha\beta}$, when evaluated on Σ, has components that characterize displacements *away from* the hypersurface. (For example, g_{tt} is such a component if Σ is a surface of constant t.) These components cannot be given meaning in terms of the geometric properties of Σ alone. To provide meaningful initial values for the spacetime metric, we must consider displacements *within* the hypersurface only. In other words, the initial values for $g_{\alpha\beta}$ can only be the six components of the *induced metric* $h_{ab} = g_{\alpha\beta}\, e_a^\alpha e_b^\beta$; the remaining four components are arbitrary, and this reflects the complete freedom in choosing the spacetime coordinates x^α.

Similarly, the initial values for the 'time derivative' of the metric must be described by a three-tensor that carries information about the derivative of the metric in the direction normal to the hypersurface. Because $K_{ab} = \frac{1}{2}(\pounds_n g_{\alpha\beta})\, e_a^\alpha e_b^\beta$, the extrinsic curvature is clearly an appropriate choice.

The *initial-value problem* of general relativity therefore consists in specifying two symmetric tensor fields, h_{ab} and K_{ab}, on a spacelike hypersurface Σ. In the complete spacetime, h_{ab} is recognized as the induced metric on the hypersurface, while K_{ab} is the extrinsic curvature. These tensors cannot be chosen freely: They must satisfy the *constraint equations* of general relativity. These are given by Eqs. (3.41) and (3.42), together with the Einstein field equations $G_{\alpha\beta} = 8\pi T_{\alpha\beta}$:

$$^3R + K^2 - K^{ab}K_{ab} = 16\pi T_{\alpha\beta}n^\alpha n^\beta \equiv 16\pi\rho \tag{3.44}$$

and

$$K^b{}_{a|b} - K_{,a} = 8\pi T_{\alpha\beta}e_a^\alpha n^\beta \equiv 8\pi j_a. \tag{3.45}$$

The remaining components of the Einstein field equations provide *evolution equations* for h_{ab} and K_{ab}; these will be considered in Chapter 4.

3.6.2 Cosmological initial values

As an example, let us solve the constraint equations for a spatially flat, isotropic, and homogeneous cosmology. To satisfy these requirements the three-metric must take the form

$$ds^2 = a^2(dx^2 + dy^2 + dz^2),$$

where a is the scale factor, which is a constant on the hypersurface. Isotropy and homogeneity also imply ρ = constant, $j_a = 0$, and

$$K_{ab} = \frac{1}{3} K h_{ab},$$

where K is a constant. The second constraint equation is therefore trivially satisfied. The first one implies

$$16\pi\rho = K^2 - K^{ab} K_{ab} = \frac{2}{3} K^2,$$

and this provides the complete solution to the initial-value problem.

To recognize the physical meaning of this last equation, we use the fact that in the complete spacetime, $K = n^\alpha{}_{;\alpha}$, where n^α is the unit normal to surfaces of constant t. The full metric is given by the Friedmann–Robertson–Walker form

$$ds^2 = -dt^2 + a^2(t)(dx^2 + dy^2 + dz^2),$$

so that $n_\alpha = -\partial_\alpha t$ and $K = 3\dot{a}/a$, where an overdot indicates differentiation with respect to t. The first constraint equation is therefore equivalent to

$$3(\dot{a}/a)^2 = 8\pi\rho,$$

which is one of the Friedmann equations governing the evolution of the scale factor.

3.6.3 Moment of time symmetry

We notice from the previous example that $K_{ab} = 0$ when $\dot{a} = 0$, that is, the extrinsic curvature vanishes when the scale factor reaches a turning point of its evolution. Because the dynamical history of the scale factor is time-symmetric about the time $t = t_0$ at which the turning point occurs, we may call this time a *moment of time symmetry* in the dynamical evolution of the spacetime. Thus, $K_{ab} = 0$ at this moment of time symmetry.

Generalizing, we shall call *any* hypersurface Σ on which $K_{ab} = 0$ a moment of time symmetry in spacetime. Because K_{ab} is essentially the 'time derivative' of the metric, a moment of time symmetry corresponds to a turning point of the metric's evolution, at which its 'time derivative' vanishes. The dynamical history of the metric is then 'time-symmetric' about Σ. From Eq. (3.45) we see that a moment of time symmetry can occur only if $j_a = 0$ on that hypersurface.

3.6.4 Stationary and static spacetimes

A spacetime is said to be *stationary* if it admits a timelike Killing vector t^α. This means that in a coordinate system (t, x^a) in which $t^\alpha \overset{*}{=} \delta^\alpha_t$, the metric does not depend on the time coordinate t: $g_{\alpha\beta,t} \overset{*}{=} 0$ (see Section 1.5). For example, a rotating star gives rise to a stationary spacetime if its mass and angular velocity do not change with time.

A stationary spacetime is also *static* if the metric does not change under a time reversal, $t \to -t$. For example, the spacetime of a rotating star is not static because a time reversal changes the direction of rotation. In the specified coordinate system, invariance of the metric under a time reversal implies $g_{ta} \overset{*}{=} 0$. This, in turn, implies that the Killing vector is proportional to a gradient: $t_\alpha \overset{*}{=} g_{tt}\partial_\alpha t$. Thus, *a spacetime is static if the timelike Killing vector field is hypersurface orthogonal.*

We may show that if a spacetime is static, then $K_{ab} = 0$ on those hypersurfaces Σ that are orthogonal to the Killing vector; these hypersurfaces therefore represent moments of time symmetry. If Σ is orthogonal to t^α, then its unit normal must be given by $n_\alpha = \mu\, t_\alpha$, where $1/\mu^2 = -t^\alpha t_\alpha$. This implies that $n_{\alpha;\beta} = \mu\, t_{\alpha;\beta} + t_\alpha \mu_{,\beta}$, and $n_{(\alpha;\beta)} = t_{(\alpha}\mu_{,\beta)}$ because t_α is a Killing vector. That $K_{ab} = 0$ follows immediately from Eq. (3.35) and the fact that t_α is orthogonal to e^α_a.

3.6.5 Spherical space, moment of time symmetry

As a second example, we solve the constraint equations for a spherically symmetric spacetime at a moment of time symmetry. The three-metric can be expressed as

$$ds^2 = \left[1 - 2m(r)/r\right]^{-1} dr^2 + r^2 d\Omega^2,$$

for some function $m(r)$; to enforce regularity of the metric at $r = 0$ we must impose $m(0) = m'(0) = 0$, with a prime denoting differentiation with respect to r. For this metric the Ricci scalar is given by $^3R = 4m'/r^2$. Because $K_{ab} = 0$ at a moment of time symmetry, Eq. (3.44) implies $16\pi\rho = {}^3R$. Solving for $m(r)$ returns

$$m(r) = \int_0^r 4\pi r'^2 \rho(r')\, dr'.$$

This states, loosely speaking, that $m(r)$ is the mass-energy contained inside a sphere of radius r, at the selected moment of time symmetry.

3.6.6 Spherical space, empty and flat

We now solve the constraint equations for a spherically symmetric space empty of matter (so that $\rho = 0 = j^a$). We assume that we can endow this space with a flat metric, so that

$$h_{ab}\, dy^a\, dy^b = dr^2 + r^2\, d\Omega^2.$$

We also assume that the hypersurface does *not* represent a moment of time symmetry. While the flat metric and $K_{ab} = 0$ make a valid solution to the constraints, this is a trivial configuration – a flat hypersurface in a flat spacetime.

Let $n_a = \partial_a r$ be a unit vector that points radially outward on the hypersurface. The fact that K_{ab} is a spherically symmetric tensor means that it can be decomposed as

$$K_{ab} = K_1(r)n_a n_b + K_2(r)(h_{ab} - n_a n_b),$$

with K_1 representing the radial component of the extrinsic curvature, and K_2 the angular components. In the usual spherical coordinates (r, θ, ϕ) we have $K^a{}_b = \mathrm{diag}(K_1, K_2, K_2)$, which is the most general expression admissible under the assumption of spherical symmetry.

Because the space is empty and flat, the first constraint equation reduces to $K^2 - K^{ab}K_{ab} = 0$, an algebraic equation for K_1 and K_2. This gives us the condition $(2K_1 + K_2)K_2 = 0$. Choosing $K_2 = 0$ would eventually return the trivial solution $K_{ab} = 0$. We choose instead $K_2 = -2K_1$ and re-express the extrinsic curvature as

$$K_{ab} = K(r)\left(\frac{2}{3} h_{ab} - n_a n_b\right),$$

where $K = -3K_1$ is the sole remaining function to be determined.

To find $K(r)$ we turn to the second constraint equation, $K^b{}_{a|b} - K_{,a} = 0$, which becomes

$$\frac{1}{3} K_{,a} + \left(K n^b{}_{|b} + K_{,b} n^b\right) n_a + K n_{a|b} n^b = 0.$$

With $K_{,a} = K' n_a$ (with a prime denoting differentiation with respect to r), $n^b{}_{|b} = 2/r$, and $n_{a|b} n^b = 0$ (because the radial curves are geodesics of the hypersurface), we arrive at $2r K' + 3K = 0$. Integration yields

$$K(r) = K_0 (r_0/r)^{3/2},$$

with K_0 denoting the value of K at the arbitrary radius r_0.

We have found a nontrivial solution to the constraint equations for a spherical space that is both empty and flat. The physical meaning of this configuration will be revealed in Section 3.13, Problem 1.

3.6.7 Conformally-flat space

A powerful technique for generating solutions to the constraint equations consists of writing the three-metric as

$$h_{ab} = \psi^4 \delta_{ab},$$

where $\psi(y^a)$ is a scalar field on the hypersurface. Such a metric is said to be conformally related to the flat metric, and the space is said to be conformally flat. For this metric the Ricci scalar is $^3R = -8\psi^{-5}\nabla^2\psi$, and Eq. (3.44) takes the form of Poisson's equation,

$$\nabla^2\psi = -2\pi\rho_{\text{eff}},$$

where

$$\rho_{\text{eff}} = \psi^5\left[\rho + \frac{1}{16\pi}\left(K^{ab}K_{ab} - K^2\right)\right]$$

is an effective mass density on the hypersurface. At a moment of time symmetry this simplifies to $\rho_{\text{eff}} = \psi^5\rho$, and one possible strategy for solving the constraint is to specify ρ_{eff}, solve for ψ, and then see what this produces for the actual mass density ρ. If $\rho = 0$ at the moment of time symmetry, then the constraint becomes Laplace's equation $\nabla^2\psi = 0$, and this admits many interesting solutions. A well-known example is Misner's (1960) solution, which describes two black holes about to undergo a head-on collision. This initial data set has been vigourously studied by numerical relativists.

3.7 Junction conditions and thin shells

The following situation sometimes presents itself: A hypersurface Σ partitions spacetime into two regions $\mathscr{V}^+$ and $\mathscr{V}^-$ (Fig. 3.5). In $\mathscr{V}^+$ the metric is $g^+_{\alpha\beta}$, and it is expressed in a system of coordinates x^α_+. In $\mathscr{V}^-$ the metric is $g^-_{\alpha\beta}$, and it is expressed in coordinates x^α_-. We ask: What conditions must be put on the metrics to ensure that $\mathscr{V}^+$ and $\mathscr{V}^-$ are joined *smoothly* at Σ, so that the union of $g^+_{\alpha\beta}$ and $g^-_{\alpha\beta}$ forms a valid solution to the Einstein field equations? To answer this question is not entirely straightforward because in practical situations, the coordinate systems $x^\alpha_\pm$ will often be different, and it may not be possible to compare the metrics directly. To circumvent this difficulty we will endeavour to formulate *junction conditions* that involve only three-tensors on Σ. In this section we will assume that Σ is either timelike or spacelike; we will return to the case of a null hypersurface in Section 3.11.

3.7.1 Notation and assumptions

We assume that the same coordinates y^a can be installed on both sides of the hypersurface, and we choose n^α, the unit normal to Σ, to point from $\mathscr{V}^-$ to $\mathscr{V}^+$. We suppose that a continuous coordinate system x^α, distinct from $x^\alpha_\pm$, can be introduced on both sides of the hypersurface. These coordinates overlap with x^α_+ in an open region of $\mathscr{V}^+$ that contains Σ, and they also overlap with x^α_- in an open region of $\mathscr{V}^-$ that contains Σ. (We introduce these coordinates for our short-term convenience only; the final formulation of the junction conditions will not involve them.)

We imagine Σ to be pierced by a congruence of geodesics that intersect it orthogonally. We take ℓ to denote proper distance (or proper time) along the geodesics, and we adjust the parameterization so that $\ell = 0$ when the geodesics cross the hypersurface; our convention is that ℓ is negative in $\mathscr{V}^-$ and positive in $\mathscr{V}^+$. We can think of ℓ as a scalar field: The point P identified by the coordinates x^α is linked to Σ by a member of the congruence, and $\ell(x^\alpha)$ is the proper distance (or proper time) from Σ to P along this geodesic. Our construction implies that a displacement away from the hypersurface along one of the geodesics is described by $dx^\alpha = n^\alpha\,d\ell$, and that

$$n_\alpha = \varepsilon\partial_\alpha\ell; \tag{3.46}$$

we also have $n^\alpha n_\alpha = \varepsilon$.

We will use the language of *distributions*. We introduce the Heaviside distribution $\Theta(\ell)$, equal to $+1$ if $\ell > 0$, 0 if $\ell < 0$, and indeterminate if $\ell = 0$. We note the following properties:

$$\Theta^2(\ell) = \Theta(\ell), \qquad \Theta(\ell)\Theta(-\ell) = 0, \qquad \frac{d}{d\ell}\Theta(\ell) = \delta(\ell),$$

where $\delta(\ell)$ is the Dirac distribution. We also note that the product $\Theta(\ell)\delta(\ell)$ is *not defined* as a distribution.

The following notation will be useful:

$$[A] \equiv A(\mathscr{V}^+)\big|_\Sigma - A(\mathscr{V}^-)\big|_\Sigma,$$

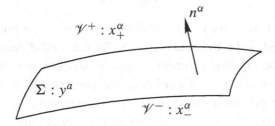

Figure 3.5 Two regions of spacetime joined at a common boundary.

where A is any tensorial quantity defined on both sides of the hypersurface; $[A]$ is therefore the *jump* of A across Σ. We note the relations

$$[n^\alpha] = [e_a^\alpha] = 0, \tag{3.47}$$

where $e_a^\alpha = \partial x^\alpha / \partial y^a$. The first follows from the relation $dx^\alpha = n^\alpha \, d\ell$ and the continuity of both ℓ and x^α across Σ; the second follows from the fact that the coordinates y^a are the same on both sides of the hypersurface.

3.7.2 First junction condition

We begin by expressing the metric $g_{\alpha\beta}$, in the coordinates x^α, as a distribution-valued tensor:

$$g_{\alpha\beta} = \Theta(\ell) \, g_{\alpha\beta}^+ + \Theta(-\ell) \, g_{\alpha\beta}^-, \tag{3.48}$$

where $g_{\alpha\beta}^\pm$ is the metric in $\mathscr{V}^\pm$ expressed in the coordinates x^α. We want to know if the metric of Eq. (3.48) makes a valid distributional solution to the Einstein field equations. To decide we must verify that geometrical quantities constructed from $g_{\alpha\beta}$, such as the Riemann tensor, are properly defined as distributions. We must then try to eliminate, or at least give an interpretation to, singular terms that might arise in these geometric quantities.

Differentiating Eq. (3.48) yields

$$g_{\alpha\beta,\gamma} = \Theta(\ell) \, g_{\alpha\beta,\gamma}^+ + \Theta(-\ell) \, g_{\alpha\beta,\gamma}^- + \varepsilon\delta(\ell)[g_{\alpha\beta}]n_\gamma,$$

where Eq. (3.46) was used. The last term is singular and it causes problems when we compute the Christoffel symbols, because it generates terms proportional to $\Theta(\ell)\delta(\ell)$. If the last term were allowed to survive, the connection would not be defined as a distribution and our program would fail. To eliminate this term we impose continuity of the metric across the hypersurface: $[g_{\alpha\beta}] = 0$. This statement holds in the coordinate system x^α only. However, we can easily turn this into a coordinate-invariant statement: $0 = [g_{\alpha\beta}]e_a^\alpha e_b^\beta = [g_{\alpha\beta}e_a^\alpha e_b^\beta]$; this last step follows by virtue of Eq. (3.47). We have obtained

$$[h_{ab}] = 0, \tag{3.49}$$

the statement that the *induced metric* must be the *same* on both sides of Σ. This is clearly required if the hypersurface is to have a well-defined geometry. Equation (3.49) will be our first junction condition, and it is expressed *independently* of the coordinates x^α or $x_\pm^\alpha$. Coordinate independence explains why Eq. (3.49) produces only six conditions while the original statement $[g_{\alpha\beta}] = 0$ contained ten: The mismatch corresponds to the four coordinate conditions $[x^\alpha] = 0$.

3.7.3 Riemann tensor

To find the second junction condition requires more work: we must calculate the distribution-valued Riemann tensor. Using the results obtained thus far, we have that the Christoffel symbols are

$$\Gamma^\alpha{}_{\beta\gamma} = \Theta(\ell)\,\Gamma^{+\alpha}_{\beta\gamma} + \Theta(-\ell)\,\Gamma^{-\alpha}_{\beta\gamma},$$

where $\Gamma^{\pm\alpha}_{\beta\gamma}$ are the Christoffel symbols constructed from $g^\pm_{\alpha\beta}$. A straightforward calculation then reveals

$$\Gamma^\alpha{}_{\beta\gamma,\delta} = \Theta(\ell)\,\Gamma^{+\alpha}_{\beta\gamma,\delta} + \Theta(-\ell)\,\Gamma^{-\alpha}_{\beta\gamma,\delta} + \varepsilon\delta(\ell)\big[\Gamma^\alpha{}_{\beta\gamma}\big]n_\delta,$$

and from this follows the Riemann tensor:

$$R^\alpha{}_{\beta\gamma\delta} = \Theta(\ell)\,R^{+\alpha}_{\beta\gamma\delta} + \Theta(-\ell)\,R^{-\alpha}_{\beta\gamma\delta} + \delta(\ell)A^\alpha{}_{\beta\gamma\delta}, \tag{3.50}$$

where

$$A^\alpha{}_{\beta\gamma\delta} = \varepsilon\big(\big[\Gamma^\alpha{}_{\beta\delta}\big]n_\gamma - \big[\Gamma^\alpha{}_{\beta\gamma}\big]n_\delta\big). \tag{3.51}$$

We see that the Riemann tensor is properly defined as a distribution, but the δ-function term represents a curvature singularity at Σ. Our second junction condition will seek to eliminate this term. Failing this, we will see that a physical interpretation can nevertheless be given to the singularity. This is our next topic.

3.7.4 Surface stress-energy tensor

Although they are constructed from Christoffel symbols, the quantities $A^\alpha{}_{\beta\gamma\delta}$ form a tensor because the *difference* between two sets of Christoffel symbols is a tensorial quantity (see Section 1.2). We would like to find an explicit expression for this tensor.

The fact that the metric is continuous across Σ in the coordinates x^α implies that its tangential derivatives also must be continuous. This means that if $g_{\alpha\beta,\gamma}$ is to be discontinuous, the discontinuity must be directed along the normal vector n^α. There must therefore exist a tensor field $\kappa_{\alpha\beta}$ such that

$$\big[g_{\alpha\beta,\gamma}\big] = \kappa_{\alpha\beta}\,n_\gamma; \tag{3.52}$$

this tensor is given explicitly by

$$\kappa_{\alpha\beta} = \varepsilon\big[g_{\alpha\beta,\gamma}\big]n^\gamma. \tag{3.53}$$

Equation (3.52) implies

$$\big[\Gamma^\alpha{}_{\beta\gamma}\big] = \frac{1}{2}\big(\kappa^\alpha{}_\beta n_\gamma + \kappa^\alpha{}_\gamma n_\beta - \kappa_{\beta\gamma}n^\alpha\big),$$

and we obtain

$$A^\alpha_{\ \beta\gamma\delta} = \frac{\varepsilon}{2}\left(\kappa^\alpha_{\ \delta}n_\beta n_\gamma - \kappa^\alpha_{\ \gamma}n_\beta n_\delta - \kappa_{\beta\delta}n^\alpha n_\gamma + \kappa_{\beta\gamma}n^\alpha n_\delta\right).$$

This is the δ-function part of the Riemann tensor.

Contracting over the first and third indices gives the δ-function part of the Ricci tensor:

$$A_{\alpha\beta} \equiv A^\mu_{\ \alpha\mu\beta} = \frac{\varepsilon}{2}\left(\kappa_{\mu\alpha}n^\mu n_\beta + \kappa_{\mu\beta}n^\mu n_\alpha - \kappa n_\alpha n_\beta - \varepsilon\kappa_{\alpha\beta}\right),$$

where $\kappa \equiv \kappa^\alpha_{\ \alpha}$. After an additional contraction we obtain the δ-function part of the Ricci scalar,

$$A \equiv A^\alpha_{\ \alpha} = \varepsilon\left(\kappa_{\mu\nu}n^\mu n^\nu - \varepsilon\kappa\right).$$

With this we form the δ-function part of the Einstein tensor, and after using the Einstein field equations we obtain an expression for the stress-energy tensor:

$$T_{\alpha\beta} = \Theta(\ell)\,T^+_{\alpha\beta} + \Theta(-\ell)\,T^-_{\alpha\beta} + \delta(\ell)S_{\alpha\beta}, \tag{3.54}$$

where $8\pi S_{\alpha\beta} \equiv A_{\alpha\beta} - \frac{1}{2}Ag_{\alpha\beta}$. On the right-hand side of Eq. (3.54) the first and second terms represent the stress-energy tensors of regions $\mathscr{V}^+$ and $\mathscr{V}^-$, respectively. The δ-function term, on the other hand, comes with a clear interpretation: It is associated with the presence of a thin distribution of matter – a *surface layer*, or a *thin shell* – at Σ; this thin shell has a surface stress-energy tensor equal to $S_{\alpha\beta}$.

3.7.5 Second junction condition

Explicitly, the surface stress-energy tensor is given by

$$16\pi\varepsilon S_{\alpha\beta} = \kappa_{\mu\alpha}n^\mu n_\beta + \kappa_{\mu\beta}n^\mu n_\alpha - \kappa n_\alpha n_\beta - \varepsilon\kappa_{\alpha\beta} - \left(\kappa_{\mu\nu}n^\mu n^\nu - \varepsilon\kappa\right)g_{\alpha\beta}.$$

From this we notice that $S_{\alpha\beta}$ is tangent to the hypersurface: $S_{\alpha\beta}n^\beta = 0$. It therefore admits the decomposition

$$S^{\alpha\beta} = S^{ab}e^\alpha_a e^\beta_b, \tag{3.55}$$

where $S_{ab} = S_{\alpha\beta}e^\alpha_a e^\beta_b$ is a symmetric three-tensor. This is evaluated as follows:

$$16\pi S_{ab} = -\kappa_{\alpha\beta}e^\alpha_a e^\beta_b - \varepsilon\left(\kappa_{\mu\nu}n^\mu n^\nu - \varepsilon\kappa\right)h_{ab}$$
$$= -\kappa_{\alpha\beta}e^\alpha_a e^\beta_b - \kappa_{\mu\nu}\left(g^{\mu\nu} - h^{mn}e^\mu_m e^\nu_n\right)h_{ab} + \kappa h_{ab}$$
$$= -\kappa_{\alpha\beta}e^\alpha_a e^\beta_b + h^{mn}\kappa_{\mu\nu}e^\mu_m e^\nu_n h_{ab}.$$

On the other hand we have

$$[n_{\alpha;\beta}] = -[\Gamma^{\gamma}{}_{\alpha\beta}]n_{\gamma}$$

$$= -\frac{1}{2}\left(\kappa_{\gamma\alpha}n_{\beta} + \kappa_{\gamma\beta}n_{\alpha} - \kappa_{\alpha\beta}n_{\gamma}\right)n^{\gamma}$$

$$= \frac{1}{2}\left(\varepsilon\kappa_{\alpha\beta} - \kappa_{\gamma\alpha}n_{\beta}n^{\gamma} - \kappa_{\gamma\beta}n_{\alpha}n^{\gamma}\right),$$

which allows us to write

$$[K_{ab}] = [n_{\alpha;\beta}]e^{\alpha}_{a}e^{\beta}_{b} = \frac{\varepsilon}{2}\kappa_{\alpha\beta}e^{\alpha}_{a}e^{\beta}_{b}.$$

Collecting these results we obtain

$$S_{ab} = -\frac{\varepsilon}{8\pi}\left([K_{ab}] - [K]h_{ab}\right), \tag{3.56}$$

which relates the surface stress-energy tensor to the jump in extrinsic curvature from one side of Σ to the other. The complete stress-energy tensor of the surface layer is

$$T^{\alpha\beta}_{\Sigma} = \delta(\ell)\,S^{ab}e^{\alpha}_{a}e^{\beta}_{b}. \tag{3.57}$$

We conclude that a smooth transition across Σ requires $[K_{ab}] = 0$ – the extrinsic curvature must be the same on both sides of the hypersurface. This requirement does more than just remove the δ-function term from the Einstein tensor: In Section 3.13, Problem 4 you will be asked to prove that $[K_{ab}] = 0$ implies $A^{\alpha}{}_{\beta\gamma\delta} = 0$, which means that the full Riemann tensor is then nonsingular at Σ.

The condition $[K_{ab}] = 0$ is our second junction condition, and it is expressed independently of the coordinates x^{α} and $x^{\alpha}_{\pm}$. If this condition is violated, then the spacetime is singular at Σ, but the singularity comes with a sound physical interpretation: a surface layer with stress-energy tensor $T^{\alpha\beta}_{\Sigma}$ is present at the hypersurface.

3.7.6 Summary

The junction conditions for a smooth joining of two metrics at a hypersurface Σ (assumed not to be null) are

$$[h_{ab}] = [K_{ab}] = 0.$$

If the extrinsic curvature is not the same on both sides of Σ, then a thin shell with surface stress-energy tensor

$$S_{ab} = -\frac{\varepsilon}{8\pi}\left([K_{ab}] - [K]h_{ab}\right)$$

is present at Σ. The complete stress-energy tensor of the surface layer is given by Eq. (3.57) in the continuous coordinates x^α. In the coordinate system $x^\alpha_\pm$ used originally in $\mathscr{V}^\pm$, it is

$$T^{\alpha\beta}_\Sigma = S^{ab} \left(\frac{\partial x^\alpha_\pm}{\partial y^a}\right)\left(\frac{\partial x^\beta_\pm}{\partial y^b}\right) \delta(\ell).$$

This follows from Eq. (3.57) by a simple coordinate transformation from x^α to $x^\alpha_\pm$; such a transformation leaves both ℓ and S^{ab} unchanged.

This formulation of the junction conditions is due to Darmois (1927) and Israel (1966). The thin-shell formalism is due to Lanczos (1922 and 1924) and Israel (1966). An extension to null hypersurfaces will be presented in Section 3.11.

3.8 Oppenheimer–Snyder collapse

In 1939, J. Robert Oppenheimer and his student Hartland Snyder published the first solution to the Einstein field equations that describes the process of gravitational collapse to a black hole. For simplicity they modelled the collapsing star as a spherical ball of pressureless matter with a uniform density. (A perfect fluid with negligible pressure is usually called *dust*.) The metric inside the dust is a Friedmann–Robertson–Walker (FRW) solution, while the metric outside is the Schwarzschild solution (Fig. 3.6). The question considered here is whether these metrics can be joined smoothly at their common boundary, the surface of the collapsing star.

The metric inside the collapsing dust (which occupies the region $\mathscr{V}^-$) is given by

$$ds^2_- = -d\tau^2 + a^2(\tau)\left(d\chi^2 + \sin^2\chi \, d\Omega^2\right), \qquad (3.58)$$

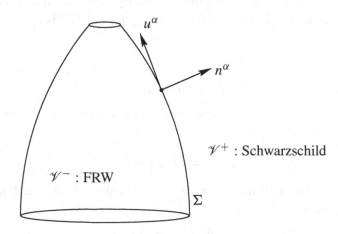

Figure 3.6   The Oppenheimer–Snyder spacetime.

where τ is proper time on comoving world lines (along which χ, θ, and ϕ are all constant), and $a(\tau)$ is the scale factor. By virtue of the Einstein field equations, this satisfies

$$\dot{a}^2 + 1 = \frac{8\pi}{3}\,\rho a^2, \tag{3.59}$$

where an overdot denotes differentiation with respect to τ. By virtue of energy-momentum conservation in the absence of pressure, the dust's mass density ρ satisfies

$$\rho a^3 = \text{constant} \equiv \frac{3}{8\pi}\,a_{\text{max}}, \tag{3.60}$$

where a_{max} is the maximum value of the scale factor. The solution to Eqs. (3.59) and (3.60) has the parametric form

$$a(\eta) = \tfrac{1}{2}a_{\text{max}}(1 + \cos\eta), \qquad \tau(\eta) = \tfrac{1}{2}a_{\text{max}}(\eta + \sin\eta);$$

the collapse begins at $\eta = 0$ when $a = a_{\text{max}}$, and it ends at $\eta = \pi$ when $a = 0$. The hypersurface Σ coincides with the surface of the collapsing star, which is located at $\chi = \chi_0$ in our comoving coordinates.

The metric outside the dust (in the region $\mathscr{V}^+$) is given by

$$ds_+^2 = -f\,dt^2 + f^{-1}\,dr^2 + r^2\,d\Omega^2, \qquad f = 1 - 2M/r, \tag{3.61}$$

where M is the gravitational mass of the collapsing star. As seen from the outside, Σ is described by the parametric equations $r = R(\tau)$, $t = T(\tau)$, where τ is proper time for observers comoving with the surface. Clearly, this is the same τ that appears in the metric of Eq. (3.58).

It is convenient to choose $y^a = (\tau, \theta, \phi)$ as coordinates on Σ. It follows that $e_\tau^\alpha = u^\alpha$, where u^α is the four-velocity of an observer comoving with the surface of the collapsing star.

We now calculate the induced metric. As seen from $\mathscr{V}^-$ the metric on Σ is

$$ds_\Sigma^2 = -d\tau^2 + a^2(\tau)\sin^2\chi_0\,d\Omega^2.$$

As seen from $\mathscr{V}^+$, on the other hand,

$$ds_\Sigma^2 = -\left(F\dot{T}^2 - F^{-1}\dot{R}^2\right)d\tau^2 + R^2(\tau)\,d\Omega^2,$$

where $F = 1 - 2M/R$. Because the induced metric must be the same on both sides of the hypersurface, we have

$$R(\tau) = a(\tau)\sin\chi_0, \qquad F\dot{T}^2 - F^{-1}\dot{R}^2 = 1. \tag{3.62}$$

The first equation determines $R(\tau)$ and the second equation can be solved for $\dot{T}$:

$$F\dot{T} = \sqrt{\dot{R}^2 + F} \equiv \beta(R, \dot{R}). \tag{3.63}$$

This equation can be integrated for $T(\tau)$ and the motion of the boundary in $\mathscr{V}^+$ is completely determined.

The unit normal to Σ can be obtained from the relations $n_\alpha u^\alpha = 0$, $n_\alpha n^\alpha = 1$. As seen from $\mathscr{V}^-$, $u^\alpha_- \partial_\alpha = \partial_\tau$ and $n^-_\alpha \, dx^\alpha = a \, d\chi$; we have chosen $n^\chi > 0$ so that n^α is directed toward $\mathscr{V}^+$. As seen from $\mathscr{V}^+$, $u^\alpha_+ \partial_\alpha = \dot{T} \partial_t + \dot{R} \partial_r$ and $n^+_\alpha \, dx^\alpha = -\dot{R} \, dt + \dot{T} \, dr$, with a consistent choice for the sign.

The extrinsic curvature is defined on either side of Σ by $K_{ab} = n_{\alpha;\beta} e^\alpha_a e^\beta_b$. The nonvanishing components are $K_{\tau\tau} = n_{\alpha;\beta} u^\alpha u^\beta = -n_\alpha u^\alpha_{;\beta} u^\beta = -a^\alpha n_\alpha$ (where a^α is the acceleration of an observer comoving with the surface), $K_{\theta\theta} = n_{\theta;\theta}$, and $K_{\phi\phi} = n_{\phi;\phi}$. A straightforward calculation reveals that as seen from $\mathscr{V}^-$,

$$K^\tau_{-\tau} = 0, \qquad K^\theta_{-\theta} = K^\phi_{-\phi} = a^{-1} \cot \chi_0; \tag{3.64}$$

the first result follows immediately from the fact that the comoving world lines of a FRW spacetime are geodesics. As seen from $\mathscr{V}^+$,

$$K^\tau_{+\tau} = \dot{\beta}/\dot{R}, \qquad K^\theta_{+\theta} = K^\phi_{+\phi} = \beta/R, \tag{3.65}$$

where $\beta(R, \dot{R})$ is defined by Eq. (3.63).

To have a smooth transition at the surface of the collapsing star, we demand that K_{ab} be the same on both sides of the hypersurface. It is therefore necessary for u^α_+ to satisfy the geodesic equation ($a^\alpha_+ = 0$) in $\mathscr{V}^+$. It is easy to check that the geodesic equation produces $\dot{R}^2 + F = \tilde{E}^2$, where $\tilde{E} = -u_t$ is the (conserved) energy parameter of the comoving observer. This relation implies $\beta = \tilde{E}$, and the fact that β is a constant enforces $K^\tau_{+\tau} = 0$, as required. On the other hand, $[K^\theta_\theta] = 0$ gives $\cot \chi_0/a = \beta/R = \tilde{E}/(a \sin \chi_0)$, or

$$\beta = \tilde{E} = \cos \chi_0. \tag{3.66}$$

We have found that the requirement for a smooth transition at Σ is that the hypersurface be generated by geodesics of both $\mathscr{V}^-$ and $\mathscr{V}^+$, and that the parameters $\tilde{E}$ and χ_0 be related by Eq. (3.66). With the help of Eqs. (3.59), (3.62), and (3.63) we may turn Eq. (3.66) into

$$M = \frac{4\pi}{3} \rho R^3, \tag{3.67}$$

which equates the gravitational mass of the collapsing star to the product of its density and volume. This relation has an immediate intuitive meaning, and it neatly summarizes the complete solution to the Oppenheimer–Snyder problem.

3.9 Thin-shell collapse

As an application of the thin-shell formalism, we consider the gravitational collapse of a thin spherical shell. We assume that spacetime is flat inside the shell (in $\mathscr{V}^-$). Outside (in $\mathscr{V}^+$), the metric is necessarily a Schwarzschild solution (by virtue of the spherical symmetry of the matter distribution). We assume also that the shell is made of pressureless matter, in the sense that its surface stress-energy tensor is constrained to have the form

$$S^{ab} = \sigma\, u^a u^b, \tag{3.68}$$

in which σ is the surface density and u^a the shell's velocity field. Our goal is to derive the shell's equations of motion under the stated conditions.

Using the results derived in the preceding section, we have

$$K^\tau_{\pm\tau} = \dot\beta_\pm / \dot R,$$

$$K^\theta_{\pm\theta} = K^\phi_{\pm\phi} = \beta_\pm / R,$$

$$\beta_+ = \sqrt{\dot R^2 + 1 - 2M/R},$$

$$\beta_- = \sqrt{\dot R^2 + 1},$$

where $R(\tau)$ is the shell's radius, and M its gravitational mass. As we did before, we use (τ, θ, ϕ) as coordinates on Σ; in these coordinates $u^a = \partial y^a / \partial \tau$. Equation (3.56) allows us to calculate the components of the surface stress-energy tensor, and we find

$$-\sigma = S^\tau_{\ \tau} = \frac{\beta_+ - \beta_-}{4\pi R}, \qquad 0 = S^\theta_{\ \theta} = \frac{\beta_+ - \beta_-}{8\pi R} + \frac{\dot\beta_+ - \dot\beta_-}{8\pi \dot R}.$$

The second equation can be integrated immediately, giving $(\beta_+ - \beta_-)R =$ constant. Substituting this into the first equation yields $4\pi R^2 \sigma = -$constant.

We have obtained

$$4\pi R^2 \sigma \equiv m = \text{constant} \tag{3.69}$$

and $\beta_- - \beta_+ = m/R$. The first equation states that m, the shell's *rest mass*, stays constant during the evolution. Squaring the second equation converts it to

$$M = m\sqrt{1 + \dot R^2} - \frac{m^2}{2R}, \tag{3.70}$$

which comes with a nice physical interpretation. The first term on the right-hand side is the shell's relativistic kinetic energy, including rest mass. The second term is the shell's binding energy, the work required to assemble the shell from its dispersed constituents. The sum of these is the total (conserved) energy, and this is equal to the shell's gravitational mass M. Equation (3.70) provides a vivid illustration of the general statement that *all forms* of energy contribute to the total gravitational mass of an isolated body.

Equations (3.69) and (3.70) are the shell's equations of motion. It is interesting to note that when $M < m$, the motion exhibits a turning point at $R = R_{max} \equiv m^2/[2(m - M)]$: An expanding shell with $M < m$ cannot escape from its own gravitational pull.

3.10 Slowly rotating shell

Our next application of the thin-shell formalism is concerned with the spacetime of a slowly rotating, spherical shell. We take the exterior metric to be the slow-rotation limit of the Kerr solution,

$$ds_+^2 = -f \, dt^2 + f^{-1} \, dr^2 + r^2 \, d\Omega^2 - \frac{4Ma}{r} \sin^2\theta \, dt \, d\phi. \qquad (3.71)$$

Here, $f = 1 - 2M/r$ with M denoting the shell's gravitational mass, and $a = J/M \ll M$, where J is the shell's angular momentum. Throughout this section we will work consistently to first order in a.

The metric of Eq. (3.71) is cut off at $r = R$, which is where the shell is located. As viewed from the exterior, the shell's induced metric is

$$ds_\Sigma^2 = -(1 - 2M/R) \, dt^2 + R^2 \, d\Omega^2 - \frac{4Ma}{R} \sin^2\theta \, dt \, d\phi.$$

It is possible to remove the off-diagonal term by going to a rotating frame of reference. We therefore introduce a new angular coordinate ψ related to ϕ by

$$\psi = \phi - \Omega t, \qquad (3.72)$$

where Ω is the angular velocity of the new frame with respect to the inertial frame of Eq. (3.71). We anticipate that Ω will be proportional to a, and this allows us to approximate $d\phi^2$ by $d\psi^2 + 2\Omega \, dt \, d\psi$. Substituting this into ds_Σ^2 returns a diagonal metric if Ω is chosen to be

$$\Omega = \frac{2Ma}{R^3}. \qquad (3.73)$$

The induced metric then becomes

$$h_{ab} \, dy^a \, dy^b = -(1 - 2M/R) \, dt^2 + R^2(d\theta^2 + \sin^2\theta \, d\psi^2). \qquad (3.74)$$

It is now clear that the shell has a spherical geometry. As Eq. (3.74) indicates, we will use the coordinates $y^a = (t, \theta, \psi)$ on the shell.

We take spacetime to be flat inside the shell, and we write the Minkowski metric in the form

$$ds_-^2 = -(1 - 2M/R)\,dt^2 + d\rho^2 + \rho^2(d\theta^2 + \sin^2\theta\,d\psi^2), \qquad (3.75)$$

where ρ is a radial coordinate. This metric must be cut off at $\rho = R$ and joined to the exterior metric of Eq. (3.71). The shell's intrinsic metric, as computed from the interior, agrees with Eq. (3.74). Continuity of the induced metric is therefore established, and we must now turn to the extrinsic curvature.

We first compute the extrinsic curvature as seen from the shell's exterior. In the metric of Eq. (3.71), the shell's unit normal is $n_\alpha = f^{-1/2}\partial_\alpha r$. The parametric equations of the hypersurface are $t = t$, $\theta = \theta$, and $\phi = \psi + \Omega t$, and they have the generic form $x^\alpha = x^\alpha(y^a)$. These allow us to compute the tangent vectors $e_a^\alpha = \partial x^\alpha/\partial y^a$ and we obtain $e_t^\alpha \partial_\alpha = \partial_t + \Omega\partial_\phi$, $e_\theta^\alpha \partial_\alpha = \partial_\theta$, and $e_\psi^\alpha \partial_\alpha = \partial_\phi$. From all this we find that the nonvanishing components of the extrinsic curvature are

$$K^t{}_t = \frac{M}{R^2\sqrt{1 - 2M/R}},$$

$$K^t{}_\psi = -\frac{3Ma\sin^2\theta}{R^2\sqrt{1 - 2M/R}},$$

$$K^\psi{}_t = \frac{3Ma}{R^4}\sqrt{1 - 2M/R},$$

$$K^\theta{}_\theta = \frac{1}{R}\sqrt{1 - 2M/R} = K^\psi{}_\psi.$$

As now seen from the shell's interior, the unit normal is $n_\alpha = \partial_\alpha\rho$ and the tangent vectors are $e_t^\alpha \partial_\alpha = \partial_t$, $e_\theta^\alpha \partial_\alpha = \partial_\theta$, and $e_\psi^\alpha \partial_\alpha = \partial_\psi$. From this we find that $K^\theta{}_\theta = 1/R = K^\psi{}_\psi$ are the only two nonvanishing components of the extrinsic curvature. This could have been obtained directly by setting $M = 0$ in our previous results.

We have a discontinuity in the extrinsic curvature, and Eq. (3.56) allows us to calculate S^{ab}, the shell's surface stress-energy tensor. After a few lines of algebra we obtain

$$S^t{}_t = -\frac{1}{4\pi R}\left(1 - \sqrt{1 - 2M/R}\right),$$

$$S^t{}_\psi = \frac{3Ma\sin^2\theta}{8\pi R^2\sqrt{1 - 2M/R}},$$

$$S^\psi{}_t = -\frac{3Ma}{8\pi R^4}\sqrt{1 - 2M/R},$$

$$S^\theta{}_\theta = \frac{1 - M/R - \sqrt{1 - 2M/R}}{8\pi R\sqrt{1 - 2M/R}} = S^\psi{}_\psi.$$

These results give us a complete description of the surface stress-energy tensor, but they are not terribly illuminating. Can we make sense of this mess?

We will attempt to cast S^{ab} in a perfect-fluid form,

$$S^{ab} = \sigma u^a u^b + p(h^{ab} + u^a u^b), \tag{3.76}$$

in terms of a velocity field u^a, a surface density σ, and a surface pressure p. How do we find these quantities? First we notice that Eq. (3.76) implies $S^a{}_b u^b = -\sigma u^a$, which shows that u^a is a normalized eigenvector of the surface stress-energy tensor, with eigenvalue $-\sigma$. This gives us three equations for three unknowns, the density and the two independent components of the velocity field. Once those have been obtained, the pressure is found by projecting S^{ab} in the directions orthogonal to u^a. The rest is just a matter of algebra.

We can save ourselves some work if we recognize that the shell must move rigidly in the ψ direction, with a uniform angular velocity ω. Its velocity vector can then be expressed as

$$u^a = \gamma(t^a + \omega \psi^a), \tag{3.77}$$

where $t^a = \partial y^a / \partial t$ and $\psi^a = \partial y^a / \partial \psi$ are Killing vectors of the induced metric h_{ab}. In Eq. (3.77), $\omega = d\psi/dt$ is the shell's angular velocity in the rotating frame of Eq. (3.72), and γ is determined by the normalization condition $h_{ab} u^a u^b = -1$. We can simplify things further if we anticipate that ω will be proportional to a. For example, neglecting $O(\omega^2)$ terms when normalizing u^a gives

$$\gamma = \frac{1}{\sqrt{1 - 2M/R}}. \tag{3.78}$$

With these assumptions, we find that the eigenvalue equation produces $\omega = -S^{\psi}_t/(-S^t{}_t + S^{\psi}{}_{\psi})$ and $\sigma = -S^t{}_t$. After simplification the first equation becomes

$$\omega = \frac{6Ma}{R^3} \frac{1 - 2M/R}{\left(1 - \sqrt{1 - 2M/R}\right)\left(1 + 3\sqrt{1 - 2M/R}\right)}, \tag{3.79}$$

and the second is

$$\sigma = \frac{1}{4\pi R}\left(1 - \sqrt{1 - 2M/R}\right). \tag{3.80}$$

We now have the surface density and the velocity field. The surface pressure can easily be obtained by projecting S^{ab} in the directions orthogonal to u^a: $p = \frac{1}{2}(h_{ab} + u_a u_b)S^{ab} = \frac{1}{2}(S + \sigma)$, where $S = h_{ab}S^{ab}$. This gives $p = S^{\theta}{}_{\theta}$, and

$$p = \frac{1 - M/R - \sqrt{1 - 2M/R}}{8\pi R\sqrt{1 - 2M/R}}. \tag{3.81}$$

The shell's material is therefore a perfect fluid of density σ, pressure p, and angular velocity ω. When R is much larger than $2M$, Eqs. (3.79)–(3.81) reduce to $\omega \simeq 3a/(2R^2)$, $\sigma \simeq M/(4\pi R^2)$, and $p \simeq M^2/(16\pi R^3)$, respectively.

The spacetime of a slowly rotating shell offers us a unique opportunity to explore the rather strange relativistic effects associated with rotation. We conclude this section with a brief description of these effects.

The metric of Eq. (3.71) is the metric outside the shell, and it is expressed in a coordinate system that goes easily into a Cartesian frame at infinity. This is the frame of the 'fixed stars,' and it is this frame which sets the standard of no rotation. The metric of Eq. (3.75), on the other hand, is the metric inside the shell, and it is expressed in a coordinate system that is rotating with respect to the frame of the fixed stars. The transformation is given by Eq. (3.72), and it shows that an observer at constant ψ moves with an angular velocity $d\phi/dt = \Omega$. Inertial observers inside the shell are therefore rotating with respect to the fixed stars, with an angular velocity $\Omega_{\text{in}} \equiv \Omega$. According to Eq. (3.73), this is

$$\Omega_{\text{in}} = \frac{2Ma}{R^3}. \tag{3.82}$$

This angular motion is induced by the rotation of the shell, and the effect is known as the *dragging of inertial frames*. It was first discovered in 1918 by Thirring and Lense.

The shell's angular velocity ω, as computed in Eq. (3.79), is measured in the rotating frame. As measured in the nonrotating frame, the shell's angular velocity is $\Omega_{\text{shell}} = d\phi/dt = d\psi/dt + \Omega = \omega + \Omega_{\text{in}}$. According to Eqs. (3.79) and (3.82), this is

$$\Omega_{\text{shell}} = \frac{2Ma}{R^3} \frac{1 + 2\sqrt{1 - 2M/R}}{\left(1 - \sqrt{1 - 2M/R}\right)\left(1 + 3\sqrt{1 - 2M/R}\right)}. \tag{3.83}$$

When R is much larger than $2M$, $\Omega_{\text{in}}/\Omega_{\text{shell}} \simeq 4M/(3R)$, and the internal observers rotate at a small fraction of the shell's angular velocity. As R approaches $2M$, however, the ratio approaches unity, and the internal observers find themselves corotating with the shell. This is a rather striking manifestation of frame dragging. (The phrase 'Mach's principle' is often attached to this phenomenon.) This spacetime, admittedly, is highly idealized, and you may wonder whether corotation could ever occur in a realistic situation. You will see in Chapter 5 that the answer is yes: A very similar phenomenon occurs in the vicinity of a rotating black hole.

3.11 Null shells

We saw in Sections 3.1 and 3.2 that the description of null hypersurfaces involves interesting subtleties, and we should not be surprised to find that the same is true of the description of null surface layers. Our purpose here, in the last section of Chapter 3, is to face these subtleties and extend the formalism of thin shells, as developed in Section 3.7, to the case of a null hypersurface. The presentation given here is adapted from Barrabès and Israel (1991).

3.11.1 Geometry

As we did in Section 3.7, we consider a hypersurface Σ that partitions spacetime into two regions $\mathcal{V}^{\pm}$ in which the metric is $g_{\alpha\beta}^{\pm}$ when expressed in coordinates $x_{\pm}^{\alpha}$. Here we assume that the hypersurface is null, and our convention is such that $\mathcal{V}^{-}$ is in the past of Σ, and $\mathcal{V}^{+}$ in its future. We assume also that the hypersurface is singular, in the sense that the Riemann tensor possesses a δ-function singularity at Σ. We will characterize the Ricci part of this singular curvature tensor, and relate it to the surface stress-energy tensor of the shell. (We shall have nothing to say about the interesting physical effects associated with the Weyl part of the singular curvature tensor.)

As we did in Section 3.1.2, we install coordinates

$$y^a = (\lambda, \theta^A)$$

on the hypersurface, and we assume that these coordinates are the same on both sides of Σ. We take λ to be an arbitrary parameter on the null generators of the hypersurface, and we use θ^A to label the generators. It is possible to choose λ to be an affine parameter on *one side* of the hypersurface. But as we shall see below, in general it is not possible to make λ an affine parameter on both sides of Σ.

As seen from $\mathcal{V}^{\pm}$, Σ is described by the parametric relations $x_{\pm}^{\alpha}(y^a)$, and using these we can introduce tangent vectors $e_{\pm a}^{\alpha} = \partial x_{\pm}^{\alpha}/\partial y^a$ on each side of the hypersurface. These are naturally segregated into a null vector $k_{\pm}^{\alpha}$ that is tangent to the generators, and two spacelike vectors $e_{\pm A}^{\alpha}$ that point in the directions transverse to the generators. Explicitly,

$$k^{\alpha} = \left(\frac{\partial x^{\alpha}}{\partial \lambda}\right)_{\theta^A} = e_{\lambda}^{\alpha}, \qquad e_A^{\alpha} = \left(\frac{\partial x^{\alpha}}{\partial \theta^A}\right)_{\lambda}. \tag{3.84}$$

(Here and below, in order to keep the notation simple, we refrain from using the '$\pm$' label in displayed equations; this should not create any confusion.) By construction, these vectors satisfy

$$k_{\alpha}k^{\alpha} = 0 = k_{\alpha}e_A^{\alpha}. \tag{3.85}$$

The remaining inner products

$$\sigma_{AB}(\lambda, \theta^C) \equiv g_{\alpha\beta} e^\alpha_A e^\beta_B \tag{3.86}$$

do not vanish, and we assume that they are the same on both sides of Σ:

$$[\sigma_{AB}] = 0. \tag{3.87}$$

We recall from Section 3.1.3 that the two-tensor σ_{AB} acts as a metric on Σ,

$$ds^2_\Sigma = \sigma_{AB}\, d\theta^A\, d\theta^B,$$

and the condition (3.87) ensures that the hypersurface possesses a well-defined intrinsic geometry.

As we did in Section 3.1.3, we complete the basis by adding an auxiliary null vector $N^\alpha_\pm$ that satisfies

$$N_\alpha N^\alpha = 0, \quad N_\alpha k^\alpha = -1, \quad N_\alpha e^\alpha_A = 0. \tag{3.88}$$

The completed basis gives us the completeness relations

$$g^{\alpha\beta} = -k^\alpha N^\beta - N^\alpha k^\beta + \sigma^{AB} e^\alpha_A e^\beta_B \tag{3.89}$$

for the inverse metric on either side of Σ (in the coordinates $x^\alpha_\pm$); σ^{AB} is the inverse of σ_{AB}, and it is the same on both sides.

To complete the geometric setup we introduce a congruence of geodesics that cross the hypersurface. In Section 3.7, in which Σ was either timelike or space-like, the congruence was selected by demanding that the geodesics intersect the hypersurface *orthogonally*: The vector field $u^\alpha_\pm$ tangent to the congruence was set equal (on Σ) to the normal vector $n^\alpha_\pm$. When the hypersurface is null, however, this requirement does not produce a unique congruence, because a vector orthogonal to k^α can still possess an arbitrary component along k^α.

We shall have to give up on the idea of adopting a *unique* congruence. An important aspect of our description of null shells is therefore that it involves an *arbitrary* congruence of *timelike* geodesics intersecting Σ. This arbitrariness comes with the lightlike nature of the singular hypersurface, and it cannot be removed. It can, however, be physically motivated: The arbitrary vector field $u^\alpha_\pm$ that enters our description of null shells can be identified with the four-velocity of a family of observers making measurements on the shell; since many different families of observers could be introduced to make such measurements, there is no reason to demand that the vector field be uniquely specified.

We therefore introduce a congruence of timelike geodesics γ that arbitrarily intersect the hypersurface. The geodesics are parameterized by proper time τ, which is adjusted so that $\tau = 0$ when a geodesic crosses Σ; thus, $\tau < 0$ in $\mathcal{V}^-$ and $\tau > 0$

in $\mathscr{V}^+$. The vector tangent to the geodesics is $u_{\pm}^{\alpha}$, and a displacement along a member of the congruence is described by

$$\mathrm{d}x^{\alpha} = u^{\alpha}\,\mathrm{d}\tau. \qquad (3.90)$$

To ensure that the congruence is smooth at the hypersurface, we demand that $u_{\pm}^{\alpha}$ be 'the same' on both sides of Σ. This means that $u_{\alpha}e_{a}^{\alpha}$, the tangential projections of the vector field, must be equal when evaluated on either side of the hypersurface:

$$\left[-u_{\alpha}k^{\alpha}\right] = 0 = \left[u_{\alpha}e_{A}^{\alpha}\right]. \qquad (3.91)$$

If, for example, u_{-}^{α} is specified in $\mathscr{V}^-$, then the three conditions (3.91) are sufficient (together with the geodesic equation) to determine the three independent components of u_{+}^{α} in $\mathscr{V}^+$. We note that $-u_{\alpha}N^{\alpha}$, the transverse projection of the four-velocity, is allowed to be discontinuous at Σ.

The proper-time parameter on the timelike geodesics can be viewed as a scalar field $\tau(x_{\pm}^{\alpha})$ defined in a neighbourhood of Σ: Select a point $x_{\pm}^{\alpha}$ off the hypersurface and locate the unique geodesic γ that connects this point to Σ; the value of the scalar field at $x_{\pm}^{\alpha}$ is equal to the proper-time parameter of this geodesic at that point. The hypersurface Σ can then be described by the statement

$$\tau(x^{\alpha}) = 0,$$

and its normal vector $k_{\alpha}^{\pm}$ will be proportional to the gradient of $\tau(x_{\pm}^{\alpha})$ evaluated at Σ. It is easy to check that the expression

$$k_{\alpha} = -\left(-k_{\mu}u^{\mu}\right)\frac{\partial\tau}{\partial x^{\alpha}} \qquad (3.92)$$

is compatible with Eq. (3.90). We recall that the factor $-k_{\mu}u^{\mu}$ in Eq. (3.92) is continuous across Σ.

3.11.2 Surface stress-energy tensor

As we did in Section 3.7, we introduce for our short-term convenience a continuous coordinate system x^{α}, distinct from $x_{\pm}^{\alpha}$, in a neighbourhood of the hypersurface; the final formulation of our null-shell formalism will be independent of these coordinates. We express the metric as a distribution-valued tensor:

$$g_{\alpha\beta} = \Theta(\tau)\,g_{\alpha\beta}^+ + \Theta(-\tau)\,g_{\alpha\beta}^-,$$

where $g_{\alpha\beta}^{\pm}(x^{\mu})$ is the metric in $\mathscr{V}^{\pm}$. We assume that in these coordinates, the metric is continuous at Σ: $[g_{\alpha\beta}] = 0$; Eq. (3.87) is compatible with this requirement. We also have $[k^{\alpha}] = [e_{A}^{\alpha}] = [N^{\alpha}] = [u^{\alpha}] = 0$. Differentiation of the metric proceeds as in Sections 3.7.2 and 3.7.3, except that we now write τ instead of ℓ, and we use Eq. (3.92) to relate the gradient of τ to the null vector k^α. We arrive at a Riemann tensor that contains a singular part given by

$$R_\Sigma{}^\alpha{}_{\beta\gamma\delta} = -(-k_\mu u^\mu)^{-1}([\Gamma^\alpha{}_{\beta\delta}]k_\gamma - [\Gamma^\alpha{}_{\beta\gamma}]k_\delta)\delta(\tau), \qquad (3.93)$$

where $[\Gamma^\alpha{}_{\beta\gamma}]$ is the jump in the Christoffel symbols across Σ.

In order to make Eq. (3.93) more explicit we must characterize the discontinuous behaviour of $g_{\alpha\beta,\gamma}$. The condition $[g_{\alpha\beta}] = 0$ guarantees that the tangential derivatives of the metric are continuous:

$$[g_{\alpha\beta,\gamma}]k^\gamma = 0 = [g_{\alpha\beta,\gamma}]e^\gamma_C.$$

The only possible discontinuity is therefore in $g_{\alpha\beta,\gamma}N^\gamma$, the transverse derivative of the metric. In view of Eq. (3.88) we conclude that there exists a tensor field $\gamma_{\alpha\beta}$ such that

$$[g_{\alpha\beta,\gamma}] = -\gamma_{\alpha\beta}k_\gamma. \qquad (3.94)$$

This tensor is given explicitly by $\gamma_{\alpha\beta} = [g_{\alpha\beta,\gamma}]N^\gamma$, and it is now easy to check that

$$[\Gamma^\alpha{}_{\beta\gamma}] = -\frac{1}{2}(\gamma^\alpha_\beta k_\gamma + \gamma^\alpha_\gamma k_\beta - \gamma_{\beta\gamma}k^\alpha). \qquad (3.95)$$

Substituting this into Eq. (3.93) gives

$$R_\Sigma{}^\alpha{}_{\beta\gamma\delta} = \frac{1}{2}(-k_\mu u^\mu)^{-1}(\gamma^\alpha_\delta k_\beta k_\gamma - \gamma_{\beta\delta}k^\alpha k_\gamma - \gamma^\alpha_\gamma k_\beta k_\delta + \gamma_{\beta\gamma}k^\alpha k_\delta)\delta(\tau), \qquad (3.96)$$

and we see that k^α and $\gamma_{\alpha\beta}$ give a complete characterization of the singular part of the Riemann tensor.

From Eq. (3.96) it is easy to form the singular part of the Einstein tensor, and the Einstein field equations then give us the singular part of the stress-energy tensor:

$$T^{\alpha\beta}_\Sigma = (-k_\mu u^\mu)^{-1}S^{\alpha\beta}\,\delta(\tau), \qquad (3.97)$$

where

$$S^{\alpha\beta} = \frac{1}{16\pi}(k^\alpha\gamma^\beta_\mu k^\mu + k^\beta\gamma^\alpha_\mu k^\mu - \gamma^\mu_\mu k^\alpha k^\beta - \gamma_{\mu\nu}k^\mu k^\nu g^{\alpha\beta})$$

is the surface stress-energy tensor of the null shell – up to a factor $-k_\mu u^\mu$ that depends on the choice of observers making measurements on the shell. Its expression can be simplified if we decompose $S^{\alpha\beta}$ in the basis $(k^\alpha, e^\alpha_A, N^\alpha)$. For this purpose we introduce the projections

$$\gamma_A \equiv \gamma_{\alpha\beta}\,e^\alpha_A k^\beta, \qquad \gamma_{AB} \equiv \gamma_{\alpha\beta}\,e^\alpha_A e^\beta_B, \qquad (3.98)$$

and we use the completeness relations (3.89) to find that the vector $\gamma^\alpha_\mu k^\mu$ admits the decomposition

$$\gamma^\alpha_\mu k^\mu = \frac{1}{2}\left(\gamma^\mu_\mu - \sigma^{AB}\gamma_{AB}\right)k^\alpha + \left(\sigma^{AB}\gamma_B\right)e^\alpha_A - \left(\gamma_{\mu\nu}k^\mu k^\nu\right)N^\alpha.$$

Substituting this into our previous expression for $S^{\alpha\beta}$ and involving once more the completeness relations, we arrive at our final expression for the surface stress-energy tensor:

$$S^{\alpha\beta} = \mu k^\alpha k^\beta + j^A\left(k^\alpha e^\beta_A + e^\alpha_A k^\beta\right) + p\,\sigma^{AB}e^\alpha_A e^\beta_B. \qquad (3.99)$$

Here,

$$\mu \equiv -\frac{1}{16\pi}\left(\sigma^{AB}\gamma_{AB}\right)$$

can be interpreted as the shell's surface density,

$$j^A \equiv \frac{1}{16\pi}\left(\sigma^{AB}\gamma_B\right)$$

as a surface current, and

$$p \equiv -\frac{1}{16\pi}\left(\gamma_{\alpha\beta}k^\alpha k^\beta\right)$$

as an isotropic surface pressure.

The surface stress-energy tensor of Eq. (3.99) is expressed in the continuous co-ordinates x^α. As a matter of fact, the derivation of Eq. (3.99) relies heavily on these coordinates: The introduction of $\gamma_{\alpha\beta}$ rests on the fact that in these coordinates, $g_{\alpha\beta}$ is continuous at Σ, so that an eventual discontinuity in the metric derivative must be directed along k^α. In the next subsection we will remove the need to involve the coordinates x^α in practical applications of the null-shell formalism. For the time being we simply note that while Eq. (3.99) is indeed expressed in the coordinates x^α, it is a *tensorial* equation involving vectors (k^α and e^α_A) and scalars (μ, j^A, and p). This equation can therefore be expressed in *any* coordinate system; in particular, when viewed from $\mathscr{V}^\pm$ the surface stress-energy tensor can be expressed in the original coordinates $x^\alpha_\pm$.

3.11.3 Intrinsic formulation

In Section 3.7, the surface stress-energy tensor of a timelike or spacelike shell was expressed in terms of *intrinsic* three-tensors – quantities that can be defined on the hypersurface only. The most important ingredients in this formulation were h_{ab}, the (continuous) induced metric, and $[K_{ab}]$, the discontinuity in the extrinsic curvature. We would like to achieve something similar here, and remove the need to involve a continuous coordinate system x^α to calculate the surface quantities μ, j^A, and p.

We can expect that the intrinsic description of the surface stress-energy tensor of a null shell will involve σ_{AB}, the nonvanishing components of the induced metric. We might also expect that it should involve the jump in the extrinsic curvature of the null hypersurface, which would be defined by $K_{ab} = k_{\alpha;\beta}\, e_a^\alpha e_b^\beta = \frac{1}{2}(\pounds_k g_{\alpha\beta})\, e_a^\alpha e_b^\beta$. Not so. The reason is that there is nothing 'transverse' about this object: In the case of a timelike or spacelike hypersurface, the normal n^α points away from the surface, and $\pounds_n g_{\alpha\beta}$ truly represents the transverse derivative of the metric; when the hypersurface is null, on the other hand, k^α is tangent to the surface, and $\pounds_k g_{\alpha\beta}$ is a *tangential* derivative. Thus, the extrinsic curvature is necessarily continuous when the hypersurface is null, and it cannot be related to the tensor $\gamma_{\alpha\beta}$ defined by Eq. (3.94).

There is, fortunately, an easy solution to this problem: We can introduce a *transverse curvature* C_{ab} that properly represents the transverse derivative of the metric. This shall be defined by $C_{ab} = \frac{1}{2}(\pounds_N g_{\alpha\beta})\, e_a^\alpha e_b^\beta = \frac{1}{2}(N_{\alpha;\beta} + N_{\beta;\alpha})e_a^\alpha e_b^\beta$, or

$$C_{ab} = -N_\alpha\, e_{a;\beta}^\alpha e_b^\beta. \tag{3.100}$$

To arrive at Eq. (3.100) we have used the fact that $N_\alpha e_a^\alpha$ is a constant, and the identity $e_{a;\beta}^\alpha e_b^\beta = e_{b;\beta}^\alpha e_a^\beta$, which states that each basis vector e_a^α is Lie transported along any other basis vector; this property ensures that C_{ab}, as defined by Eq. (3.100), is a *symmetric* three-tensor.

In the continuous coordinates x^α, the jump in the transverse curvature is given by

$$\begin{aligned}
[C_{ab}] &= [N_{\alpha;\beta}]\, e_a^\alpha e_b^\alpha \\
&= -[\Gamma^\gamma{}_{\alpha\beta}]N_\gamma e_a^\alpha e_b^\alpha \\
&= \frac{1}{2}\, \gamma_{\alpha\beta}\, e_a^\alpha e_b^\alpha,
\end{aligned}$$

where we have used Eq. (3.95) and the fact that k^α is orthogonal to e_a^α. We therefore have $[C_{\lambda\lambda}] = \frac{1}{2}\gamma_{\alpha\beta}k^\alpha k^\beta$, $[C_{A\lambda}] = \frac{1}{2}\gamma_{\alpha\beta}e_A^\alpha k^\beta \equiv \frac{1}{2}\gamma_A$, and $[C_{AB}] = \frac{1}{2}\gamma_{\alpha\beta}e_A^\alpha e_B^\beta \equiv \frac{1}{2}\gamma_{AB}$, where we have involved Eq. (3.98). Finally, we find that the surface quantities can be expressed as

$$\mu = -\frac{1}{8\pi}\, \sigma^{AB}[C_{AB}], \quad j^A = \frac{1}{8\pi}\, \sigma^{AB}[C_{\lambda B}], \quad p = -\frac{1}{8\pi}\,[C_{\lambda\lambda}]. \tag{3.101}$$

We have established that the shell's surface quantities can all be related to the induced metric σ_{AB} and the jump of the transverse curvature C_{ab}. This completes the intrinsic formulation of our null-shell formalism.

3.11.4 Summary

A singular null hypersurface Σ possesses a surface stress-energy tensor characterized by tangent vectors $k_{\pm}^{\alpha}$ and $e_{\pm A}^{\alpha}$, as well as a surface density μ, a surface current j^{A}, and an isotropic surface pressure p. The surface quantities can all be related to a discontinuity in the surface's transverse curvature,

$$C_{ab} = -N_{\alpha}\, e_{a;\beta}^{\alpha}\, e_{b}^{\beta},$$

which is defined on either side of Σ in the appropriate coordinate system $x_{\pm}^{\alpha}$. The relations are

$$\mu = -\frac{1}{8\pi}\,\sigma^{AB}[C_{AB}], \quad j^{A} = \frac{1}{8\pi}\,\sigma^{AB}[C_{\lambda B}], \quad p = -\frac{1}{8\pi}\,[C_{\lambda\lambda}].$$

The surface stress-energy tensor is given by

$$S^{\alpha\beta} = \mu k^{\alpha} k^{\beta} + j^{A}\left(k^{\alpha} e_{A}^{\beta} + e_{A}^{\alpha} k^{\beta}\right) + p\,\sigma^{AB} e_{A}^{\alpha} e_{B}^{\beta},$$

and the complete stress-energy tensor of the surface layer is

$$T_{\Sigma}^{\alpha\beta} = \left(-k_{\mu} u^{\mu}\right)^{-1} S^{\alpha\beta}\,\delta(\tau).$$

In this expression, the factor $\left(-k_{\mu} u^{\mu}\right)^{-1}$ is continuous at Σ, and the vector field $u_{\pm}^{\alpha}$ is tangent to an arbitrary congruence of timelike geodesics parameterized by proper time τ (the congruence represents a family of observers making measurements on the shell). The presence of this factor implies that μ, j^{A}, and p are not truly the surface quantities that would be measured by the observers. The physically-measured surface quantities are given instead by

$$\mu_{\text{physical}} = \left(-k_{\mu} u^{\mu}\right)^{-1}\mu, \quad j_{\text{physical}}^{A} = \left(-k_{\mu} u^{\mu}\right)^{-1} j^{A},$$

$$p_{\text{physical}} = \left(-k_{\mu} u^{\mu}\right)^{-1} p.$$

The arbitrariness associated with the choice of congruence is thus limited to a single multiplicative factor. The 'bare' quantities μ, j^{A}, and p are independent of this choice, and it is often more convenient to work in terms of those.

3.11.5 Parameterization of the null generators

Our null-shell formalism is now complete, and it is ready to be involved in applications. We will consider a few in the following subsections, but we first return to a statement made earlier, that in general λ cannot be an affine parameter on both sides of the hypersurface. We shall justify this here, and also consider what happens to μ, j^{A}, and p when the parameterization of the null generators is altered.

Whether or not λ is an affine parameter can be decided by computing $\kappa_\pm$, the 'acceleration' of the null vector $k^\alpha_\pm$. This is defined on either side of the hypersurface by (Section 1.3)

$$k^\alpha_{;\beta} k^\beta = \kappa k^\alpha,$$

and λ will be an affine parameter on the $\mathscr{V}^\pm$ side of Σ if $\kappa_\pm = 0$. According to Eq. (3.88), $\kappa = -N_\alpha k^\alpha_{;\beta} k^\beta = -N_\alpha e^\alpha_{\lambda;\beta} e^\beta_\lambda = C_{\lambda\lambda}$, where we have also used Eqs. (3.85) and (3.100). Equation (3.101) then relates the discontinuity in the acceleration to the surface pressure:

$$[\kappa] = -8\pi\, p. \qquad (3.102)$$

We conclude that λ can be an affine parameter on both sides of Σ only when the null shell has a vanishing surface pressure. When $p \neq 0$, λ can be chosen to be an affine parameter on *one side* of the hypersurface, but it will not be an affine parameter on the other side.

Additional insight into this matter can be gained from Raychaudhuri's equation, which describes the transverse evolution of a congruence of null geodesics (Section 2.4). In Section 2.6, Problem 8, Raychaudhuri's equation was written in terms of an arbitrary parameterization of the null geodesics. When the congruence is hypersurface orthogonal, it reads

$$\frac{d\theta}{d\lambda} + \frac{1}{2}\theta^2 + \sigma^{\alpha\beta}\sigma_{\alpha\beta} = \kappa\,\theta - 8\pi T_{\alpha\beta} k^\alpha k^\beta,$$

where θ and $\sigma_{\alpha\beta}$ are the expansion and shear of the congruence, respectively; the equation holds on either side of Σ. Because it depends only on the intrinsic geometry of the hypersurface, the left-hand side of Raychaudhuri's equation is guaranteed to be continuous across the shell. Continuity of the right-hand side therefore implies

$$[\kappa]\theta = 8\pi \lfloor T_{\alpha\beta} k^\alpha k^\beta \rfloor. \qquad (3.103)$$

This relation shows that $[\kappa] \neq 0$ (and therefore $p \neq 0$) whenever the component $T_{\alpha\beta} k^\alpha k^\beta$ of the stress-energy tensor is discontinuous at the shell. We conclude that λ cannot be an affine parameter on both sides of Σ when $[T_{\alpha\beta} k^\alpha k^\beta] \neq 0$. (Notice that this conclusion breaks down when $\theta = 0$, that is, when the shell is stationary.)

Recalling (Section 2.4.8) that the expansion θ is equal to the fractional rate of change of the congruence's cross-sectional area, we find that with the help of Eq. (3.102), Eq. (3.103) can be expressed as

$$p\,\frac{d}{d\lambda} dS + [T_{\alpha\beta} k^\alpha k^\beta] dS = 0, \qquad (3.104)$$

where $dS = \sqrt{\sigma}\, d^2\theta$ is an element of cross-sectional area on the shell (Section 3.2.2). This equation has a simple interpretation: The first term represents the work done by the shell as it expands or contracts, while the second term is the energy absorbed by the shell from its surroundings; Eq. (3.104) therefore states that all of the absorbed energy goes into work.

Having established that λ cannot, in general, be an affine parameter on both sides of the hypersurface, let us now investigate how a change of parameterization might affect the surface density μ, surface current j^A, and surface pressure p of the null shell. Because each generator can be reparameterized independently of any other generator, we must consider transformations of the form

$$\lambda \to \bar{\lambda}(\lambda, \theta^A). \tag{3.105}$$

The question before us is: How do the surface quantities change under such a transformation?

To answer this we need to work out how the transformation of Eq. (3.105) affects the vectors k^α, e^α_A, and N^α. We first note that the differential form of Eq. (3.105) is

$$d\bar{\lambda} = e^\beta\, d\lambda + c_A\, d\theta^A, \tag{3.106}$$

where

$$e^\beta \equiv \left(\frac{\partial\bar{\lambda}}{\partial\lambda}\right)_{\theta^A}, \qquad c_A \equiv \left(\frac{\partial\bar{\lambda}}{\partial\theta^A}\right)_\lambda; \tag{3.107}$$

both e^β and c_A depend on $y^a = (\lambda, \theta^A)$, but because they depend on the intrinsic coordinates only, we have that $[e^\beta] = 0 = [c_A]$. A displacement within the hypersurface can then be described either by

$$dx^\alpha = k^\alpha\, d\lambda + e^\alpha_A\, d\theta^A,$$

where $k^\alpha = (\partial x^\alpha/\partial\lambda)_{\theta_A}$ and $e^\alpha_A = (\partial x^\alpha/\partial\theta^A)_\lambda$, or by

$$dx^\alpha = \bar{k}^\alpha\, d\bar{\lambda} + \bar{e}^\alpha_A\, d\theta^A,$$

where $\bar{k}^\alpha = (\partial x^\alpha/\partial\bar{\lambda})_{\theta_A}$ and $\bar{e}^\alpha_A = (\partial x^\alpha/\partial\theta^A)_{\bar{\lambda}}$; these relations hold on either side of Σ, in the relevant coordinate system $x^\alpha_\pm$. Using Eq. (3.106), it is easy to see that the tangent vectors transform as

$$\bar{k}^\alpha = e^{-\beta} k^\alpha, \qquad \bar{e}^\alpha_A = e^\alpha_A - c_A e^{-\beta} k^\alpha \tag{3.108}$$

under the reparameterization of Eq. (3.105). It may be checked that the new basis vectors satisfy the orthogonality relations (3.85), and that the induced metric σ_{AB} is invariant under this transformation: $\bar{\sigma}_{AB} \equiv g_{\alpha\beta}\bar{e}^\alpha_A\bar{e}^\beta_B = g_{\alpha\beta}e^\alpha_A e^\beta_B \equiv \sigma_{AB}$.

To preserve the relations (3.88) we let the new auxiliary null vector be

$$\bar{N}^\alpha = e^\beta N^\alpha + \frac{1}{2} c^A c_A e^{-\beta} k^\alpha - c^A e_A^\alpha, \tag{3.109}$$

where $c^A = \sigma^{AB} c_B$. This ensures that the completeness relations (3.89) take the same form in the new basis.

It is a straightforward (but slightly tedious) task to compute how the transverse curvature C_{ab} changes under a reparameterization of the generators, and to then compute how the surface quantities transform. You will be asked to go through this calculation in Section 3.13, Problem 8. The answer is that under the reparameterization of Eq. (3.105), the surface quantities transform as

$$\bar{\mu} = e^\beta \mu + 2c_A j^A + c^A c_A e^{-\beta} p,$$
$$\bar{j}^A = j^A + c^A e^{-\beta} p, \tag{3.110}$$
$$\bar{p} = e^{-\beta} p.$$

These transformations, together with Eq. (3.108), imply that the surface stress-energy tensor becomes $\bar{S}^{\alpha\beta} = e^{-\beta} S^{\alpha\beta}$. We also have $(-\bar{k}_\mu u^\mu)^{-1} = e^\beta (-k_\mu u^\mu)$, and these results reveal that the combination $(-k_\mu u^\mu)^{-1} S^{\alpha\beta}$ is *invariant* under the reparameterization. This, finally, establishes the invariance of $T_\Sigma^{\alpha\beta}$, the full stress-energy tensor of the surface layer.

As a final remark, we note that under the reparameterization of Eq. (3.105), the physically-measured surface quantities transform as

$$\bar{\mu}_{\text{physical}} = e^{2\beta} \mu_{\text{physical}} + 2c_A e^\beta j^A_{\text{physical}} + c^A c_A p_{\text{physical}},$$
$$\bar{j}^A_{\text{physical}} = e^\beta j^A_{\text{physical}} + c^A p_{\text{physical}}, \tag{3.111}$$
$$\bar{p}_{\text{physical}} = p_{\text{physical}};$$

we see in particular that the physically-measured surface pressure is an invariant.

3.11.6 Imploding spherical shell

For our first application of the null-shell formalism, we take another look at the gravitational collapse of a thin spherical shell, a problem that was first formulated in Section 3.9. Here we imagine that the collapse proceeds at the speed of light, and that the thin shell lies on a null hypersurface Σ. We take spacetime to be flat inside the shell (in $\mathscr{V}^-$), and write the metric there as

$$ds_-^2 = -dt_-^2 + dr^2 + r^2 d\Omega^2,$$

in terms of spatial coordinates (r, θ, ϕ) and a time coordinate t_-. The metric outside the shell (in $\mathscr{V}^+$) is the Schwarzschild solution,

$$ds_+^2 = -f\, dt_+^2 + f^{-1}\, dr^2 + r^2\, d\Omega^2,$$

which is expressed in the same spatial coordinates but in terms of a distinct time t_+; here, $f = 1 - 2M/r$ and M denotes the gravitational mass of the collapsing shell.

As seen from $\mathscr{V}^-$, the null hypersurface Σ is described by the equation $t_- + r \equiv v_- = \text{constant}$, which means that the induced metric on Σ is given by $ds_\Sigma^2 = r^2\, d\Omega^2$. As seen from $\mathscr{V}^+$, on the other hand, the hypersurface is described by $t_+ + r^*(r) \equiv v_+ = \text{constant}$, where $r^*(r) = \int f^{-1}\, dr = r + 2M \ln(r/2M - 1)$, and this gives rise to the same induced metric. From these considerations we see that it was permissible to express the metrics of $\mathscr{V}^\pm$ in terms of the same spatial coordinates (r, θ, ϕ), but that t_+ cannot be equal to t_-. The induced metric on the shell is

$$\sigma_{AB}\, d\theta^A d\theta^B = \lambda^2 \big(d\theta^2 + \sin^2 \theta\, d\phi^2\big),$$

where we have set $\theta^A = (\theta, \phi)$ and identified $-r$ with the parameter λ on the null generators of the hypersurface; we shall see that here, λ is an affine parameter on both sides of Σ.

As seen from $\mathscr{V}^-$, the parametric equations $x_-^\alpha = x_-^\alpha(\lambda, \theta^A)$ that describe the hypersurface have the explicit form $t_- = v_- + \lambda$, $r = -\lambda$, $\theta = \theta$, and $\phi = \phi$. These give us the tangent vectors $k^\alpha \partial_\alpha = \partial_t - \partial_r$, $e_\theta^\alpha \partial_\alpha = \partial_\theta$, and $e_\phi^\alpha \partial_\alpha = \partial_\phi$, and the basis is completed by $N_\alpha\, dx^\alpha = -\frac{1}{2}(dt - dr)$. From all this and Eq. (3.100) we find that the nonvanishing components of the transverse curvature are

$$C_{AB}^- = \frac{1}{2r}\, \sigma_{AB}.$$

The fact that $C_{\lambda\lambda}^- = 0$ confirms that $\lambda \equiv -r$ is an affine parameter on the $\mathscr{V}^-$ side of Σ.

As seen from $\mathscr{V}^+$, the parametric equations are $t_+ = v_+ - r^*(-\lambda)$, $r = -\lambda$, $\theta = \theta$, and $\phi = \phi$. The basis vectors are $k^\alpha \partial_\alpha = f^{-1}\partial_t - \partial_r$, $e_\theta^\alpha \partial_\alpha = \partial_\theta$, $e_\phi^\alpha \partial_\alpha = \partial_\phi$, and $N_\alpha\, dx^\alpha = -\frac{1}{2}(f\, dt - dr)$. The nonvanishing components of the transverse curvature are now

$$C_{AB}^+ = \frac{f}{2r}\, \sigma_{AB}.$$

The fact that $C_{\lambda\lambda}^+ = 0$ confirms that $\lambda \equiv -r$ is an affine parameter on the $\mathscr{V}^+$ side of Σ; λ is therefore an affine parameter on both sides.

The angular components of the transverse curvature are discontinuous across the shell: $[C_{AB}] = -(M/r^2)\sigma_{AB}$. According to Eq. (3.101), this means that the shell has a vanishing surface current j^A and a vanishing surface pressure p, but that its surface density is

$$\mu = \frac{M}{4\pi r^2}.$$

We have obtained the very sensible result that the surface density of a collapsing null shell is equal to its gravitational mass divided by its (ever decreasing) surface area. Notice that $\mu_{\text{physical}} = \mu$ for observers at rest in $\mathscr{V}^-$. Because of the focusing action of the null shell, however, these observers do not remain at rest after crossing over to the $\mathscr{V}^+$ side: A simple calculation, based on Eq. (3.91), reveals that an observer at rest before crossing the shell will move according to $dr/d\tau = -(\tilde{E}^2 - f)^{1/2}$ after crossing the shell; the energy parameter $\tilde{E}$ varies from observer to observer, and is related by $\tilde{E} = 1 - M/r_\Sigma$ to the radius r_Σ at which a given observer crosses the hypersurface.

3.11.7 Accreting black hole

Our second application of the null-shell formalism features a nonrotating black hole of mass $(M - m)$ which suddenly acquires additional material of mass m and angular momentum $J \equiv aM$. We suppose that the accretion process is virtually instantaneous, that the material falls in with the speed of light, and that $J \ll M^2$. We idealize the accreting material as a singular matter distribution supported on a null hypersurface Σ.

The spacetime in the future of Σ (in $\mathscr{V}^+$) is that of a slowly rotating black hole of mass M and (small) angular momentum aM. We write the metric in $\mathscr{V}^+$ as in Eq. (3.71),

$$ds_+^2 = -f\,dt^2 + f^{-1}\,dr^2 + r^2\,d\Omega^2 - \frac{4Ma}{r}\sin^2\theta\,dt\,d\psi,$$

where $f = 1 - 2M/r$; this is the slow-rotation limit of the Kerr metric, and throughout this subsection we will work consistently to first order in the small parameter a.

As seen from $\mathscr{V}^+$, the null hypersurface Σ is described by $v \equiv t + r^* = 0$, where $r^* = \int f^{-1}\,dr = r + 2M\ln(r/2M - 1)$; you may check that in the slow-rotation limit, every surface $v = $ constant is null. It follows that the vector $k^\alpha = g^{\alpha\beta}(-\partial_\beta v)$ is normal to Σ and tangent to its null generators. We have

$$k^\alpha \partial_\alpha = \frac{1}{f}\,\partial_t - \partial_r + \frac{2Ma}{r^3 f}\,\partial_\phi,$$

and from this expression we deduce four important properties of the generators. First, the generators are affinely parameterized by $\lambda \equiv -r$. Second, as measured by inertial observers at infinity, the generators move with an (ever increasing) angular velocity

$$\frac{d\phi}{dt} \equiv \Omega_{\text{generators}} = \frac{2Ma}{r^3}.$$

Third, θ is constant on each generator. And fourth, integration of $d\phi/(-dr) = 2Ma/(r^3 f)$ reveals that

$$\psi \equiv \phi + \frac{a}{r}\left(1 + \frac{r}{2M} \ln f\right)$$

also is constant on the generators.

We shall use $y^a = (\lambda \equiv -r, \theta, \psi)$ as coordinates on Σ; as we have just seen, these coordinates are well adapted to the generators, and this property is required by the null-shell formalism. Remembering that $dt = -dr/f$ and $d\phi = d\psi - (2Ma/r^3 f)\,dr$ on Σ, we find that the induced metric is

$$\sigma_{AB}\,d\theta^A d\theta^B = r^2\big(d\theta^2 + \sin^2\theta\,d\psi^2\big),$$

and that the hypersurface is intrinsically spherical.

The parametric description of Σ, as seen from $\mathscr{V}^+$, is $x^\alpha(-r, \theta, \psi)$, and from this we form the tangent vectors $e_\lambda^\alpha = k^\alpha$, $e_\theta^\alpha = \delta_\theta^\alpha$, and $e_\psi^\alpha = \delta_\phi^\alpha$. The basis is completed by $N_\alpha\,dx^\alpha = \frac{1}{2}(-f\,dt + dr)$. From Eq. (3.100) we obtain

$$C_{\lambda\psi}^+ = \frac{3Ma}{r^2}\sin^2\theta, \qquad C_{AB}^+ = \frac{f}{2r}\sigma_{AB}$$

for the nonvanishing components of the transverse curvature.

The spacetime in the past of Σ (in $\mathscr{V}^-$) is that of a nonrotating black hole of mass $(M - m)$. Here we write the metric as

$$ds_-^2 = -F\,d\bar{t}^2 + F^{-1}\,dr^2 + r^2\big(d\theta^2 + \sin^2\theta\,d\psi^2\big),$$

in terms of a distinct time coordinate $\bar{t}$ and the angles θ and ψ; we also have $F \equiv 1 - 2(M - m)/r$. This choice of angular coordinates implies that inertial observers within $\mathscr{V}^-$ corotate with the shell's null generators; this is another manifestation of the dragging of inertial frames, a phenomenon already encountered in Section 3.10. As we shall see presently, this choice of coordinates is dictated by continuity of the induced metric at Σ.

The mathematical description of the hypersurface, as seen from $\mathscr{V}^-$, is identical to its external description provided that we make the substitutions $t \to \bar{t}$, $\phi \to \psi$, $M \to M - m$, and $a \to 0$. According to this, the induced metric on Σ is still given by $ds_\Sigma^2 = r^2(d\theta^2 + \sin^2\theta\,d\psi^2)$, as required. The basis vectors are now $k^\alpha\partial_\alpha =$

$F^{-1}\partial_{\bar{t}} - \partial_r$, $e^\alpha_\theta \partial_\alpha = \partial_\theta$, $e^\alpha_\psi \partial_\alpha = \partial_\psi$, and $N_\alpha \, dx^\alpha = \frac{1}{2}(-F \, d\bar{t} + dr)$. This gives us

$$C^-_{AB} = \frac{F}{2r} \sigma_{AB}$$

for the nonvanishing components of the transverse curvature.

The transverse curvature is discontinuous at Σ, and Eqs. (3.101) allow us to compute the shell's surface quantities. Because the generators are affinely parameterized by $-r$ on both sides of the shell, we have that $p = 0$ – the shell has a vanishing surface pressure. On the other hand, its surface density is given by

$$\mu = \frac{m}{4\pi r^2},$$

the ratio of the shell's gravitational mass m to its (ever decreasing) surface area $4\pi r^2$. Thus far our results are virtually identical to those obtained in the preceding subsection. What is new in this context is the presence of a surface current j^A, whose sole component is

$$j^\psi = \frac{3Ma}{8\pi r^4}.$$

This comes from the shell's rotation, and the fact that the situation is not entirely spherically symmetric.

To better understand the physical significance of the surface current, we express the shell's surface stress-energy tensor,

$$S^{\alpha\beta} = \mu k^\alpha k^\beta + j^\psi \left(k^\alpha e^\beta_\psi + e^\alpha_\psi k^\beta \right),$$

in terms of the vector $\ell^\alpha \equiv k^\alpha + (j^\psi/\mu)\, e^\alpha_\psi$. This vector is null (when we appropriately discard terms of order a^2 in the calculation of $g_{\alpha\beta}\ell^\alpha \ell^\beta$), and it has the components

$$\ell^\alpha \partial_\alpha = \frac{1}{f}\,\partial_t - \partial_r + \frac{1}{f}\,\Omega_{\text{fluid}}\partial_\phi$$

in the coordinates $x^\alpha = (t, r, \theta, \phi)$ used in $\mathscr{V}^+$; we have set

$$\Omega_{\text{fluid}} \equiv \frac{2Ma}{r^3} + \frac{3Ma}{2mr}\,f.$$

The shell's surface stress-energy tensor is now given by the simple expression

$$S^{\alpha\beta} = \mu\, \ell^\alpha \ell^\beta,$$

which corresponds to a pressureless fluid of density μ moving with a four-velocity ℓ^α. We see that the fluid is moving along null curves (not geodesics!) that do not coincide with the shell's null generators. The motion across generators is created by a mismatch between Ω_{fluid}, the fluid's angular velocity, and $\Omega_{\text{generators}}$, the angular velocity of the generators. The mismatch is directly related to j^A:

$$\Omega_{\text{relative}} \equiv \Omega_{\text{fluid}} - \Omega_{\text{generators}} = \frac{j^\psi}{\mu} = \frac{3Ma}{2mr}\, f.$$

Notice that the fluid rotates faster than the generators, which share their angular velocity with inertial observers within $\mathscr{V}^-$; such a phenomenon was encountered before, in the context of the stationary rotating shell of Section 3.10. But notice also that Ω_{relative} decreases to zero as r approaches $2M$: The fluid ends up corotating with the generators when the shell crosses the black-hole horizon.

3.11.8 Cosmological phase transition

In this third (and final) application of the formalism, we consider an intriguing (but entirely artificial) cosmological scenario according to which the universe was initially expanding in two directions only, but was then made to expand isotropically by a sudden explosive event.

The $\mathscr{V}^-$ region of spacetime is the one in which the universe is expanding in the x and y directions only. Its metric is

$$ds_-^2 = -dt^2 + a^2(t)\big(dx^2 + dy^2\big) + dz_-^2,$$

and the scale factor is assumed to be given by $a(t) \propto t^{1/2}$. The cosmological fluid moves with a four-velocity $u^\alpha = \partial x^\alpha / \partial t$, and it has a density and (isotropic) pressure given by $\rho_- = p_- = 1/(32\pi t^2)$, respectively.

In the $\mathscr{V}^+$ region of spacetime, the universe expands uniformly in all three directions. Here the metric is

$$ds_+^2 = -dt^2 + a^2(t)\big(dx^2 + dy^2 + dz_+^2\big),$$

with the same scale factor $a(t)$ as in $\mathscr{V}^-$, and the cosmological fluid has a density and pressure given by $\rho_+ = 3p_+ = 3/(32\pi t^2)$, respectively; this corresponds to a radiation-dominated universe.

The history of the explosive event that changes the metric from $g_{\alpha\beta}^-$ to $g_{\alpha\beta}^+$ traces a null hypersurface Σ in spacetime. This surface moves in the positive $z_\pm$ direction and as we shall see, it supports a singular stress-energy tensor. The 'agent' that alters the course of the universe's expansion is therefore a null shell.

As seen from $\mathscr{V}^-$, the hypersurface is described by $t = z_- + \text{constant}$, and the vector $k^\alpha \partial_\alpha = \partial_t + \partial_z$ is tangent to the null generators, which are parameterized by t. In fact, because $k^\alpha_{\;;\beta} k^\beta = 0$ we have that t is an affine parameter on this side of the hypersurface. The coordinates x and y are constant on the generators, and we use them, together with t, as intrinsic coordinates on Σ. We therefore have

$y^a = (t, \theta^A)$, $\theta^A = (x, y)$, and the shell's induced metric is

$$\sigma_{AB} \, d\theta^A \, d\theta^B = a^2(t)(dx^2 + dy^2).$$

The remaining basis vectors are $e_x^\alpha \partial_\alpha = \partial_x$, $e_y^\alpha \partial_\alpha = \partial_y$, and $N_\alpha \, dx^\alpha = -\frac{1}{2}(dt + dz_-)$. The nonvanishing components of the transverse curvature are

$$C_{AB}^- = \frac{1}{4t} \sigma_{AB}.$$

We note that on the $\mathscr{V}^-$ side of Σ, the null generators have an expansion given by $\theta = k^\alpha_{;\alpha} = 1/t$, and that $T_{\alpha\beta} k^\alpha k^\beta = \rho_- + p_- = 1/(16\pi t^2)$, where $T^{\alpha\beta}$ is the stress-energy tensor of the cosmological fluid.

As seen from $\mathscr{V}^+$, the description of the hypersurface is obtained by integrating $dt = a(t) \, dz_+$, and $k^\alpha \partial_\alpha = \partial_t + a^{-1} \partial_z$ is tangent to the null generators. We note that t is *not* an affine parameter on this side of the hypersurface: we have that $k^\alpha_{;\beta} k^\beta = (2t)^{-1} k^\alpha$. The remaining basis vectors are $e_x^\alpha \partial_\alpha = \partial_x$, $e_y^\alpha \partial_\alpha = \partial_y$, $N_\alpha \, dx^\alpha = -\frac{1}{2}(dt + a \, dz_+)$, and the nonvanishing components of the transverse curvature are now

$$C_{tt}^+ = \frac{1}{2t}, \qquad C_{AB}^+ = \frac{1}{4t} \sigma_{AB}.$$

On this side of Σ, the generators have an expansion also given by $\theta = 1/t$ (since continuity of θ is implied by continuity of the induced metric), and $T_{\alpha\beta} k^\alpha k^\beta = \rho_+ + p_+ = 1/(8\pi t^2)$.

The fact that t is an affine parameter on one side of the hypersurface only tells us that the shell must possess a surface pressure. In fact, continuity of C_{AB} across the shell implies that p is the *only* nonvanishing surface quantity. It is given by

$$p = -\frac{1}{16\pi t},$$

the negative sign indicating that this surface quantity would be better described as a *tension*, not a pressure. The shell's surface stress-energy tensor is $S^{\alpha\beta} = p \sigma^{AB} e_A^\alpha e_B^\alpha$. If we select observers comoving with the cosmological fluid as our preferred observers to make measurements on the shell, then $-k_\alpha u^\alpha = 1$ and the full stress-energy tensor of the singular hypersurface is $T_\Sigma^{\alpha\beta} = S^{\alpha\beta} \delta(t - t_\Sigma)$, with t_Σ denoting the time at which a given observer crosses the shell. We see that for these observers, $-p$ is the physically-measured surface tension.

Finally, we note that the expressions $-p = 1/(16\pi t)$, $\theta = 1/t$, and $[T_{\alpha\beta} k^\alpha k^\beta] = 1/(16\pi t^2)$ are compatible with the general relation $-p\theta = [T_{\alpha\beta} k^\alpha k^\beta]$ derived in Section 3.11.5. This shows that the energy released by the shell as it expands is absorbed by the cosmological fluid, whose density increases by a factor of $\rho_+/\rho_- = 3$; this energy is provided by the shell's surface tension.

3.12 Bibliographical notes

During the preparation of this chapter I have relied on the following references: Barrabès and Israel (1991); Barrabès and Hogan (1998); de la Cruz and Israel (1968); Israel (1966); Misner, Thorne, and Wheeler (1973); Musgrave and Lake (1997); and Wald (1984).

More specifically:

Sections 3.1, 3.2, and 3.3 are based partially on unpublished lecture notes by Werner Israel. Sections 3.4, 3.5, and 3.9 are based on Israel's paper. Section 3.6 is based on Section 10.2 of Wald. Sections 3.7 and 3.11 (as well as Problem 9 below) are based on Barrabès and Israel. Section 3.8 is based on Exercise 32.4 of Misner, Thorne, and Wheeler. Section 3.10 is based on de la Cruz and Israel. Finally, the examples of Sections 3.11.7 and 3.11.8 are adapted from Musgrave and Lake, and Barrabès and Hogan, respectively.

Suggestions for further reading:

Solving the initial-value problem of general relativity is an important aspect of numerical relativity, and a lot of effort is currently devoted to finding initial data that involve compact bodies in astrophysically realistic situations. The situation is reviewed by Greg Cook in his 2000 *Living Reviews* article.

Could our four-dimensional universe be a singular hypersurface in an extended five-dimensional world? This intriguing idea, a variation on the old Kaluza–Klein scenario, was proposed recently by Randall and Sundrum (1999a and 1999b). The intense scientific activity that followed the publication of their papers is reviewed by Brax and van de Bruck (2003).

3.13 Problems

Warning: The results derived in Problem 1 are used in later portions of this book.

1. We consider a hypersurface $T = $ constant in Schwarzschild spacetime, where

$$T = t + 4M \left[\sqrt{r/2M} + \frac{1}{2} \ln\left(\frac{\sqrt{r/2M} - 1}{\sqrt{r/2M} + 1} \right) \right].$$

We use (r, θ, ϕ) as coordinates on the hypersurface.

(a) Calculate the unit normal n_α and find parametric equations that describe the hypersurface.

(b) Calculate the induced metric h_{ab}.

(c) Calculate the extrinsic curvature K_{ab}. Verify that your results agree with those of Section 3.6.6, and show that K is equal to the expansion of the congruence considered in Section 2.3.7.

(d) Prove that when it is expressed in terms of the coordinates (T, r, θ, ϕ), the Schwarzschild metric takes the form

$$ds^2 = -dT^2 + \left(dr + \sqrt{2M/r}\, dT\right)^2 + r^2\, d\Omega^2.$$

This shows very clearly that the sections $T = \text{constant}$ are intrinsically flat. [This coordinate system was discovered independently by Painlevé (1921) and Gullstrand (1922). It is presented in some detail in a 2001 paper by Martel and Poisson.]

2. A four-dimensional hypersurface is embedded in a flat, five-dimensional spacetime. We use coordinates z^A in the five-dimensional world, and express the metric as

$$ds^2 = \eta_{AB}\, dz^A\, dz^B = -(dz^0)^2 + (dz^1)^2 + (dz^2)^2 + (dz^3)^2 + (dz^4)^2;$$

we let uppercase Latin indices run from 0 to 4. In the four-dimensional world we use coordinates $x^\alpha = (t, \chi, \theta, \phi)$. The hypersurface is defined by parametric relations $z^A(x^\alpha)$. Explicitly,

$$z^0 = a\,\sinh(t/a), \quad z^1 = a\,\cosh(t/a)\cos\chi, \quad z^2 = a\,\cosh(t/a)\sin\chi\cos\theta,$$

$$z^3 = a\,\cosh(t/a)\sin\chi\sin\theta\cos\phi, \quad z^4 = a\,\cosh(t/a)\sin\chi\sin\theta\sin\phi,$$

where a is a constant.

(a) Compute the unit normal n^A and the tangent vectors $e_\alpha^A = \partial z^A/\partial x^\alpha$ to the hypersurface.

(b) Compute the induced metric $g_{\alpha\beta}$. What is the physical significance of this four-dimensional metric? Does it satisfy the Einstein field equations?

(c) Compute the extrinsic curvature $K_{\alpha\beta}$. Use the Gauss–Codazzi equations to prove that the induced Riemann tensor can be expressed as

$$R_{\alpha\beta\gamma\delta} = \frac{1}{a^2}\left(g_{\alpha\gamma}g_{\beta\delta} - g_{\alpha\delta}g_{\beta\gamma}\right).$$

This implies that the four-dimensional hypersurface is a spacetime of constant Ricci curvature.

3. In this problem we consider a spherically symmetric space at a moment of time symmetry. We write the three-metric as

$$ds^2 = d\ell^2 + r^2(\ell)\, d\Omega^2,$$

where ℓ is proper distance from the centre.

(a) Show that in these coordinates, the mass function introduced in Section 3.6.5 is given by

$$m(r) = \frac{r}{2}\left[1 - (dr/d\ell)^2\right].$$

(b) Solve the constraint equations for a uniform mass density ρ on the hypersurface. Make sure to impose the asymptotic condition $r(\ell \to 0) \to \ell$ to force the three-metric to be regular at the centre.

(c) Prove that $r(\ell)$ can be no larger than $r_{\max} = \sqrt{3/(8\pi\rho)}$.

(d) Prove that $2m(r_{\max}) = r_{\max}$, and that $m(r_{\max})$ is the maximum value of the mass function.

(e) What happens when $\ell \to \pi r_{\max}$?

4. Prove the statement made near the end of Section 3.7.5, that $[K_{ab}] = 0$ is a sufficient condition for the regularity of the full Riemann tensor at the hypersurface Σ.

5. Prove that the surface stress-energy tensor of a thin shell satisfies the conservation equation

$$S^{ab}_{\ \ |b} = -\varepsilon[j^a],$$

where $j_a \equiv T_{\alpha\beta}e^\alpha_a n^\beta$. Interpret this equation physically. (Consider the case where the shell is timelike.)

6. The metric

$$ds^2 = -dt^2 + d\ell^2 + r^2(\ell)\,d\Omega^2,$$

where $r(\ell) = \ell$ when $0 < \ell < \ell_0$ and $r(\ell) = 2\ell_0 - \ell$ when $\ell_0 < \ell < 2\ell_0$, describes a spacetime with closed spatial sections. (What is the volume of a hypersurface $t = $ constant?) The spacetime is flat in both $\mathscr{V}^-$ ($\ell < \ell_0$) and $\mathscr{V}^+$ ($\ell > \ell_0$), but it contains a surface layer at $\ell = \ell_0$.

(a) Calculate the surface stress-energy tensor of the thin shell. Express this in terms of a velocity field u^a, a density σ, and a surface pressure p.

(b) Consider a congruence of outgoing null geodesics in this spacetime, with its tangent vector $k_\alpha = -\partial_\alpha(t - \ell)$. Calculate θ, the expansion of this congruence. Show that it abruptly changes sign (from positive to negative) at $\ell = \ell_0$. The surface layer therefore produces a strong focusing of the null geodesics.

(c) Use Raychaudhuri's equation to prove that the discontinuity in $d\theta/d\ell$ is precisely accounted for by the surface stress-energy tensor.

7. Two Schwarzschild solutions, one with mass parameter m_-, the other with mass parameter m_+, are joined at a radius $r = R(\tau)$ by means of a spherical thin shell; τ denotes proper time for an observer comoving with the shell. It is assumed that m_- is the interior mass (m_+ is the exterior mass), that $m_+ > m_-$, and that $R(\tau) > 2m_+$ for all values of τ. The shell's surface stress-energy tensor is given by

$$S^{ab} = (\sigma + p)u^a u^b + ph^{ab},$$

where u^a is the fluid's velocity field, $\sigma(\tau)$ the surface density, $p(\tau)$ the surface pressure, and h_{ab} the induced metric.

(a) Derive, and interpret physically, the equation

$$\frac{\mathrm{d}}{\mathrm{d}\tau}(\sigma R^2) + p \frac{\mathrm{d}}{\mathrm{d}\tau}(R^2) = 0.$$

(b) Find the values of σ and p which produce a static configuration: $R(\tau) = R_0 = \text{constant}$. Verify that both σ and p are positive. [The stability of these static configurations was examined by Brady, Louko, and Poisson (1991).]

8. Derive the relations (3.109).

9. Let spacetime be partitioned into two regions $\mathscr{V}^{\pm}$ with metrics

$$\mathrm{d}s_{\pm}^2 = -f_{\pm}\,\mathrm{d}v^2 + 2\,\mathrm{d}v\mathrm{d}r + r^2\,\mathrm{d}\Omega^2.$$

We assume that the coordinate system (v, r, θ, ϕ) is common to both $\mathscr{V}^{-}$ and $\mathscr{V}^{+}$. (In each region we could introduce a conventional time coordinate $t_{\pm}$ defined by $\mathrm{d}t_{\pm} = \mathrm{d}v - \mathrm{d}r/f_{\pm}$, but it is much more convenient to work with the original system.) In $\mathscr{V}^{-}$ we set $f_{-} = 1 - r_0/r$, so that the metric is a Schwarzschild solution with mass parameter $M \equiv \frac{1}{2}r_0$. In $\mathscr{V}^{+}$ we set $f_{+} = 1 - (r/r_0)^2$, so that the metric is a de Sitter solution with cosmological constant $\Lambda \equiv 3/r_0^2$. (This metric is a solution to the modified Einstein field equations, $G_{\alpha\beta} + \Lambda g_{\alpha\beta} = 0$.) The boundary Σ between the two regions is the null surface $r = r_0$, the common horizon of the Schwarzschild and de Sitter spacetimes.

Using $y^a = (v, \theta, \phi)$ as coordinates on Σ, calculate the surface quantities μ, j^A, and p associated with the null shell. Explain whether your results are compatible with the general relation $p\,\theta = [T_{\alpha\beta}k^{\alpha}k^{\beta}]$ derived in Section 3.11.5.

4

Lagrangian and Hamiltonian formulations
of general relativity

Variational principles play a fundamental role in virtually all areas of physics, and general relativity is no exception. This chapter is devoted to a general discussion of the Lagrangian and Hamiltonian formulations of field theories in curved spacetime, with a special focus on general relativity.

The Lagrangian formulation of a field theory (Section 4.1) begins with the introduction of an action functional, which is usually defined as an integral of a Lagrangian density over a finite region $\mathcal{V}$ of spacetime. As we shall see, general relativity is peculiar in this respect, as its action involves also an integration over $\partial \mathcal{V}$, the boundary of the region $\mathcal{V}$; this is necessary for the well-posedness of the variational principle. We will, in this chapter, provide a systematic treatment of the boundary terms in the gravitational action.

The Hamiltonian formulation of a field theory (Section 4.2) involves a decomposition of spacetime into space and time. Geometrically, this corresponds to a foliation of spacetime by nonintersecting spacelike hypersurfaces Σ. In this $3 + 1$ decomposition, the spacetime metric $g_{\alpha\beta}$ is broken down into an induced metric h_{ab}, a shift vector N^a, and a lapse scalar N; while the induced metric is concerned with displacements within Σ, the lapse and shift are concerned with displacements away from the hypersurface. The Hamiltonian is a functional of the field configuration and its conjugate momentum on Σ. In general relativity, the Hamiltonian is a functional of h_{ab} and its conjugate momentum p^{ab}, which is closely related to the extrinsic curvature of the hypersurface Σ; the lapse and shift are freely specifiable, and they do not appear in the Hamiltonian as dynamical variables. The gravitational Hamiltonian inherits boundary terms from the action functional; those are defined on the two-surface S formed by the intersection of $\partial \mathcal{V}$ and Σ.

There is a close connection between the gravitational Hamiltonian and the total mass M and angular momentum J of an asymptotically-flat spacetime; this connection is explored in Section 4.3. We will see that the value of the gravitational Hamiltonian for a solution to the Einstein field equations depends only on the conditions at the two-dimensional boundary S. When the spacetime is asymptotically flat and S is pushed to infinity, the Hamiltonian becomes M if the lapse and shift are chosen so as to correspond to an asymptotic time translation. For an alternative choice of lapse and shift, corresponding to an asymptotic rotation about an axis, the Hamiltonian becomes J, the component of the angular-momentum vector along this axis. These Hamiltonian definitions for mass and angular momentum form the starting point of a rather broad review of the different notions of mass and angular momentum in general relativity.

4.1 Lagrangian formulation

4.1.1 Mechanics

In the Lagrangian formulation of Newtonian mechanics, one is given a *Lagrangian* $L(q, \dot{q})$, a function of the generalized coordinate q and its velocity $\dot{q} \equiv dq/dt$. One then forms an *action functional* $S[q]$,

$$S[q] = \int_{t_1}^{t_2} L(q, \dot{q}) \, dt, \tag{4.1}$$

by integrating the Lagrangian over a selected path $q(t)$. The path that satisfies the equations of motion is the one about which $S[q]$ is stationary: Under a variation $\delta q(t)$ of this path, restricted by

$$\delta q(t_1) = \delta q(t_2) = 0 \tag{4.2}$$

but otherwise arbitrary in the interval $t_1 < t < t_2$, the action does not change, $\delta S = 0$.

The change in the action is given by

$$\delta S = \int_{t_1}^{t_2} \delta L \, dt$$

$$= \int_{t_1}^{t_2} \left(\frac{\partial L}{\partial q} \delta q + \frac{\partial L}{\partial \dot{q}} \delta \dot{q} \right) dt$$

$$= \frac{\partial L}{\partial \dot{q}} \delta q \bigg|_{t_1}^{t_2} + \int_{t_1}^{t_2} \left(\frac{\partial L}{\partial q} - \frac{d}{dt} \frac{\partial L}{\partial \dot{q}} \right) \delta q \, dt,$$

where, in the last step, we have used $\delta \dot{q} = d(\delta q)/dt$ and integrated by parts. The boundary terms vanish by virtue of Eq. (4.2). Because the variation is arbitrary between t_1 and t_2,

$$\delta S = 0 \quad \Rightarrow \quad \frac{d}{dt} \frac{\partial L}{\partial \dot{q}} - \frac{\partial L}{\partial q} = 0. \tag{4.3}$$

This is the *Euler–Lagrange* equation for a one-dimensional mechanical system. Generalization to higher dimensions is immediate.

4.1.2 Field theory

We now consider the dynamics of a field $q(x^\alpha)$ in curved spacetime. Although this field could be of any type (scalar, vector, tensor, spinor), for simplicity we shall restrict our attention to the case of a scalar field.

In the Lagrangian formulation of a field theory, one is given an arbitrary region $\mathcal{V}$ of the spacetime manifold, bounded by a closed hypersurface $\partial\mathcal{V}$. One is also given a *Lagrangian density* $\mathscr{L}(q, q_{,\alpha})$, a scalar function of the field and its first derivatives. The action functional is then

$$S[q] = \int_{\mathcal{V}} \mathscr{L}(q, q_{,\alpha})\sqrt{-g}\,\mathrm{d}^4 x. \qquad (4.4)$$

Dynamical equations for q are obtained by introducing a variation $\delta q(x^\alpha)$ that is arbitrary within $\mathcal{V}$ but vanishes everywhere on $\partial\mathcal{V}$,

$$\delta q\big|_{\partial\mathcal{V}} = 0, \qquad (4.5)$$

and by demanding that δS vanish if the variation is about the actual path $q(x^\alpha)$. Equation (4.5) is the field-theoretical counterpart to Eq. (4.2).

Upon such a variation (we use the notation $\mathscr{L}' \equiv \partial\mathscr{L}/\partial q$, $\mathscr{L}^\alpha \equiv \partial\mathscr{L}/\partial q_{,\alpha}$),

$$\begin{aligned}
\delta S &= \int_{\mathcal{V}} (\mathscr{L}'\delta q + \mathscr{L}^\alpha \delta q_{,\alpha})\sqrt{-g}\,\mathrm{d}^4 x \\
&= \int_{\mathcal{V}} [\mathscr{L}'\delta q + (\mathscr{L}^\alpha \delta q)_{;\alpha} - \mathscr{L}^\alpha_{\;;\alpha}\delta q]\sqrt{-g}\,\mathrm{d}^4 x \\
&= \int_{\mathcal{V}} (\mathscr{L}' - \mathscr{L}^\alpha_{\;;\alpha})\delta q\sqrt{-g}\,\mathrm{d}^4 x + \oint_{\partial\mathcal{V}} \mathscr{L}^\alpha \delta q\,\mathrm{d}\Sigma_\alpha,
\end{aligned}$$

where Gauss' theorem (Section 3.3) was used in the last step. The surface integral vanishes by virtue of Eq. (4.5), and because δq is arbitrary within $\mathcal{V}$ we obtain

$$\delta S = 0 \quad \Rightarrow \quad \nabla_\alpha \frac{\partial\mathscr{L}}{\partial q_{,\alpha}} - \frac{\partial\mathscr{L}}{\partial q} = 0. \qquad (4.6)$$

This is the Euler–Lagrange equation for a single scalar field q. Generalization to a collection of fields is immediate, and the procedure can be taken over to fields of arbitrary tensorial or spinorial types.

As a concrete example, let us consider a Klein–Gordon field ψ with Lagrangian density

$$\mathscr{L} = -\frac{1}{2}\left(g^{\mu\nu}\psi_{,\mu}\psi_{,\nu} + m^2\psi^2\right).$$

We have $\mathscr{L}^\alpha = -g^{\alpha\beta}\psi_{,\beta}$, $\mathscr{L}^\alpha{}_{;\alpha} = -g^{\alpha\beta}\psi_{;\alpha\beta}$, and $\mathscr{L}' = -m^2\psi$. The Euler–Lagrange equation becomes

$$g^{\alpha\beta}\psi_{;\alpha\beta} - m^2\psi = 0,$$

which is the curved-spacetime version of the Klein–Gordon equation.

4.1.3 General relativity

The action functional for general relativity contains a contribution $S_G[g]$ from the gravitational field $g_{\alpha\beta}$ and a contribution $S_M[\phi; g]$ from the matter fields, which we collectively denote ϕ.

The gravitational action contains a Hilbert term $S_H[g]$, a boundary term $S_B[g]$, and a nondynamical term S_0 that affects the numerical value of the action but not the equations of motion. More explicitly,

$$S_G[g] = S_H[g] + S_B[g] - S_0, \tag{4.7}$$

where

$$S_H[g] = \frac{1}{16\pi}\int_{\mathscr{V}} R\sqrt{-g}\,\mathrm{d}^4x, \tag{4.8}$$

$$S_B[g] = \frac{1}{8\pi}\oint_{\partial\mathscr{V}} \varepsilon K|h|^{1/2}\,\mathrm{d}^3y, \tag{4.9}$$

$$S_0 = \frac{1}{8\pi}\oint_{\partial\mathscr{V}} \varepsilon K_0|h|^{1/2}\,\mathrm{d}^3y. \tag{4.10}$$

Here, R is the Ricci scalar in $\mathscr{V}$, K is the trace of the extrinsic curvature of $\partial\mathscr{V}$, ε is equal to $+1$ where $\partial\mathscr{V}$ is timelike and -1 where $\partial\mathscr{V}$ is spacelike (it is assumed that $\partial\mathscr{V}$ is nowhere null), and h is the determinant of the induced metric on $\partial\mathscr{V}$. Coordinates x^α are used in $\mathscr{V}$, and coordinates y^a are used on $\partial\mathscr{V}$. The role of $S_B[g]$ in the variational principle will be elucidated below. The presence of S_0 in the action will also be explained, and this explanation will come with a precise definition for the quantity K_0.

The matter action is taken to be of the form

$$S_M[\phi; g] = \int_{\mathscr{V}} \mathscr{L}(\phi, \phi_{,\alpha}; g_{\alpha\beta})\sqrt{-g}\,\mathrm{d}^4x, \tag{4.11}$$

for some Lagrangian density $\mathscr{L}$. As Eq. (4.11) indicates, it is assumed that only $g_{\alpha\beta}$, and none of its derivatives, appears in the matter action. This assumption is made for simplicity and it could easily be removed.

The complete action functional is therefore

$$S[g; \phi] = \int_{\mathscr{V}} \left(\frac{R}{16\pi} + \mathscr{L} \right) \sqrt{-g}\, \mathrm{d}^4 x + \frac{1}{8\pi} \oint_{\partial\mathscr{V}} \varepsilon (K - K_0) |h|^{1/2}\, \mathrm{d}^3 y. \quad (4.12)$$

The Einstein field equations, $G_{\alpha\beta} = 8\pi T_{\alpha\beta}$, are recovered by varying $S[g, \phi]$ with respect to $g_{\alpha\beta}$. The variation is subjected to the condition

$$\delta g_{\alpha\beta}\big|_{\partial\mathscr{V}} = 0. \quad (4.13)$$

This implies that $h_{ab} = g_{\alpha\beta}\, e^\alpha_a e^\beta_b$, the induced metric on $\partial\mathscr{V}$, is *held fixed* during the variation.

4.1.4 Variation of the Hilbert term

It is convenient to use the variations $\delta g^{\alpha\beta}$ instead of $\delta g_{\alpha\beta}$. These are of course not independent: the relations $g^{\alpha\mu} g_{\mu\beta} = \delta^\alpha_{\ \beta}$ imply

$$\delta g_{\alpha\beta} = -g_{\alpha\mu} g_{\beta\nu}\, \delta g^{\mu\nu}. \quad (4.14)$$

We recall (from Section 1.7) that the variation of the metric determinant is given by $\delta \ln |g| = g^{\alpha\beta} \delta g_{\alpha\beta} = -g_{\alpha\beta} \delta g^{\alpha\beta}$, which implies

$$\delta\sqrt{-g} = -\frac{1}{2}\sqrt{-g}\, g_{\alpha\beta}\, \delta g^{\alpha\beta}. \quad (4.15)$$

We also recall (from Section 1.2) that although $\Gamma^\alpha_{\ \beta\gamma}$ is not a tensor, the *difference* between two sets of Christoffel symbols *is* a tensor; the variation $\delta\Gamma^\alpha_{\ \beta\gamma}$ is therefore a tensor.

We now proceed with the variation of the Hilbert term in the gravitational action:

$$(16\pi)\delta S_H = \int_{\mathscr{V}} \delta\left(g^{\alpha\beta} R_{\alpha\beta} \sqrt{-g}\right) \mathrm{d}^4 x$$

$$= \int_{\mathscr{V}} \left(R_{\alpha\beta}\sqrt{-g}\, \delta g^{\alpha\beta} + g^{\alpha\beta}\sqrt{-g}\, \delta R_{\alpha\beta} + R\, \delta\sqrt{-g}\right) \mathrm{d}^4 x$$

$$= \int_{\mathscr{V}} \left(R_{\alpha\beta} - \frac{1}{2} R g_{\alpha\beta}\right) \delta g^{\alpha\beta} \sqrt{-g}\, \mathrm{d}^4 x + \int_{\mathscr{V}} g^{\alpha\beta} \delta R_{\alpha\beta} \sqrt{-g}\, \mathrm{d}^4 x.$$

In the last step we have used Eq. (4.15). The first integral seems to give us what we need for the left-hand side of the Einstein field equations, but we must still account for the second integral.

Let us work on this integral. We begin with $\delta R_{\alpha\beta}$, which we calculate in a local Lorentz frame at a point P:

$$\delta R_{\alpha\beta} \overset{*}{=} \delta\big(\Gamma^\mu{}_{\alpha\beta,\mu} - \Gamma^\mu{}_{\alpha\mu,\beta}\big)$$

$$\overset{*}{=} \big(\delta\Gamma^\mu{}_{\alpha\beta}\big)_{,\mu} - \big(\delta\Gamma^\mu{}_{\alpha\mu}\big)_{,\beta}$$

$$\overset{*}{=} \big(\delta\Gamma^\mu{}_{\alpha\beta}\big)_{;\mu} - \big(\delta\Gamma^\mu{}_{\alpha\mu}\big)_{;\beta}.$$

Here, covariant differentiation is defined with respect to the reference metric $g_{\alpha\beta}$, about which the variation is taken. We notice that the last expression is tensorial; it is therefore valid in any coordinate system. We have found

$$g^{\alpha\beta}\delta R_{\alpha\beta} = \delta v^\mu{}_{;\mu}, \qquad \delta v^\mu = g^{\alpha\beta}\delta\Gamma^\mu{}_{\alpha\beta} - g^{\alpha\mu}\delta\Gamma^\beta{}_{\alpha\beta}. \tag{4.16}$$

We use the 'slash' notation δv^μ to emphasize the fact that δv^μ is not the variation of some quantity v^μ. Using Eq. (4.16), the second integral in δS_H becomes

$$\int_{\mathscr{V}} g^{\alpha\beta}\delta R_{\alpha\beta}\sqrt{-g}\,\mathrm{d}^4x = \int \delta v^\mu{}_{;\mu}\sqrt{-g}\,\mathrm{d}^4x$$

$$= \oint_{\partial\mathscr{V}} \delta v^\mu\,\mathrm{d}\Sigma_\mu$$

$$= \oint_{\partial\mathscr{V}} \varepsilon\,\delta v^\mu n_\mu |h|^{1/2}\,\mathrm{d}^3y,$$

where n_μ is the unit normal to $\partial\mathscr{V}$ and $\varepsilon \equiv n^\mu n_\mu = \pm 1$.

We must now evaluate $\delta v^\mu n_\mu$, keeping in mind that on $\partial\mathscr{V}$, $\delta g_{\alpha\beta} = 0 = \delta g^{\alpha\beta}$. Under these conditions,

$$\delta\Gamma^\mu{}_{\alpha\beta}\big|_{\partial\mathscr{V}} = \frac{1}{2}g^{\mu\nu}\big(\delta g_{\nu\alpha,\beta} + \delta g_{\nu\beta,\alpha} - \delta g_{\alpha\beta,\nu}\big),$$

and substituting this into Eq. (4.16) yields $\delta v_\mu = g^{\alpha\beta}(\delta g_{\mu\beta,\alpha} - \delta g_{\alpha\beta,\mu})$, so that

$$n^\mu \delta v_\mu\big|_{\partial\mathscr{V}} = n^\mu(\varepsilon n^\alpha n^\beta + h^{\alpha\beta})(\delta g_{\mu\beta,\alpha} - \delta g_{\alpha\beta,\mu})$$

$$= n^\mu h^{\alpha\beta}(\delta g_{\mu\beta,\alpha} - \delta g_{\alpha\beta,\mu}).$$

In the first line we have inserted the completeness relations $g^{\alpha\beta} = \varepsilon n^\alpha n^\beta + h^{\alpha\beta}$, where $h^{\alpha\beta} \equiv h^{ab}e^\alpha_a e^\beta_b$ (see Section 3.1). To obtain the second line we have multiplied $n^\alpha n^\mu$ by the antisymmetric quantity within the brackets. Proceeding, we observe that because $\delta g_{\alpha\beta}$ vanishes everywhere on $\partial\mathscr{V}$, its tangential derivatives must vanish also: $\delta g_{\alpha\beta,\gamma}e^\gamma_c = 0$. It follows that $h^{\alpha\beta}\delta g_{\mu\beta,\alpha} = 0$ and we finally obtain

$$n^\mu \delta v_\mu\big|_{\partial\mathscr{V}} = -h^{\alpha\beta}\delta g_{\alpha\beta,\mu}n^\mu. \tag{4.17}$$

This is nonzero because the *normal* derivative of $\delta g_{\alpha\beta}$ is not required to vanish on the hypersurface.

Gathering the results we obtain

$$(16\pi)\delta S_H = \int_{\mathscr{V}} G_{\alpha\beta}\delta g^{\alpha\beta}\sqrt{-g}\,\mathrm{d}^4x - \oint_{\partial\mathscr{V}} \varepsilon h^{\alpha\beta}\delta g_{\alpha\beta,\mu}n^\mu|h|^{1/2}\,\mathrm{d}^3y. \qquad (4.18)$$

The boundary term in Eq. (4.18) will be cancelled by the variation of $S_B[g]$: this is the reason for including a boundary term in the gravitational action. That a boundary term is needed is due to the fact that R, the gravitational Lagrangian density, contains *second derivatives* of the metric tensor. This is a nontypical feature of field theories, which are usually formulated in terms of Lagrangians that involve q and $q_{,\alpha}$ only.

4.1.5 Variation of the boundary term

We now work on the variation of $S_B[g]$, as given by Eq. (4.9). Because the induced metric is fixed on $\partial\mathscr{V}$, the only quantity to be varied is K, the trace of the extrinsic curvature. We recall from Section 3.4 that

$$K = n^\alpha{}_{;\alpha}$$
$$= (\varepsilon n^\alpha n^\beta + h^{\alpha\beta})n_{\alpha;\beta}$$
$$= h^{\alpha\beta}n_{\alpha;\beta}$$
$$= h^{\alpha\beta}(n_{\alpha,\beta} - \Gamma^\gamma{}_{\alpha\beta}n_\gamma),$$

so that its variation is

$$\delta K = -h^{\alpha\beta}\delta\Gamma^\gamma{}_{\alpha\beta}n_\gamma$$
$$= -\frac{1}{2}h^{\alpha\beta}\left(\delta g_{\mu\alpha,\beta} + \delta g_{\mu\beta,\alpha} - \delta g_{\alpha\beta,\mu}\right)n^\mu$$
$$= \frac{1}{2}h^{\alpha\beta}\delta g_{\alpha\beta,\mu}n^\mu;$$

we have used the fact that the tangential derivatives of $\delta g_{\alpha\beta}$ vanish on $\partial\mathscr{V}$. We have obtained

$$(16\pi)\delta S_B = \oint_{\partial\mathscr{V}} \varepsilon h^{\alpha\beta}\delta g_{\alpha\beta,\mu}n^\mu|h|^{1/2}\,\mathrm{d}^3y, \qquad (4.19)$$

and we see that this indeed cancels out the second integral on the right-hand side of Eq. (4.18). Because $\delta S_0 \equiv 0$, the complete variation of the gravitational action is

$$\delta S_G = \frac{1}{16\pi}\int_{\mathscr{V}} G_{\alpha\beta}\,\delta g^{\alpha\beta}\sqrt{-g}\,\mathrm{d}^4x. \qquad (4.20)$$

This produces the correct left-hand side to the Einstein field equations.

4.1.6 Variation of the matter action

Variation of $S_M[\phi; g]$, as given by Eq. (4.11), yields

$$\delta S_M = \int_{\mathscr{V}} \delta(\mathscr{L}\sqrt{-g})\, \mathrm{d}^4 x$$

$$= \int_{\mathscr{V}} \left(\frac{\partial \mathscr{L}}{\partial g^{\alpha\beta}} \delta g^{\alpha\beta} \sqrt{-g} + \mathscr{L}\, \delta\sqrt{-g} \right) \mathrm{d}^4 x$$

$$= \int_{\mathscr{V}} \left(\frac{\partial \mathscr{L}}{\partial g^{\alpha\beta}} - \frac{1}{2} \mathscr{L} g_{\alpha\beta} \right) \delta g^{\alpha\beta} \sqrt{-g}\, \mathrm{d}^4 x.$$

If we *define* the stress-energy tensor by

$$T_{\alpha\beta} \equiv -2\frac{\partial \mathscr{L}}{\partial g^{\alpha\beta}} + \mathscr{L} g_{\alpha\beta}, \qquad (4.21)$$

then

$$\delta S_M = -\frac{1}{2} \int_{\mathscr{V}} T_{\alpha\beta}\, \delta g^{\alpha\beta} \sqrt{-g}\, \mathrm{d}^4 x, \qquad (4.22)$$

and this produces the correct right-hand side to the Einstein field equations.

We have obtained

$$\delta(S_G + S_M) = 0 \quad \Rightarrow \quad G_{\alpha\beta} = 8\pi T_{\alpha\beta}, \qquad (4.23)$$

because the variation $\delta g^{\alpha\beta}$ is arbitrary within $\mathscr{V}$. The Einstein field equations therefore follow from a variational principle, and the action functional for the theory is given by Eq. (4.12).

To see that Eq. (4.21) gives a reasonable definition for the stress-energy tensor, let us consider once more a Klein–Gordon field ψ with Lagrangian density

$$\mathscr{L} = -\frac{1}{2}\left(g^{\mu\nu} \psi_{,\mu} \psi_{,\nu} + m^2 \psi^2 \right).$$

It is easy to check that for this, Eq. (4.21) becomes

$$T_{\alpha\beta} = \psi_{,\alpha}\psi_{,\beta} - \frac{1}{2}\left(\psi^{,\mu}\psi_{,\mu} + m^2\psi^2 \right) g_{\alpha\beta}.$$

This is the correct expression for the Klein–Gordon stress-energy tensor. You may look into the consistency of this result by checking that the statement of energy-momentum conservation, $T^{\alpha\beta}{}_{;\beta} = 0$, implies the Klein–Gordon equation.

4.1.7 Nondynamical term

What is the role of

$$S_0 = \frac{1}{8\pi} \oint_{\partial \mathcal{V}} \varepsilon K_0 |h|^{1/2} \, \mathrm{d}^3 y$$

in the gravitational action? Because S_0 depends only on the induced metric h_{ab} (through the factor $|h|^{1/2}$ in the integrand), its variation with respect to $g_{\alpha\beta}$ gives zero, and the presence of S_0 cannot affect the equations of motion. Its purpose can only be to change the numerical value of the gravitational action.

Let us first assume that $g_{\alpha\beta}$ is a solution to the *vacuum* field equations. Then $R = 0$ and the numerical value of the gravitational action is

$$S_G = \frac{1}{8\pi} \oint_{\partial \mathcal{V}} \varepsilon K |h|^{1/2} \, \mathrm{d}^3 y,$$

where we omit the subtraction term K_0 for the time being. Let us evaluate this for flat spacetime. We choose $\partial \mathcal{V}$ to consist of two hypersurfaces $t = \text{constant}$ and a large three-cylinder at $r = R$ (Fig. 4.1). It is easy to check that $K = 0$ on the hypersurfaces of constant time. On the three-cylinder, the induced metric is $\mathrm{d}s^2 = -\mathrm{d}t^2 + R^2 \, \mathrm{d}\Omega^2$, so that $|h|^{1/2} = R^2 \sin\theta$. The unit normal is $n_\alpha = \partial_\alpha r$, so that $\varepsilon = 1$ and $K = n^\alpha{}_{;\alpha} = 2/R$. We then have

$$\oint_{\partial \mathcal{V}} \varepsilon K |h|^{1/2} \, \mathrm{d}^3 y = 8\pi R(t_2 - t_1),$$

and this *diverges* when $R \to \infty$, that is, when the spatial boundary is pushed all the way to infinity. The gravitational action of flat spacetime is therefore infinite, even when $\mathcal{V}$ is bounded by two hypersurfaces of constant time. Because this problem does not go away when the spacetime is curved, this would imply that the gravitational action is not a well-defined quantity for asymptotically-flat spacetimes. (Of course, the problem goes away if the spacetime manifold is compact.)

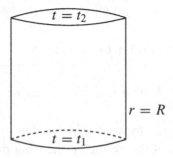

Figure 4.1   The boundary of a region $\mathcal{V}$ of flat spacetime.

This problem is remedied by S_0. This term is chosen to be equal to the gravitational action of flat spacetime, as regularized by the procedure adopted before. The difference $S_B - S_0$ is then well defined in the limit $R \to \infty$, and there is no longer a difficulty in defining a gravitational action for asymptotically-flat spacetimes. (The subtraction term is irrelevant for compact manifolds.) In other words, the choice

$$K_0 = \text{extrinsic curvature of } \partial \mathscr{V} \text{ embedded in flat spacetime} \qquad (4.24)$$

cures the divergence of the gravitational action, which is then well defined when the spacetime is asymptotically flat. In particular, $S_G = 0$ for flat spacetime.

4.1.8 Bianchi identities

The Lagrangian formulation of general relativity provides us with an elegant derivation of the contracted Bianchi identities,

$$G^{\alpha\beta}{}_{;\beta} = 0. \qquad (4.25)$$

In this approach, Eq. (4.25) comes as a consequence of the invariance of $S_G[g]$ under a change of coordinates in $\mathscr{V}$.

To prove this it is sufficient to consider infinitesimal transformations,

$$x^\alpha \to x'^\alpha = x^\alpha + \epsilon^\alpha, \qquad (4.26)$$

where ϵ^α is an infinitesimal vector field, arbitrary within $\mathscr{V}$ but constrained to vanish on $\partial \mathscr{V}$. The variation of the metric under such a transformation is

$$
\begin{aligned}
\delta g_{\alpha\beta} &\equiv g'_{\alpha\beta}(x) - g_{\alpha\beta}(x) \\
&= g'_{\alpha\beta}(x') - g_{\alpha\beta}(x) + g'_{\alpha\beta}(x) - g'_{\alpha\beta}(x') \\
&= \frac{\partial x^\mu}{\partial x'^\alpha} \frac{\partial x^\nu}{\partial x'^\beta} g_{\mu\nu}(x) - g_{\alpha\beta}(x) + g'_{\alpha\beta}(x) - g'_{\alpha\beta}(x+\epsilon) \\
&= (\delta^\mu_\alpha - \epsilon^\mu{}_{,\alpha})(\delta^\nu_\beta - \epsilon^\nu{}_{,\beta}) g_{\mu\nu}(x) - g_{\alpha\beta}(x) - g_{\alpha\beta,\mu}(x)\epsilon^\mu \\
&= -\epsilon^\mu{}_{,\alpha} g_{\mu\beta} - \epsilon^\mu{}_{,\beta} g_{\alpha\mu} - g_{\alpha\beta,\mu}\epsilon^\mu \\
&= -\pounds_\epsilon g_{\alpha\beta},
\end{aligned}
$$

discarding all terms of the second order in ϵ^α. Using Eq. (4.14) we find that the metric variation is

$$\delta g^{\alpha\beta} = \epsilon^{\alpha;\beta} + \epsilon^{\beta;\alpha}. \qquad (4.27)$$

Substituting this into Eq. (4.20), we find

$$(8\pi)\delta S_G = \int_{\mathscr{V}} G^{\alpha\beta} \epsilon_{\alpha;\beta} \sqrt{-g}\, \mathrm{d}^4 x$$

$$= -\int_{\mathscr{V}} G^{\alpha\beta}{}_{;\beta} \epsilon_\alpha \sqrt{-g}\, \mathrm{d}^4 x + \oint_{\partial\mathscr{V}} G^{\alpha\beta} \epsilon_\alpha \, \mathrm{d}\Sigma_\beta.$$

With ϵ^α arbitrary within $\mathscr{V}$ but vanishing on $\partial\mathscr{V}$, the contracted Bianchi identities follow from the requirement that $\delta S_G = 0$ under the variation of Eq. (4.27).

4.2 Hamiltonian formulation

4.2.1 Mechanics

The Hamiltonian formulation of Newtonian mechanics begins with the introduction of the canonical momentum p, defined by

$$p = \frac{\partial L}{\partial \dot{q}}. \tag{4.28}$$

It is assumed that this relation can be inverted to give $\dot{q}$ as a function of p and q. The *Hamiltonian* is then

$$H(p, q) = p\dot{q} - L. \tag{4.29}$$

Hamilton's form of the equations of motion can be derived from a variational principle. Here, the action is varied with respect to p and q *independently*, with the restriction that δq must vanish at the endpoints. Thus,

$$\delta S = \int_{t_1}^{t_2} \delta(p\dot{q} - H)\, \mathrm{d}t$$

$$= \int_{t_1}^{t_2} \left(p\,\delta\dot{q} + \dot{q}\,\delta p - \frac{\partial H}{\partial p}\,\delta p - \frac{\partial H}{\partial q}\,\delta q \right) \mathrm{d}t$$

$$= p\,\delta q \Big|_{t_1}^{t_2} + \int_{t_1}^{t_2} \left[-\left(\dot{p} + \frac{\partial H}{\partial q} \right)\delta q + \left(\dot{q} - \frac{\partial H}{\partial p} \right)\delta p \right] \mathrm{d}t.$$

Because the variations are arbitrary between t_1 and t_2, but $\delta q(t_1) = \delta q(t_2) = 0$, we have

$$\delta S = 0 \quad \Rightarrow \quad \dot{p} = -\frac{\partial H}{\partial q}, \qquad \dot{q} = \frac{\partial H}{\partial p}. \tag{4.30}$$

These are *Hamilton's equations*. They are equivalent to the Euler–Lagrange equation (4.3).

4.2.2 3 + 1 decomposition

The Hamiltonian formulation of a field theory is more involved. Here, the Hamiltonian $H[p, q]$ is a *functional* of q, the field configuration, and p, the canonical momentum, on a spacelike hypersurface Σ. To express the action in terms of the Hamiltonian it is necessary to foliate $\mathcal{V}$ with a family of spacelike hypersurfaces, one for each 'instant of time.' This is the purpose of the $3 + 1$ decomposition.

To effect this decomposition we introduce a scalar field $t(x^\alpha)$ such that $t =$ constant describes a family of nonintersecting spacelike hypersurfaces Σ_t. This 'time function' is completely arbitrary; the only requirements are that t be a single-valued function of x^α, and that $n_\alpha \propto \partial_\alpha t$, the unit normal to the hypersurfaces, be a future-directed timelike vector field.

On each of the hypersurfaces Σ_t we install coordinates y^a. A priori, the coordinates on one hypersurface need not be related to the coordinates on another hypersurface. It is, however, convenient to introduce a relationship, as follows (Fig. 4.2). Consider a congruence of curves γ intersecting the hypersurfaces Σ_t. We do not assume that these curves are geodesics, nor that they intersect the hypersurfaces orthogonally. We use t as a parameter on the curves, and the vector t^α is tangent to the congruence. It is easy to check that the relation

$$t^\alpha \partial_\alpha t = 1 \tag{4.31}$$

follows from the construction. A particular curve γ_P from the congruence defines a mapping from a point P on Σ_t to a point P' on $\Sigma_{t'}$, and then to a point P'' on $\Sigma_{t''}$, and so on. To fix the coordinates of P' and P'', given $y^a(P)$ on Σ_t, we simply impose $y^a(P'') = y^a(P') = y^a(P)$. Thus, y^a is held constant on each member of the congruence.

This construction defines a coordinate system (t, y^a) in $\mathcal{V}$. There exists a transformation between this and the system x^α originally in use: $x^\alpha = x^\alpha(t, y^a)$. We

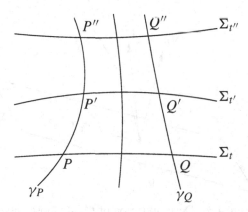

Figure 4.2 Foliation of spacetime by spacelike hypersurfaces.

have

$$t^\alpha = \left(\frac{\partial x^\alpha}{\partial t}\right)_{y^a},$$

(4.32)

and we define

$$e_a^\alpha = \left(\frac{\partial x^\alpha}{\partial y^a}\right)_t$$

(4.33)

to be tangent vectors on Σ_t. These relations imply that in the coordinates (t, y^a), $t^\alpha \overset{*}{=} \delta_t^\alpha$ and $e_a^\alpha \overset{*}{=} \delta_a^\alpha$. We also have

$$\pounds_t e_a^\alpha = 0,$$

(4.34)

which holds in any coordinate system.

We now introduce the unit normal to the hypersurfaces:

$$n_\alpha = -N\partial_\alpha t, \qquad n_\alpha e_a^\alpha = 0,$$

(4.35)

where the scalar function N, called the *lapse*, ensures that n_α is properly normalized. Because the curves γ do not intersect Σ_t orthogonally, t^α is not parallel to n^α. We may decompose t^α in the basis provided by the normal and tangent vectors (Fig. 4.3):

$$t^\alpha = Nn^\alpha + N^a e_a^\alpha;$$

(4.36)

the three-vector N^a is called the *shift*. It is easy to check that Eq. (4.36) is compatible with Eq. (4.31).

We can use the coordinate transformation $x^\alpha = x^\alpha(t, y^a)$ to express the metric in the coordinates (t, y^a). We start by writing

$$dx^\alpha = t^\alpha \, dt + e_a^\alpha \, dy^a$$

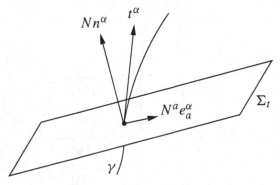

Figure 4.3   Decomposition of t^α into lapse and shift.

$$= (N \, dt) n^\alpha + (dy^a + N^a \, dt) e_a^\alpha,$$

which follows at once from Eqs. (4.32), (4.33), and (4.36). The line element is then given by $ds^2 = g_{\alpha\beta} dx^\alpha dx^\beta$, or

$$ds^2 = -N^2 dt^2 + h_{ab} (dy^a + N^a \, dt)(dy^b + N^b \, dt), \qquad (4.37)$$

where $h_{ab} = g_{\alpha\beta} \, e_a^\alpha e_b^\beta$ is the induced metric on Σ_t.

We may now express the metric determinant g in terms of $h \equiv \det[h_{ab}]$ and the lapse function. We recall that $g^{tt} = \text{cofactor}(g_{tt})/g = h/g$, as follows from Eq. (4.37). But $g^{tt} = g^{\alpha\beta} t_{,\alpha} t_{,\beta} = N^{-2} g^{\alpha\beta} n_\alpha n_\beta = -N^{-2}$, where Eq. (4.35) was used. The desired expression is therefore

$$\sqrt{-g} = N\sqrt{h}. \qquad (4.38)$$

Equations (4.36), (4.37), and (4.38) are the fundamental results of the $3 + 1$ decomposition.

4.2.3 Field theory

We now return to the Hamiltonian formulation of a field theory. For simplicity we will assume that the field is a scalar, but the procedure can easily be applied to fields of other tensorial types. We begin by defining the 'time derivative' of q to be its Lie derivative along the flow vector t^α,

$$\dot{q} \equiv \pounds_t \, q. \qquad (4.39)$$

In the coordinates (t, y^a), $\pounds_t q \overset{*}{=} \partial q / \partial t$, and $\dot{q}$ reduces to the ordinary time derivative. We also introduce the spatial derivatives, $q_{,a} \equiv q_{,\alpha} e_a^\alpha$. The field's Lagrangian density can then be expressed as $\mathscr{L}(q, \dot{q}, q_{,a})$.

The field's canonical momentum p is defined by

$$p = \frac{\partial}{\partial \dot{q}} \left(\sqrt{-g} \, \mathscr{L} \right). \qquad (4.40)$$

It is assumed that this relation can be inverted to give $\dot{q}$ in terms of q, $q_{,a}$, and p. The *Hamiltonian density* is then

$$\mathscr{H}(p, q, q_{,a}) = p \dot{q} - \sqrt{-g} \, \mathscr{L}. \qquad (4.41)$$

The presence of $\sqrt{-g}$ in Eqs. (4.40) and (4.41) implies that the Hamiltonian density is not a scalar with respect to transformations $y^a \to y^{a'}$. We might introduce

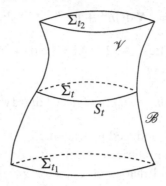

Figure 4.4   The region $\mathscr{V}$, its boundary $\partial\mathscr{V}$, and their foliations.

a scalarized version $\mathscr{H}_{\text{scalar}}$ defined by $\mathscr{H} = \sqrt{h}\,\mathscr{H}_{\text{scalar}} = \sqrt{-g}\,\mathscr{H}_{\text{scalar}}/N$, but such an object would turn out not to be as useful as the original, nonscalar, Hamiltonian density. The *Hamiltonian functional* is defined by

$$H[p,q] = \int_{\Sigma_t} \mathscr{H}\,(p,q,q_{,a})\,\mathrm{d}^3 y. \tag{4.42}$$

The Hamiltonian functional is an ordinary (nonscalar) function of time t.

We consider a region $\mathscr{V}$ of spacetime foliated by spacelike hypersurfaces Σ_t bounded by closed two-surfaces S_t (Fig. 4.4); $\mathscr{V}$ itself is bounded by the hypersurfaces Σ_{t_1}, Σ_{t_2}, and $\mathscr{B}$, the union of all two-surfaces S_t. To obtain the Hamilton form of the field equations, we will vary the action with respect to q and p, treating the variations δq and δp as independent. We will demand that δq vanish on the boundaries Σ_{t_1}, Σ_{t_2}, and $\mathscr{B}$.

The action functional is given by

$$S = \int_{t_1}^{t_2} \mathrm{d}t \int_{\Sigma_t} \left(p\,\dot{q} - \mathscr{H} \right) \mathrm{d}^3 y,$$

and variation yields

$$\delta S = \int_{t_1}^{t_2} \mathrm{d}t \int_{\Sigma_t} \left(p\,\delta\dot{q} + \dot{q}\,\delta p - \frac{\partial\mathscr{H}}{\partial p}\,\delta p - \frac{\partial\mathscr{H}}{\partial q}\,\delta q - \frac{\partial\mathscr{H}}{\partial q_{,a}}\,\delta q_{,a} \right) \mathrm{d}^3 y.$$

The first term may be integrated by parts:

$$\int_{t_1}^{t_2} \mathrm{d}t \int_{\Sigma_t} p\,\delta\dot{q}\,\mathrm{d}^3 y = \int_{t_1}^{t_2} \mathrm{d}t\,\frac{\mathrm{d}}{\mathrm{d}t} \int_{\Sigma_t} p\,\delta q\,\mathrm{d}^3 y - \int_{t_1}^{t_2} \mathrm{d}t \int_{\Sigma_t} \dot{p}\,\delta q\,\mathrm{d}^3 y$$

$$= \int_{\Sigma_{t_2}} p\,\delta q\,\mathrm{d}^3 y - \int_{\Sigma_{t_1}} p\,\delta q\,\mathrm{d}^3 y - \int_{t_1}^{t_2} \mathrm{d}t \int_{\Sigma_t} \dot{p}\,\delta q\,\mathrm{d}^3 y$$

$$= -\int_{t_1}^{t_2} \mathrm{d}t \int_{\Sigma_t} \dot{p}\,\delta q\,\mathrm{d}^3 y,$$

because $\delta q = 0$ on Σ_{t_1} and Σ_{t_2}. We treat the last term similarly:

$$-\int_{t_1}^{t_2} dt \int_{\Sigma_t} \frac{\partial \mathcal{H}}{\partial q_{,a}} \delta q_{,a} \, d^3 y = -\int_{t_1}^{t_2} dt \int_{\Sigma_t} \frac{\partial \mathcal{H}_{\text{scalar}}}{\partial q_{,a}} \delta q_{,a} \sqrt{h} \, d^3 y$$

$$= -\int_{t_1}^{t_2} dt \oint_{S_t} \frac{\partial \mathcal{H}_{\text{scalar}}}{\partial q_{,a}} \delta q \, dS_a$$

$$+ \int_{t_1}^{t_2} dt \int_{\Sigma_t} \left(\frac{\partial \mathcal{H}_{\text{scalar}}}{\partial q_{,a}} \right)_{|a} \delta q \sqrt{h} \, d^3 y$$

$$= \int_{t_1}^{t_2} dt \int_{\Sigma_t} \left(\frac{\partial \mathcal{H}}{\partial q_{,a}} \right)_{,a} \delta q \, d^3 y.$$

In the second line we have used the three-dimensional version of Gauss' theorem, with dS_a denoting the surface element on S_t. In the third line we have used the divergence formula $A^a{}_{|a} = h^{-1/2}(h^{1/2} A^a)_{,a}$ and the fact that δq vanishes on S_t.

Gathering the results, we have

$$\delta S = \int_{t_1}^{t_2} dt \int_{\Sigma_t} \left\{ -\left[\dot{p} + \frac{\partial \mathcal{H}}{\partial q} - \left(\frac{\partial \mathcal{H}}{\partial q_{,a}} \right)_{,a} \right] \delta q + \left[\dot{q} - \frac{\partial \mathcal{H}}{\partial p} \right] \delta p \right\} d^3 y,$$

and

$$\delta S = 0 \quad \Rightarrow \quad \dot{p} = -\frac{\partial \mathcal{H}}{\partial q} + \left(\frac{\partial \mathcal{H}}{\partial q_{,a}} \right)_{,a}, \qquad \dot{q} = \frac{\partial \mathcal{H}}{\partial p}. \qquad (4.43)$$

These are Hamilton's equations for a scalar field q and its canonical momentum p.

As a concrete example we consider once again a Klein–Gordon field ψ with its Lagrangian density

$$\mathcal{L} = -\frac{1}{2} \left(g^{\mu\nu} \psi_{,\mu} \psi_{,\nu} + m^2 \psi^2 \right).$$

For simplicity we choose our foliation to be such that $N^a = 0$. This implies $g^{tt} = 1/g_{tt}$, $g^{ta} = 0$, and $g^{ab} = h^{ab}$. Then $\mathcal{L} = -\frac{1}{2}(g^{tt} \dot{\psi}^2 + h^{ab} \psi_{,a} \psi_{,b} + m^2 \psi^2)$, $p = -\sqrt{-g} \, g^{tt} \dot{\psi}$, and Eq. (4.41) gives

$$\mathcal{H} = -\frac{p^2}{2\sqrt{-g} \, g^{tt}} + \frac{1}{2} \sqrt{-g} \left(h^{ab} \psi_{,a} \psi_{,b} + m^2 \psi^2 \right).$$

The equations of motion are

$$\dot{\psi} = -\frac{p}{\sqrt{-g} \, g^{tt}}, \qquad \dot{p} = -\sqrt{-g} \, m^2 \psi + \left(\sqrt{-g} \, h^{ab} \psi_{,b} \right)_{,a}.$$

It is easy to check that from these follow the Klein–Gordon equation, $g^{\alpha\beta} \psi_{;\alpha\beta} - m^2 \psi = 0$, in the selected foliation.

4.2.4 Foliation of the boundary

Before tackling the case of the gravitational field, we need to provide additional details regarding the foliation of $\mathscr{B}$, the timelike boundary of $\mathscr{V}$, by the two-surfaces S_t. (Refer back to Fig. 4.4.)

The closed two-surface S_t is the boundary of the spacelike hypersurface Σ_t, on which we have coordinates y^a, tangent vectors e_a^α, and an induced metric h_{ab}. It is described by an equation of the form $\Phi(y^a) = 0$, or by parametric relations $y^a(\theta^A)$, where θ^A are coordinates on S_t. We use r_a to denote the unit normal to S_t, and we define an associated four-vector r^α by

$$r^\alpha = r^a e_a^\alpha. \tag{4.44}$$

This satisfies the relations $r^\alpha r_\alpha = 1$ and $r^\alpha n_\alpha = 0$, where n^α is the normal to Σ_t. The three-vectors $e_A^a = \partial y^a / \partial \theta^A$ are tangent to S_t, so that $r_a e_A^a = 0$. This implies $r_\alpha e_A^\alpha = 0$, where

$$e_A^\alpha \equiv e_a^\alpha e_A^a = \frac{\partial x^\alpha}{\partial \theta^A}. \tag{4.45}$$

In this equation it is understood that x^α stands for the functions $x^\alpha(y^a(\theta^A))$, where $x^\alpha(y^a)$ are the parametric relations describing Σ_t.

The induced metric on S_t is given by

$$ds^2 = \sigma_{AB}\, d\theta^A\, d\theta^B, \tag{4.46}$$

where $\sigma_{AB} = h_{ab}\, e_A^a e_B^b = (g_{\alpha\beta}\, e_a^\alpha e_b^\beta) e_A^a e_B^b$ or, using Eq. (4.45),

$$\sigma_{AB} = g_{\alpha\beta}\, e_A^\alpha e_B^\beta. \tag{4.47}$$

Its inverse is denoted σ^{AB}. The three-dimensional completeness relations, $h^{ab} = r^a r^b + \sigma^{AB} e_A^a e_B^b$, are easily established (see Section 3.1). It follows that the four-dimensional relations, $g^{\alpha\beta} = -n^\alpha n^\beta + h^{ab} e_a^\alpha e_b^\beta$, can be expressed as

$$g^{\alpha\beta} = -n^\alpha n^\beta + r^\alpha r^\beta + \sigma^{AB} e_A^\alpha e_B^\beta. \tag{4.48}$$

This can be verified by computing all inner products between the vectors n^α, r^α, and e_A^α.

The extrinsic curvature of S_t embedded in Σ_t is defined by $k_{AB} = r_{a|b} e_A^a e_B^b$ (see Section 3.4), or

$$k_{AB} = r_{\alpha;\beta}\, e_A^\alpha e_B^\beta. \tag{4.49}$$

We use k to denote its trace: $k = \sigma^{AB} k_{AB}$.

A priori, the coordinates θ^A on a given two-surface ($S_{t'}$, say) are not related to the coordinates on another two-surface ($S_{t''}$, say). To introduce a relationship we consider a congruence of curves β running on $\mathscr{B}$, intersecting the two-surfaces S_t *orthogonally*, and therefore having n^α as their tangent vectors. We demand that if the curve β_P intersects $S_{t'}$ at the point P' labelled by θ^A, then the same coordinates will designate the point P'' at which β_P intersects $S_{t''}$. Because θ^A does not vary along these curves, and because t can be chosen as a parameter on the curves, we have

$$n^\alpha = N^{-1}\left(\frac{\partial x^\alpha}{\partial t}\right)_{\theta^A},$$

(4.50)

where the factor N^{-1} comes from Eq. (4.35) and the normalization condition $n^\alpha n_\alpha = -1$. The construction ensures that n^α and e_A^α are everywhere orthogonal.

The hypersurface $\mathscr{B}$ is foliated by the two-surfaces S_t. We place coordinates z^i on $\mathscr{B}$, and introduce the tangent vectors $e_i^\alpha = \partial x^\alpha/\partial z^i$. The induced metric on $\mathscr{B}$ is then given by

$$\gamma_{ij} = g_{\alpha\beta}\, e_i^\alpha e_j^\beta.$$

(4.51)

Its inverse is γ^{ij}, and the completeness relations take the form

$$g^{\alpha\beta} = r^\alpha r^\beta + \gamma^{ij}\, e_i^\alpha e_j^\beta.$$

(4.52)

While the coordinates z^i are a priori arbitrary, the choice $z^i = (t, \theta^A)$ is clearly convenient. In these coordinates a displacement on $\mathscr{B}$ is described by

$$
\begin{aligned}
\mathrm{d}x^\alpha &= \left(\frac{\partial x^\alpha}{\partial t}\right)_{\theta^A} \mathrm{d}t + \left(\frac{\partial x^\alpha}{\partial \theta^A}\right)_t \mathrm{d}\theta^A \\
&= Nn^\alpha\, \mathrm{d}t + e_A^\alpha\, \mathrm{d}\theta^A,
\end{aligned}
$$

where Eqs. (4.45) and (4.50) were used. The line element is

$$
\begin{aligned}
\mathrm{d}s_{\mathscr{B}}^2 &= g_{\alpha\beta}\mathrm{d}x^\alpha\, \mathrm{d}x^\beta \\
&= g_{\alpha\beta}\left(Nn^\alpha\, \mathrm{d}t + e_A^\alpha\, \mathrm{d}\theta^A\right)\left(Nn^\beta\, \mathrm{d}t + e_B^\beta\, \mathrm{d}\theta^B\right) \\
&= \left(g_{\alpha\beta}n^\alpha n^\beta\right)N^2\, \mathrm{d}t^2 + \left(g_{\alpha\beta}e_A^\alpha e_B^\beta\right)\mathrm{d}\theta^A\, \mathrm{d}\theta^B,
\end{aligned}
$$

where the relation $n_\alpha e_A^\alpha = 0$ was used. We have obtained

$$\gamma_{ij}\, \mathrm{d}z^i\, \mathrm{d}z^j = -N^2\, \mathrm{d}t^2 + \sigma_{AB}\, \mathrm{d}\theta^A\, \mathrm{d}\theta^B.$$

(4.53)

This implies $\sqrt{-\gamma} = N\sqrt{\sigma}$, where γ and σ are the determinants of γ_{ij} and σ_{AB}, respectively.

Finally, we let $\mathscr{K}_{ij}$ be the extrinsic curvature of $\mathscr{B}$ embedded in the four-dimensional spacetime. This is given by

$$\mathscr{K}_{ij} = r_{\alpha;\beta}\, e_i^\alpha e_j^\beta,$$

(4.54)

Table 4.1 *Geometric quantities of Σ_t, S_t, and $\mathscr{B}$.*

Surface	Σ_t	S_t	$\mathscr{B}$
Unit normal	n^α	r^α	r^α
Coordinates	y^a	θ^A	z^i
Tangent vectors	e^α_a	e^α_A	e^α_i
Induced metric	h_{ab}	σ_{AB}	γ_{ij}
Extrinsic curvature	K_{ab}	k_{AB}	$\mathscr{K}_{ij}$

because r_α, the unit normal to the two-surfaces S_t, is also normal to $\mathscr{B}$. We will use $\mathscr{K}$ to denote its trace: $\mathscr{K} = \gamma^{ij}\mathscr{K}_{ij}$.

Table 4.1 provides a list of the various geometric quantities introduced in this subsection.

4.2.5 Gravitational action

As a first step toward constructing the gravitational Hamiltonian, we must subject the gravitational action S_G to the $3+1$ decomposition described in Section 4.2.2. Our starting point is Eq. (4.12),

$$(16\pi)S_G = \int_{\mathscr{V}} R\sqrt{-g}\,\mathrm{d}^4x + 2\oint_{\partial\mathscr{V}} \varepsilon K|h|^{1/2}\,\mathrm{d}^3y,$$

where the subtraction term S_0 is omitted for the time being; it will be re-instated at the end of the calculation. Here, $\partial\mathscr{V}$ is the closed hypersurface bounding the four-dimensional region $\mathscr{V}$, y^a are coordinates on $\partial\mathscr{V}$, h_{ab} is the induced metric, K is the trace of the extrinsic curvature, and $\varepsilon = n^\alpha n_\alpha$, where n^α is the outward normal to $\partial\mathscr{V}$.

Throughout this section the quantities n^α, y^a, h_{ab}, and K_{ab} have referred specifically to the spacelike hypersurfaces Σ_t, and we need to be more precise with our notation. We have seen that $\mathscr{V}$'s boundary is the union of two spacelike hypersurfaces Σ_{t_1} and Σ_{t_2} with a timelike hypersurface $\mathscr{B}$ (Fig. 4.4):

$$\partial\mathscr{V} = \Sigma_{t_2} \cup (-\Sigma_{t_1}) \cup \mathscr{B}.$$

The minus sign in front of Σ_{t_1} serves to remind us that while the normal to $\partial\mathscr{V}$ must be directed outward, the normal to Σ_{t_1} is future-directed and therefore points inward. With the notation introduced in the preceding subsection, the gravitational action takes the form

$$(16\pi)S_G = \int_{\mathscr{V}} R\sqrt{-g}\,d^4x - 2\int_{\Sigma_{t_2}} K\sqrt{h}\,d^3y + 2\int_{\Sigma_{t_1}} K\sqrt{h}\,d^3y$$

$$+2\int_{\mathscr{B}} \mathscr{K}\sqrt{-\gamma}\,d^3z,$$

and the integration over Σ_{t_1} incorporates the extra minus sign just discussed.

The region $\mathscr{V}$ is foliated by spacelike hypersurfaces Σ_t on which the Ricci scalar is given by (Section 3.5.3)

$$R = {}^3R + K^{ab}K_{ab} - K^2 - 2\left(n^\alpha_{\ ;\beta}n^\beta - n^\alpha n^\beta_{\ ;\beta}\right)_{;\alpha},$$

where ${}^3R$ is the Ricci scalar constructed from h_{ab}. Using Eq. (4.38), which we write as $\sqrt{-g}\,d^4x = N\sqrt{h}\,dt\,d^3y$, we have that

$$\int_{\mathscr{V}} R\sqrt{-g}\,d^4x = \int_{t_1}^{t_2} dt \int_{\Sigma_t} \left({}^3R + K^{ab}K_{ab} - K^2\right)N\sqrt{h}\,d^3y$$

$$-2\oint_{\partial\mathscr{V}} \left(n^\alpha_{\ ;\beta}n^\beta - n^\alpha n^\beta_{\ ;\beta}\right)d\Sigma_\alpha.$$

The new boundary term can be broken down into integrals over Σ_{t_1}, Σ_{t_2}, and $\mathscr{B}$. On Σ_{t_1}, $d\Sigma_\alpha = n_\alpha\sqrt{h}\,d^3y$ – this also incorporates an extra minus sign – and

$$-2\int_{\Sigma_{t_1}} \left(n^\alpha_{\ ;\beta}n^\beta - n^\alpha n^\beta_{\ ;\beta}\right)d\Sigma_\alpha = -2\int_{\Sigma_{t_1}} n^\beta_{\ ;\beta}\sqrt{h}\,d^3y = -2\int_{\Sigma_{t_1}} K\sqrt{h}\,d^3y.$$

We see that this term cancels out the other integral over Σ_{t_1} coming from the original boundary term in the gravitational action. The integrals over Σ_{t_2} cancel out also. There remains a contribution from $\mathscr{B}$, on which $d\Sigma_\alpha = r_\alpha\sqrt{-\gamma}\,d^3z$, giving

$$-2\int_{\mathscr{B}} \left(n^\alpha_{\ ;\beta}n^\beta - n^\alpha n^\beta_{\ ;\beta}\right)d\Sigma_\alpha = -2\int_{\mathscr{B}} n^\alpha_{\ ;\beta}n^\beta r_\alpha\sqrt{-\gamma}\,d^3z$$

$$= 2\int_{\mathscr{B}} r_{\alpha;\beta}n^\alpha n^\beta\sqrt{-\gamma}\,d^3z,$$

where we have used $n^\alpha r_\alpha = 0$.

Collecting the results, we have

$$(16\pi)S_G = \int_{t_1}^{t_2} dt \int_{\Sigma_t} \left({}^3R + K^{ab}K_{ab} - K^2\right)N\sqrt{h}\,d^3y$$

$$+2\int_{\mathscr{B}} \left(\mathscr{K} + r_{\alpha;\beta}n^\alpha n^\beta\right)\sqrt{-\gamma}\,d^3z.$$

We now use the fact that $\mathcal{B}$ is foliated by the closed two-surfaces S_t. We substitute $\sqrt{-\gamma}\,\mathrm{d}^3 z = N\sqrt{\sigma}\,\mathrm{d}t\,\mathrm{d}^2\theta$ and express $\mathcal{K}$ as

$$
\begin{aligned}
\mathcal{K} &= \gamma^{ij}\mathcal{K}_{ij} \\
&= \gamma^{ij}\left(r_{\alpha;\beta}\,e_i^\alpha e_j^\beta\right) \\
&= r_{\alpha;\beta}\left(\gamma^{ij} e_i^\alpha e_j^\beta\right) \\
&= r_{\alpha;\beta}(g^{\alpha\beta} - r^\alpha r^\beta),
\end{aligned}
$$

so that the integrand becomes

$$
\begin{aligned}
\mathcal{K} + r_{\alpha;\beta} n^\alpha n^\beta &= r_{\alpha;\beta}(g^{\alpha\beta} - r^\alpha r^\beta + n^\alpha n^\beta) \\
&= r_{\alpha;\beta}\left(\sigma^{AB} e_A^\alpha e_B^\beta\right) \\
&= \sigma^{AB}\left(r_{\alpha;\beta} e_A^\alpha e_B^\beta\right) \\
&= \sigma^{AB} k_{AB} \\
&= k.
\end{aligned}
$$

We have used Eqs. (4.48) and (4.52) in these manipulations. Substituting this into our previous expression for the gravitational action, we arrive at

$$
S_G = \frac{1}{16\pi} \int_{t_1}^{t_2} \mathrm{d}t \left\{ \int_{\Sigma_t} \left({}^3R + K^{ab} K_{ab} - K^2\right) N\sqrt{h}\,\mathrm{d}^3 y \right.
$$

$$
\left. + 2\oint_{S_t} (k - k_0) N\sqrt{\sigma}\,\mathrm{d}^2\theta \right\}. \tag{4.55}
$$

We have re-instated the subtraction term, by inserting k_0 into the integral over S_t. This is justified by the fact that the integral over Σ_t vanishes for flat spacetime, so that the sole contribution to S_G comes from the boundary integral; the k_0 term prevents this integral from diverging in the limit $S_t \to \infty$, and it ensures that S_G vanishes identically for flat spacetime. Thus,

$$
k_0 = \text{extrinsic curvature of } S_t \text{ embedded in flat space.}
$$

The k_0 term makes the gravitational action well defined for any asymptotically-flat spacetime. For compact spacetime manifolds, this term is irrelevant.

   The matter action should also be subjected to the 3 + 1 decomposition. Because the procedure is straightforward, and because we would do well to keep things as simple as possible, we shall omit this step here. In the remainder of this section we will consider pure gravity only, and set the matter action to zero.

4.2.6 Gravitational Hamiltonian

To construct the Hamiltonian we must express S_G in terms of

$$\dot{h}_{ab} \equiv \pounds_t h_{ab}, \tag{4.56}$$

where t^α is the timelike vector field defined by Eq. (4.36). We calculate this as follows. First we recall the definition of the induced metric and write

$$\dot{h}_{ab} = \pounds_t \left(g_{\alpha\beta} e_a^\alpha e_b^\beta \right) = \left(\pounds_t g_{\alpha\beta} \right) e_a^\alpha e_b^\beta,$$

where we have used Eq. (4.34). Equation (4.36) implies that the Lie derivative of the metric is given by

$$\begin{aligned}
\pounds_t g_{\alpha\beta} &= t_{\alpha;\beta} + t_{\beta;\alpha} \\
&= (N n_\alpha + N_\alpha)_{;\beta} + (N n_\beta + N_\beta)_{;\alpha} \\
&= n_\alpha N_{,\beta} + N_{,\alpha} n_\beta + N(n_{\alpha;\beta} + n_{\beta;\alpha}) + N_{\alpha;\beta} + N_{\beta;\alpha},
\end{aligned}$$

where $N^\alpha = N^a e_a^\alpha$. Finally, projecting this along $e_a^\alpha e_b^\beta$ gives

$$\dot{h}_{ab} = 2N K_{ab} + N_{a|b} + N_{b|a},$$

where we have used the definitions of extrinsic curvature and intrinsic covariant differentiation found in Section 3.4.

We have obtained

$$K_{ab} = \frac{1}{2N} \left(\dot{h}_{ab} - N_{a|b} - N_{b|a} \right). \tag{4.57}$$

The gravitational action therefore depends on $\dot{h}_{ab}$ through the extrinsic curvature. Notice that the action does *not* involve $\dot{N}$ nor $\dot{N}^a$, so that momenta conjugate to N and N^a are *not defined*. This means that unlike h_{ab}, the lapse and the shift are not *dynamical* variables. This was to be expected: N and N^a only serve to specify the foliation of $\mathcal{V}$ into the spacelike hypersurfaces Σ_t; because this foliation is arbitrary, we are completely free in our choice of lapse and shift.

The momentum conjugate to h_{ab} is defined by

$$p^{ab} = \frac{\partial}{\partial \dot{h}_{ab}} \left(\sqrt{-g}\, \mathscr{L}_G \right), \tag{4.58}$$

where $\mathscr{L}_G$ is the 'volume part' of the gravitational Lagrangian. (The 'boundary part' is independent of $\dot{h}_{ab}$.) Because $\mathscr{L}_G$ is expressed in terms of K_{ab}, it is convenient to write Eq. (4.58) in the form

$$(16\pi) p^{ab} = \frac{\partial K_{mn}}{\partial \dot{h}_{ab}} \frac{\partial}{\partial K_{mn}} \left(16\pi \sqrt{-g}\, \mathscr{L}_G \right),$$

where

$$16\pi \sqrt{-g}\,\mathscr{L}_G = [{}^3R + (h^{ac}h^{bd} - h^{ab}h^{cd})K_{ab}K_{cd}]N\sqrt{h}$$

follows from Eq. (4.55). Evaluating the partial derivatives gives

$$(16\pi)p^{ab} = \sqrt{h}\,(K^{ab} - Kh^{ab}),\tag{4.59}$$

and we see that the canonical momentum is closely related to the extrinsic curvature.

The 'volume part' of the Hamiltonian density is

$$\mathscr{H}_G = p^{ab}\dot{h}_{ab} - \sqrt{-g}\,\mathscr{L}_G.\tag{4.60}$$

Using our previous results, we have

$$\begin{aligned}(16\pi)\mathscr{H}_G = {}&\sqrt{h}\,(K^{ab} - Kh^{ab})(2NK_{ab} + N_{a|b} + N_{b|a})\\&- ({}^3R + K^{ab}K_{ab} - K^2)N\sqrt{h}\\={}&(K^{ab}K_{ab} - K^2 - {}^3R)N\sqrt{h} + 2(K^{ab} - Kh^{ab})N_{a|b}\sqrt{h}\\={}&(K^{ab}K_{ab} - K^2 - {}^3R)N\sqrt{h} + 2\big[(K^{ab} - Kh^{ab})N_a\big]_{|b}\sqrt{h}\\&- 2(K^{ab} - Kh^{ab})_{|b}N_a\sqrt{h}.\end{aligned}$$

The full Hamiltonian is obtained by integrating $\mathscr{H}_G$ over Σ_t and adding the boundary terms:

$$\begin{aligned}(16\pi)H_G ={}& \int_{\Sigma_t} 16\pi\,\mathscr{H}_G\,\mathrm{d}^3y - 2\oint_{S_t}(k - k_0)N\sqrt{\sigma}\,\mathrm{d}^2\theta\\={}&\int_{\Sigma_t}\Big[N(K^{ab}K_{ab} - K^2 - {}^3R) - 2N_a(K^{ab} - Kh^{ab})_{|b}\Big]\sqrt{h}\,\mathrm{d}^3y\\&+ 2\oint_{S_t}(K^{ab} - Kh^{ab})N_a\,\mathrm{d}S_b - 2\oint_{S_t}(k - k_0)N\sqrt{\sigma}\,\mathrm{d}^2\theta.\end{aligned}$$

Writing $\mathrm{d}S_b = r_b\sqrt{\sigma}\,\mathrm{d}^2\theta$, the gravitational Hamiltonian becomes

$$\begin{aligned}(16\pi)H_G ={}&\int_{\Sigma_t}\Big[N(K^{ab}K_{ab} - K^2 - {}^3R) - 2N_a(K^{ab} - Kh^{ab})_{|b}\Big]\sqrt{h}\,\mathrm{d}^3y\\&- 2\oint_{S_t}\Big[N(k - k_0) - N_a(K^{ab} - Kh^{ab})r_b\Big]\sqrt{\sigma}\,\mathrm{d}^2\theta.\end{aligned}\tag{4.61}$$

It is understood that here, K_{ab} stands for the function of h_{ab} and p^{ab} defined by Eq. (4.59); this is given explicitly by

$$\sqrt{h}\,K^{ab} = 16\pi\left(p^{ab} - \frac{1}{2}p\,h^{ab}\right),\tag{4.62}$$

where $p \equiv h_{ab}\,p^{ab}$.

4.2.7 Variation of the Hamiltonian

The equations of motion for the gravitational field are obtained by varying the action of Eq. (4.55) with respect to N, N^a, h_{ab}, and p^{ab}, which are all treated as independent variables. The variation is restricted by the conditions

$$\delta N = \delta N^a = \delta h_{ab} = 0 \qquad \text{on } S_t, \tag{4.63}$$

but there is no requirement that δp^{ab} vanish on the boundary. As a preliminary step toward calculating δS_G, we will now carry out the variation of H_G. The computations presented here are rather formidable; the punch line is delivered in Eq. (4.73) below.

We begin with a variation with respect to both N and N^a. Taking Eq. (4.63) into account, Eq. (4.61) gives

$$(16\pi)\delta_N H_G = \int_{\Sigma_t} (-\hat{C}\,\delta N - 2\hat{C}_a\,\delta N^a)\sqrt{h}\,d^3y, \tag{4.64}$$

where

$$\hat{C} \equiv {}^3R + K^2 - K^{ab}K_{ab}, \qquad \hat{C}_a \equiv (K_a{}^b - K\delta_a{}^b)_{|b}. \tag{4.65}$$

This was easy; the remaining variations will require a much larger effort.

To carry out a variation with respect to h_{ab} or p^{ab}, we must express H_G in terms of these variables, instead of h_{ab} and K^{ab} as was done in Eq. (4.61). Using Eq. (4.62), a few steps of algebra give

$$(16\pi)H_G = \hat{H}_\Sigma + \hat{H}_S, \tag{4.66}$$

where

$$\hat{H}_\Sigma = \int_{\Sigma_t} \left[Nh^{-1/2}(\hat{p}^{ab}\hat{p}_{ab} - \tfrac{1}{2}\hat{p}^2) - Nh^{1/2}\,{}^3R - 2N_a h^{1/2}(h^{-1/2}\,\hat{p}^{ab})_{|b} \right] d^3y \tag{4.67}$$

is the 'volume' term, while

$$\hat{H}_S = -2\oint_{S_t} \left[N(k - k_0) - N_a h^{-1/2}\,\hat{p}^{ab}r_b \right] \sqrt{\sigma}\,d^2\theta \tag{4.68}$$

is the 'boundary' term. We have introduced the notation $\hat{H}_\Sigma \equiv (16\pi)H_\Sigma$, $\hat{p}^{ab} \equiv (16\pi)p^{ab}$, and so on; this usage was anticipated in Eqs. (4.65).

We first vary $\hat{H}_G$ with respect to $\hat{p}^{ab}$. From Eq. (4.67) we have

$$\delta_p \hat{H}_\Sigma = \int_{\Sigma_t} Nh^{-1/2}\delta_p(\hat{p}^{ab}\hat{p}_{ab} - \tfrac{1}{2}\hat{p}^2)\,d^3y - 2\delta_p \int_{\Sigma_t} N_a(h^{-1/2}\,\hat{p}^{ab})_{|b}h^{1/2}\,d^3y.$$

We substitute

$$\delta_p(\hat{p}^{ab}\hat{p}_{ab} - \tfrac{1}{2}\hat{p}^2) = 2(\hat{p}_{ab} - \tfrac{1}{2}\hat{p}\,h_{ab})\delta\hat{p}^{ab}$$

inside the first integral, and we integrate the second by parts. This gives

$$\delta_p \hat{H}_\Sigma = \int_{\Sigma_t} 2\Big[Nh^{-1/2}(\hat{p}_{ab} - \tfrac{1}{2}\hat{p}\, h_{ab}) + N_{(a|b)}\Big]\delta\hat{p}^{ab}\, d^3y$$

$$- 2\oint_{S_t} N_a h^{-1/2}\delta\hat{p}^{ab}\, r_b\sqrt{\sigma}\, d^2\theta.$$

The boundary term is precisely equal to (minus) the variation of $\hat{H}_S$. We therefore have obtained

$$\delta_p \hat{H}_G = \int_{\Sigma_t} \mathcal{H}_{ab}\, \delta\hat{p}^{ab}\, d^3y, \qquad (4.69)$$

where

$$\mathcal{H}_{ab} = 2Nh^{-1/2}\Big(\hat{p}_{ab} - \frac{1}{2}\hat{p}\, h_{ab}\Big) + 2N_{(a|b)}. \qquad (4.70)$$

To vary $\hat{H}_G$ with respect to h_{ab} is more labourious, and we will rely on computations already presented in Section 4.1. We begin with the volume term:

$$\delta_h \hat{H}_\Sigma = \int_{\Sigma_t} \Big[-Nh^{-1}(\hat{p}^{ab}\hat{p}_{ab} - \tfrac{1}{2}\hat{p}^2)\delta_h h^{1/2} + Nh^{-1/2}\delta_h(\hat{p}^{ab}\hat{p}_{ab} - \tfrac{1}{2}\hat{p}^2)$$

$$- N\delta_h(h^{1/2}\, {}^3R)\Big]d^3y - 2\delta_h\oint_{S_t} N_a h^{-1/2}\,\hat{p}^{ab}\, r_b\sqrt{\sigma}\, d^2\theta$$

$$+ 2\delta_h \int_{\Sigma_t} N_{a|b}\, \hat{p}^{ab}\, d^3y,$$

in which the last term on the right-hand side of Eq. (4.67) was integrated by parts. The variation of the integral over S_t vanishes because h_{ab} is fixed on the boundary. In the first term within the integral over Σ_t we substitute

$$\delta_h h^{1/2} = \frac{1}{2}\, h^{1/2}h^{ab}\delta h_{ab},$$

while in the second term,

$$\delta_h(\hat{p}^{ab}\hat{p}_{ab} - \tfrac{1}{2}\hat{p}^2) = 2(\hat{p}^a{}_c\, \hat{p}^{cb} - \tfrac{1}{2}\hat{p}\,\hat{p}^{ab})\delta h_{ab}.$$

In the third term, we use the three-dimensional version of Eq. (4.16),

$$\delta_h(h^{1/2}\, {}^3R) = -h^{1/2}G^{ab}\delta h_{ab} + h^{1/2}\, \delta v^c{}_{|c},$$

where $G^{ab} = R^{ab} - \frac{1}{2}\, {}^3R\, h^{ab}$ is the three-dimensional Einstein tensor and $\delta v^c = h^{ab}\delta\Gamma^c{}_{ab} - h^{ac}\delta\Gamma^b{}_{ab}$. Finally, in the last term we substitute

$$\delta_h N_{a|b} = N^c{}_{|b}\delta h_{ac} + h_{ac}N^d\delta\Gamma^c{}_{bd}.$$

After a few steps of algebra we obtain

$$\delta_h \hat{H}_\Sigma = \int_{\Sigma_t} \left[-\tfrac{1}{2} N h^{-1/2} \left(\hat{p}^{cd} \hat{p}_{cd} - \tfrac{1}{2} \hat{p}^2 \right) h^{ab} + 2 N h^{-1/2} \left(\hat{p}^a_{\ c} \hat{p}^{bc} - \tfrac{1}{2} \hat{p} \, \hat{p}^{ab} \right) \right.$$

$$\left. + N h^{1/2} G^{ab} + 2 \hat{p}^{c(a} N^{b)}_{\ |c} \right] \delta h_{ab} \, \mathrm{d}^3 y$$

$$+ \int_{\Sigma_t} \left[-N h^{1/2} \delta v^c_{\ |c} + 2 \hat{p}^b_{\ c} N^d \delta \Gamma^c_{\ bd} \right] \mathrm{d}^3 y.$$

We now leave the first integral alone and set to work on the second integral, beginning with the first term. After integrating by parts,

$$-\int_{\Sigma_t} N h^{1/2} \delta v^c_{\ |c} \, \mathrm{d}^3 y = \int_{\Sigma_t} N_{,c} \delta v^c h^{1/2} \, \mathrm{d}^3 y - \oint_{S_t} N \delta v^c r_c \sqrt{\sigma} \, \mathrm{d}^2 \theta$$

$$= \int_{\Sigma_t} N_{,c} \delta v^c h^{1/2} \, \mathrm{d}^3 y + \oint_{S_t} N h^{ab} \delta h_{ab,c} r^c \sqrt{\sigma} \, \mathrm{d}^2 \theta,$$

where the three-dimensional version of Eq. (4.17) was used. To express the first integral in terms of δh_{ab} we use the relation

$$\delta \Gamma^c_{\ ab} = \frac{1}{2} h^{cd} \left[(\delta h_{da})_{|b} + (\delta h_{db})_{|a} - (\delta h_{ab})_{|d} \right],$$

which is easy to establish. (Note that the covariant derivative is defined with respect to the reference metric h_{ab}, about which the variation is taken.) We have

$$\delta v^c = \frac{1}{2} \left(h^{ab} h^{cd} - h^{ac} h^{bd} \right) \left[(\delta h_{da})_{|b} + (\delta h_{db})_{|a} - (\delta h_{ab})_{|d} \right]$$

and then

$$N_{,c} \delta v^c = \frac{1}{2} \left(h^{ab} N^{,d} - N^{,a} h^{bd} \right) \left[(\delta h_{da})_{|b} + (\delta h_{db})_{|a} - (\delta h_{ab})_{|d} \right]$$

$$= -\left(h^{ab} N^{,d} - N^{,a} h^{bd} \right) (\delta h_{ab})_{|d};$$

the second line follows by virtue of the antisymmetry in a and d of the first factor. After another integration by parts we obtain

$$-\int_{\Sigma_t} N h^{1/2} \delta v^c_{\ |c} \, \mathrm{d}^3 y = \int_{\Sigma_t} \left(h^{ab} N^{|d}_{\ d} - N^{|ab} \right) \delta h_{ab} h^{1/2} \, \mathrm{d}^3 y$$

$$+ \oint_{S_t} N h^{ab} \delta h_{ab,c} r^c \sqrt{\sigma} \, \mathrm{d}^2 \theta,$$

where we have used the fact that δh_{ab} vanishes on S_t. All this takes care of the first term inside the second integral for $\delta_h \hat{H}_\Sigma$. We now turn to the second term inside the same integral. We have

$$\int_{\Sigma_t} 2\hat{p}^b_{\ c} N^d \delta\Gamma^c_{bd}\, \mathrm{d}^3 y = \int_{\Sigma_t} \hat{p}^{ab} N^d \big[(\delta h_{ab})_{|d} + (\delta h_{ad})_{|b} - (\delta h_{bd})_{|a} \big]\, \mathrm{d}^3 y$$

$$= \int_{\Sigma_t} h^{-1/2}\, \hat{p}^{ab} N^d (\delta h_{ab})_{|d} h^{1/2}\, \mathrm{d}^3 y$$

$$= -\int_{\Sigma_t} (h^{-1/2}\, \hat{p}^{ab} N^d)_{|d}\, \delta h_{ab} h^{1/2}\, \mathrm{d}^3 y,$$

where we have integrated by parts and put $\delta h_{ab} = 0$ on S_t.

Gathering the results, we find that the variation of the volume term is

$$\delta_h \hat{H}_\Sigma = \int_{\Sigma_t} \hat{\mathcal{P}}^{ab} \delta h_{ab}\, \mathrm{d}^3 y + \oint_{S_t} N h^{ab} \delta h_{ab,c} r^c \sqrt{\sigma}\, \mathrm{d}^2\theta,$$

where $\hat{\mathcal{P}}^{ab}$ will be written in full below. On the other hand, variation of the boundary term gives

$$\delta_h \hat{H}_S = -2 \oint_{S_t} N \delta k \sqrt{\sigma}\, \mathrm{d}^2\theta,$$

and $\delta k = \frac{1}{2} h^{ab} \delta h_{ab,c} r^c$ is the three-dimensional analogue of a result previously derived in Section 4.1.5. Thus,

$$\delta_h \hat{H}_S = -\oint_{S_t} N h^{ab} \delta h_{ab,c} r^c \sqrt{-\sigma}\, \mathrm{d}^2\theta,$$

and this cancels out the boundary integral in $\delta_h \hat{H}_\Sigma$. The variation of the full Hamiltonian is therefore

$$\delta_h \hat{H}_G = \int_{\Sigma_t} \hat{\mathcal{P}}^{ab} \delta h_{ab}\, \mathrm{d}^3 y, \qquad (4.71)$$

where

$$\hat{\mathcal{P}}^{ab} = N h^{1/2} G^{ab} - \tfrac{1}{2} N h^{-1/2} \big(\hat{p}^{cd} \hat{p}_{cd} - \tfrac{1}{2} \hat{p}^2 \big) h^{ab} + 2N h^{-1/2} \big(\hat{p}^a_{\ c} \hat{p}^{bc} - \tfrac{1}{2} \hat{p}\, \hat{p}^{ab} \big)$$

$$- h^{1/2} \big(N^{|ab} - h^{ab} N^{|c}_{\ c} \big) - h^{1/2} \big(h^{-1/2}\, \hat{p}^{ab} N^c \big)_{|c} + 2\hat{p}^{c(a} N^{b)}_{\ |c}. \qquad (4.72)$$

Here, as before, $G^{ab} = R^{ab} - \tfrac{1}{2}\,{}^3R\, h^{ab}$ is the three-dimensional Einstein tensor.

Combining Eqs. (4.64), (4.69), and (4.71) we find that the complete variation of the gravitational Hamiltonian, under the conditions of Eq. (4.63), is given by

$$\delta H_G = \int_{\Sigma_t} \big(\mathcal{P}^{ab} \delta h_{ab} + \mathcal{H}_{ab} \delta p^{ab} - \mathcal{C}\, \delta N - 2\mathcal{C}_a\, \delta N^a \big)\, \mathrm{d}^3 y, \qquad (4.73)$$

where $\mathcal{P}^{ab} \equiv \hat{\mathcal{P}}^{ab}/(16\pi)$ is given by Eq. (4.72), $\mathcal{H}_{ab}$ by Eq. (4.70), while $\mathcal{C} \equiv \hat{\mathcal{C}}/(16\pi)$ and $\mathcal{C}^a \equiv \hat{\mathcal{C}}^a/(16\pi)$ are given by Eq. (4.65).

4.2.8 Hamilton's equations

The equations of motion are obtained by varying the gravitational action, expressed as

$$S_G = \int_{t_1}^{t_2} dt \left[\int_{\Sigma_t} p^{ab} \dot{h}_{ab} \, d^3y - H_G \right],$$

with respect to the independent variables N, N^a, h_{ab}, and p^{ab}. Variation yields

$$\delta S_G = \int_{t_1}^{t_2} dt \left[\int_{\Sigma_t} \left(p^{ab} \delta \dot{h}_{ab} + \dot{h}_{ab} \delta p^{ab} \right) d^3y - \delta H_G \right],$$

where δH_G is given by Eq. (4.73). After integrating the first term by parts we obtain

$$\delta S_G = \int_{t_1}^{t_2} dt \int_{\Sigma_t} \left[\left(\dot{h}_{ab} - \mathcal{H}_{ab} \right) \delta p^{ab} - \left(\dot{p}^{ab} + \mathcal{P}^{ab} \right) \delta h_{ab} \right.$$

$$\left. + \mathcal{C} \, \delta N + 2 \mathcal{C}_a \, \delta N^a \right] d^3y. \tag{4.74}$$

Demanding that the action be stationary implies

$$\dot{h}_{ab} = \mathcal{H}_{ab}, \quad \dot{p}^{ab} = -\mathcal{P}^{ab}, \quad \mathcal{C} = 0, \quad \mathcal{C}_a = 0. \tag{4.75}$$

These are the vacuum Einstein field equations in Hamilton form. The first two govern the evolution of the conjugate variables h_{ab} and p^{ab}; it is easy to check that $\dot{h}_{ab} = \mathcal{H}_{ab}$ just reproduces the relation between $\dot{h}_{ab}$ and p^{ab} implied by Eqs. (4.57) and (4.62). The last two are the constraints equations first derived in Section 3.6; the relations $\mathcal{C} = 0$ and $\mathcal{C}_a = 0$ are usually referred to as the *Hamiltonian* and *momentum constraints* of general relativity, respectively.

The Hamiltonian formulation of general relativity suggests the following strategy for solving the Einstein field equations. First, select a foliation of spacetime by specifying the lapse N and the shift N^a as functions of $x^\alpha = (t, y^a)$; the choice of foliation is completely arbitrary. Defining h_{ab} to be the induced metric on the spacelike hypersurfaces, the full spacetime metric is given by Eq. (4.37):

$$ds^2 = -N^2 \, dt^2 + h_{ab}(dy^a + N^a \, dt)(dy^b + N^b \, dt). \tag{4.76}$$

Next, choose initial values for the tensor fields h_{ab} and K_{ab}, where K_{ab} is the extrinsic curvature of the spacelike hypersurfaces. This choice is not entirely arbitrary because the constraint equations must be satisfied: The initial values must be solutions to

$$^3R + K^2 - K^{ab} K_{ab} = 0, \quad (K^{ab} - K h^{ab})_{|b} = 0, \tag{4.77}$$

where ^3R is the Ricci scalar associated with h_{ab}, and $K = h^{ab} K_{ab}$. Finally, evolve these initial values using Hamilton's equations, $\dot{h}_{ab} = \mathcal{H}_{ab}$ and $\dot{p}^{ab} = -\mathcal{P}^{ab}$, which may be written in the form (Section 4.5, Problem 4)

$$\dot{h}_{ab} = 2N K_{ab} + \pounds_N h_{ab} \tag{4.78}$$

and

$$\dot{K}_{ab} = N_{|ab} - N\left(R_{ab} + K K_{ab} - 2K^c{}_a K_{bc}\right) + \pounds_N K_{ab}. \tag{4.79}$$

In these equations, the Lie derivatives are directed along N^a, the shift vector. This formulation of the vacuum field equations, usually referred to as their $3 + 1$ decomposition, is the usual starting point of numerical relativity.

4.2.9 Value of the Hamiltonian for solutions

We now return to Eq. (4.61) and ask: What is the value of the gravitational Hamiltonian when the fields h_{ab} and K_{ab} satisfy the vacuum field equations (4.77)–(4.79)? The answer is that by virtue of the constraint equations, only the boundary term contributes to the solution-valued Hamiltonian:

$$H_G^{\text{solution}} = -\frac{1}{8\pi} \oint_{S_t} \left[N(k - k_0) - N_a \left(K^{ab} - K h^{ab} \right) r_b \right] \sqrt{\sigma}\, d^2\theta. \tag{4.80}$$

As was discussed previously, this boundary term is relevant only when the space-time manifold is noncompact. For compact manifolds, $H_G^{\text{solution}} \equiv 0$. The physical significance of H_G^{solution} for asymptotically-flat spacetimes will be examined in the next section.

4.3 Mass and angular momentum

4.3.1 Hamiltonian definitions

It is natural to expect that the gravitational mass of an asymptotically-flat space-time – its total energy – should be related to the value of the gravitational Hamiltonian for this spacetime. We will explore this relation in this section, and motivate another between the Hamiltonian and the spacetime's total angular momentum.

The solution-valued Hamiltonian, H_G^{solution} given by Eq. (4.80), depends on the asymptotic behaviour of the spacelike hypersurface Σ_t, and on the asymptotic behaviour of the lapse and shift. While the lapse and shift are always arbitrary, the fact that the spacetime is asymptotically flat gives us a preferred behaviour for the hypersurfaces. We shall demand that Σ_t asymptotically coincide with a surface of constant time in Minkowski spacetime: If $(\bar{t}, \bar{x}, \bar{y}, \bar{z})$ is a Lorentzian frame at infinity, then the asymptotic portion of Σ_t must coincide with a sur-face $\bar{t}$ = constant. In this portion of Σ_t, the (arbitrary) coordinates y^a are re-lated to the spatial Minkowski coordinates, and we have the asymptotic relations $y^a \to y^a(\bar{x}, \bar{y}, \bar{z})$; similarly, $x^\alpha \to x^\alpha(\bar{t}, \bar{x}, \bar{y}, \bar{z})$. We note that $\bar{t}$ is proper time for an observer at rest in the asymptotic region, and infer that this observer moves with a four-velocity $u^\alpha = \partial x^\alpha / \partial \bar{t}$. Because this vector is normalized and orthogonal to the surfaces $\bar{t}$ = constant, it must coincide with the normal vector n^α, and we have another asymptotic relation, $n^\alpha \to \partial x^\alpha / \partial \bar{t}$. Substituting this into Eq. (4.36) gives us

$$t^\alpha \to N \left(\frac{\partial x^\alpha}{\partial \bar{t}} \right)_{y^a} + N^a \left(\frac{\partial x^\alpha}{\partial y^a} \right)_{\bar{t}},$$

an asymptotic relation for the flow vector. This shows that once the asymptotic behaviour of Σ_t has been specified, there is a one-to-one correspondence between a choice of lapse and shift and a choice of flow vector. The solution-valued Hamil-tonian can then be regarded either as a functional of N and N^a, or as a functional of t^α.

We shall define M, the gravitational mass of an asymptotically-flat spacetime, to be the limit of H_G^{solution} when S_t is a two-sphere at spatial infinity, evaluated with the following choice of lapse and shift: $N = 1$ and $N^a = 0$. From Eq. (4.80),

$$M = -\frac{1}{8\pi} \lim_{S_t \to \infty} \oint_{S_t} (k - k_0) \sqrt{\sigma} \, d^2\theta. \tag{4.81}$$

Here, σ_{AB} is the metric on S_t, $k = \sigma^{AB} k_{AB}$ is the extrinsic curvature of S_t em-bedded in Σ_t, and k_0 is the extrinsic curvature of S_t embedded in flat space. The quantity defined by Eq. (4.81) is called the *ADM mass* of the asymptotically-flat spacetime; the name refers to the seminal work by Arnowitt, Deser, and Misner.

The choice $N = 1$, $N^a = 0$ implies that asymptotically, $t^\alpha \to \partial x^\alpha / \partial \bar{t}$, so that the flow vector generates an *asymptotic time translation*. The ADM mass is then just the gravitational Hamiltonian for this choice of flow vector, and we have made a formal connection between total energy and time translations. This connection is both deep and compelling, and it can be adapted to give a definition of total angular momentum. Indeed, the gravitational Hamiltonian should provide a sim-ilar connection between angular momentum and *asymptotic rotations*, which are characterized by $t^\alpha \to \phi^\alpha \equiv \partial x^\alpha / \partial \phi$, where ϕ is a rotation angle defined in the asymptotic region in terms of the Cartesian frame $(\bar{x}, \bar{y}, \bar{z})$. This corresponds to the choice $N = 0$, $N^a = \phi^a \equiv \partial y^a / \partial \phi$ of lapse and shift.

We shall define J, the angular momentum of an asymptotically-flat spacetime, to be (minus) the limit of H_G^{solution} when S_t is a two-sphere at spatial infinity, evaluated with $N = 0$ and $N^a = \phi^a$. From Eq. (4.80),

$$J = -\frac{1}{8\pi} \lim_{S_t \to \infty} \oint_{S_t} (K_{ab} - K h_{ab}) \phi^a r^b \sqrt{\sigma} \, \mathrm{d}^2\theta. \qquad (4.82)$$

The minus sign was inserted to recover the usual right-hand rule for the angular momentum. Notice that this definition of angular momentum refers to a specific choice of rotation axis, and ϕ is the angle around this axis.

4.3.2 Mass and angular momentum for stationary, axially symmetric spacetimes

To show that these definitions are in fact reasonable, we shall calculate M and J for an asymptotically-flat spacetime that is both stationary and axially symmetric. In the asymptotic region $r \gg m$, the metric of such a spacetime can be expressed as

$$\mathrm{d}s^2 = -\left(1 - \frac{2m}{r}\right) \mathrm{d}t^2 + \left(1 + \frac{2m}{r}\right) (\mathrm{d}r^2 + r^2 \, \mathrm{d}\Omega^2) - \frac{4j \sin^2\theta}{r} \, \mathrm{d}t \, \mathrm{d}\phi, \quad (4.83)$$

where m and j are the spacetime's mass and angular-momentum parameters, respectively. We will show that $M = m$ and $J = j$, and thus confirm that the Hamiltonian definitions are well founded. We note that the validity of this metric in the asymptotic region could always be used to define mass and angular momentum. Our Hamiltonian definitions are more powerful, however, because they do not involve a particular coordinate system, and they stay meaningful even when the spacetime is not stationary or axially symmetric.

We choose the hypersurfaces Σ_t to be surfaces of constant t, and $n_\alpha = -(1 - m/r)\partial_\alpha t$ is the unit normal. (Throughout this calculation we work consistently to first order in m/r.) The induced metric on Σ_t is given by

$$h_{ab} \, \mathrm{d}y^a \, \mathrm{d}y^b = \left(1 + \frac{2m}{r}\right) (\mathrm{d}r^2 + r^2 \, \mathrm{d}\Omega^2).$$

The boundary S_t is the two-sphere $r = R$, and the limit $R \to \infty$ will be taken at the end of the calculation. Its unit normal is $r_a = (1 + m/r)\partial_a r$ and

$$\sigma_{AB} \, \mathrm{d}\theta^A \, \mathrm{d}\theta^B = \left(1 + \frac{2m}{R}\right) R^2 \, \mathrm{d}\Omega^2$$

gives the two-metric on S_t.

To evaluate M we must first calculate k. This is given by $k = r^a{}_{|a}$ and a brief calculation yields $k = 2(1 - 2m/R)/R$. To this we must subtract k_0, the extrinsic curvature of a two-surface of *identical* intrinsic geometry, but embedded in flat space. On this surface,

$$\sigma^0_{AB} \, d\theta^A \, d\theta^B = R'^2 \, d\Omega^2,$$

where $R' \equiv R(1 + m/R)$ so that $\sigma^0_{AB} = \sigma_{AB}$. We have $k_0 = 2/R' = 2(1 - m/R)/R$ and simple algebra yields $k - k_0 = -2m/R^2$. On the other hand, $\sqrt{\sigma} \, d^2\theta = R^2(1 + 2m/R)\sin\theta \, d\theta \, d\phi$, and substitution into Eq. (4.81) yields

$$M = m, \tag{4.84}$$

the expected result.

To evaluate J we must first calculate $K_{ab}\phi^a r^b = K_{\phi r}(1 - m/r)$, where $K_{ab} = n_{\alpha;\beta} e^\alpha_a e^\beta_b$. (The second term in the integrand, $Kh_{ab}\phi^a r^b$, vanishes because the vectors r^a and ϕ^a are orthogonal.) The relevant component of the extrinsic curvature is $K_{\phi r} = (1 - m/r)\Gamma^t_{\phi r}$. Using

$$g^{tt} = -\left(1 + \frac{2m}{r}\right), \qquad g^{t\phi} = -\frac{2j}{r^3},$$

we find that $\Gamma^t_{\phi r} = -3j \sin^2\theta/r^2$ and this gives $K_{\phi r} = -3j \sin^2\theta/R^2$. Substituting this into Eq. (4.82) yields $J = (3j/4)\int_0^\pi \sin^3\theta \, d\theta$, or

$$J = j, \tag{4.85}$$

the expected result.

4.3.3 Komar formulae

An appealing feature of the Hamiltonian definitions for mass and angular momentum is that they do not involve a specific choice of coordinates. Alternative definitions that share this property can be produced for stationary and axially symmetric spacetimes. These are known as the *Komar formulae*, and they are

$$M = -\frac{1}{8\pi} \lim_{S_t \to \infty} \oint_{S_t} \nabla^\alpha \xi^\beta_{(t)} \, dS_{\alpha\beta} \tag{4.86}$$

and

$$J = \frac{1}{16\pi} \lim_{S_t \to \infty} \oint_{S_t} \nabla^\alpha \xi^\beta_{(\phi)} \, dS_{\alpha\beta}. \tag{4.87}$$

Here, $\xi^\alpha_{(t)}$ is the spacetime's timelike Killing vector and $\xi^\alpha_{(\phi)}$ is the rotational Killing vector; they both satisfy Killing's equation, $\xi_{\alpha;\beta} + \xi_{\beta;\alpha} = 0$. The surface element is given by (Section 3.2.3)

$$dS_{\alpha\beta} = -2n_{[\alpha}r_{\beta]}\sqrt{\sigma} \, d^2\theta, \tag{4.88}$$

where n_α and r_α are the timelike and spacelike normals to S_t, respectively.

To establish that these formulae do indeed give $M = m$ and $J = j$, we must prove that for the spacetime of Eq. (4.83), the relations

$$-2\nabla^\alpha \xi_{(t)}^\beta n_\alpha r_\beta = -2m/r^2 = k - k_0, \qquad \nabla^\alpha \xi_{(\phi)}^\beta n_\alpha r_\beta = K_{ab}\phi^a r^b$$

hold in the limit $r \to \infty$. We begin with the first relation:

$$\begin{aligned}
-2\nabla^\alpha \xi_{(t)}^\beta n_\alpha r_\beta &= 2\xi_{(t);\beta}^\alpha n_\alpha r^\beta \\
&= 2\Gamma^\alpha_{\beta\gamma} n_\alpha r^\beta \xi_{(t)}^\gamma \\
&= -2(1 - 2m/r)\Gamma^t_{rt} \\
&= -2m/r^2,
\end{aligned}$$

as required. We have used Killing's equation in the first line, and inserted $n_\alpha = -(1 - m/r)\partial_\alpha t$, $r^\alpha = r^a e_a^\alpha = (1 - m/r)\partial x^\alpha/\partial r$, and $\Gamma^t_{rt} = m(1 + 2m/r)/r^2$ in the following steps. To establish the second relation requires less work:

$$\begin{aligned}
\nabla^\alpha \xi_{(\phi)}^\beta n_\alpha r_\beta &= -\xi_{(\phi);\beta}^\alpha n_\alpha r^\beta \\
&= \xi_{(\phi)}^\alpha n_{\alpha;\beta} r^\beta \\
&= n_{\alpha;\beta}\left(\phi^a\, e_a^\alpha\right)\left(r^b\, e_b^\beta\right) \\
&= \left(n_{\alpha;\beta}\, e_a^\alpha e_b^\beta\right)\phi^a r^b \\
&= K_{ab}\phi^a r^b.
\end{aligned}$$

Once again, Killing's equation was in the first line. The second line follows from the fact that the Killing vector is orthogonal to n^α. In the third line the vectors $\xi_{(\phi)}^\alpha$ and r^β were decomposed into the basis e_a^α. Finally, the last line follows from the definition of the extrinsic curvature. These computations prove that the definitions of Eq. (4.86) and (4.87) do indeed imply $M = m$ and $J = j$. We see that for stationary and axially symmetric spacetimes, the Komar formulae are equivalent to our Hamiltonian definitions for mass and angular momentum.

The Komar formulae can be turned into hypersurface integrals by invoking Stokes' theorem (Section 3.3.3),

$$\oint_S B^{\alpha\beta}\, dS_{\alpha\beta} = 2\int_\Sigma B^{\alpha\beta}_{\;\;\;;\beta}\, d\Sigma_\alpha,$$

where $B^{\alpha\beta}$ is an antisymmetric tensor field and S is the two-dimensional boundary of the hypersurface Σ. This is possible because when ξ^α is a Killing vector, the tensor $B^{\alpha\beta} = \nabla^\alpha \xi^\beta$ is necessarily antisymmetric. We have

$$B^{\alpha\beta}_{\;\;\;;\beta} = (\nabla^\alpha \xi^\beta)_{;\beta} = -(\nabla^\beta \xi^\alpha)_{;\beta} = -\Box\xi^\alpha,$$

where $\Box \equiv \nabla^\alpha \nabla_\alpha$. Using the fact that all Killing vectors satisfy $\Box \xi^\alpha = -R^\alpha{}_\beta \xi^\beta$ (Section 1.13, Problem 9), we have established the identity

$$\oint_S \nabla^\alpha \xi^\beta \, dS_{\alpha\beta} = 2 \int_\Sigma R^\alpha{}_\beta \xi^\beta \, d\Sigma_\alpha.$$

Because the hypersurface Σ is spacelike, we have that $d\Sigma_\alpha = -n_\alpha \sqrt{h} \, d^3 y$. Using the Einstein field equations we then obtain

$$\oint_S \nabla^\alpha \xi^\beta \, dS_{\alpha\beta} = -16\pi \int_\Sigma \left(T_{\alpha\beta} - \frac{1}{2} T g_{\alpha\beta} \right) n^\alpha \xi^\beta \sqrt{h} \, d^3 y.$$

Finally, combining this with Eqs. (4.86) and (4.87), we arrive at

$$M = 2 \int_\Sigma \left(T_{\alpha\beta} - \frac{1}{2} T g_{\alpha\beta} \right) n^\alpha \xi^\beta_{(t)} \sqrt{h} \, d^3 y \qquad (4.89)$$

and

$$J = -\int_\Sigma \left(T_{\alpha\beta} - \frac{1}{2} T g_{\alpha\beta} \right) n^\alpha \xi^\beta_{(\phi)} \sqrt{h} \, d^3 y. \qquad (4.90)$$

In these equations, Σ stands for any spacelike hypersurface that extends to spatial infinity. If Σ had two boundaries instead of just one, then an additional contribution from the inner boundary would appear on the right-hand side of Eqs. (4.89) and (4.90). Such a situation arises when the stationary, axially symmetric spacetime contains a black hole (see Section 5.5.3).

It is a remarkable fact that M and J are defined fundamentally in terms of integrals over a closed two-surface at infinity. These quantities should therefore be thought of as properties of the asymptotic structure of spacetime. It is only in the case of stationary, axially symmetric spacetimes that M and J can be defined as hypersurface integrals.

4.3.4 Bondi–Sachs mass

The ADM mass was constructed in Section 4.3.1 by selecting a closed two-surface $S_t \equiv S(t, r)$, integrating $k - k_0$ over this surface, and then taking the limit $r \to \infty$. Thus,

$$M_{\text{ADM}}(t) = -\frac{1}{8\pi} \oint_{S(t, r \to \infty)} (k - k_0) \sqrt{\sigma} \, d^2 \theta. \qquad (4.91)$$

Here, $S(t, r)$ denotes a surface of constant t and r which becomes a round two-sphere of area $4\pi r^2$ as $r \to \infty$. This limit, which is taken while keeping t fixed, is what defines 'spatial infinity.'

There exists another way of reaching infinity, and to this new limiting procedure corresponds a distinct notion of mass. This is the *Bondi–Sachs* mass, which is obtained by taking $S(t, r)$ to 'null infinity' instead of spatial infinity. To define this we introduce the null coordinates $u = t - r$ (retarded time) and $v = t + r$ (advanced time). In these coordinates, a two-surface of constant t and r becomes a surface of constant u and v, which we denote $S(u, v)$. Null infinity corresponds to the limit $v \to \infty$ keeping u fixed, and the Bondi–Sachs mass is defined by

$$M_{\mathrm{BS}}(u) = -\frac{1}{8\pi} \oint_{S(u,v\to\infty)} (k - k_0)\sqrt{\sigma}\, \mathrm{d}^2\theta. \qquad (4.92)$$

The physical importance of the Bondi–Sachs mass comes from the fact that when an isolated body emits radiation (in the form, say, of electromagnetic or gravitational waves), the rate of change of $M_{\mathrm{BS}}(u)$ is directly related to the outward flux of radiated energy. If F denotes this flux, then the Bondi–Sachs mass satisfies

$$\frac{\mathrm{d}M_{\mathrm{BS}}}{\mathrm{d}u} = -\oint_{S(u,v\to\infty)} F\sqrt{\sigma}\, \mathrm{d}^2\theta. \qquad (4.93)$$

Thus, the mass of a radiating body decreases as the radiation escapes to infinity. The proof of this statement is rather involved; it can be found in the original papers by Bondi, Sachs, and their collaborators.

4.3.5 Distinction between ADM and Bondi–Sachs masses: Vaidya spacetime

For stationary spacetimes, the ADM and Bondi–Sachs masses are identical: there is no distinction. For the dynamical spacetime of an isolated body emitting gravitational (or other types of) radiation, the two notions of mass are distinct. For such a system the Bondi–Sachs mass decreases according to Eq. (4.93), while the ADM mass stays constant.

The metric of a radiating spacetime is difficult to write down; usually it is expressed as a messy expansion in powers of $1/r$. We shall not attempt to deal with these complications here. For the purpose of illustrating the difference between the ADM and Bondi–Sachs masses, we shall instead adopt a simple spherically-symmetric model. Consider the Schwarzschild metric expressed in terms of the null coordinate $u = t - r - 2M \ln(r/2M - 1)$, and allow the mass parameter M to become a function of retarded time: $M \to m(u)$. This new metric is given by

$$\mathrm{d}s^2 = -f\, \mathrm{d}u^2 - 2\, \mathrm{d}u\, \mathrm{d}r + r^2\, \mathrm{d}\Omega^2, \qquad f = 1 - 2m(u)/r, \qquad (4.94)$$

and it is a good candidate to represent a radiating spacetime. To see if it makes a sensible solution to the Einstein field equations, let us examine the Einstein tensor, whose only nonvanishing component is $G_{uu} = -(2/r^2)(\mathrm{d}m/\mathrm{d}u)$. This means that the stress-energy tensor must be of the form

$$T_{\alpha\beta} = -\frac{dm/du}{4\pi r^2} l_\alpha l_\beta, \tag{4.95}$$

where $l_\alpha = -\partial_\alpha u$ is tangent to radial, outgoing null geodesics. This stress-energy tensor describes a pressureless fluid with energy density $\rho = (-dm/du)/(4\pi r^2)$ moving with a four-velocity l^α. Such a fluid is usually referred to as *null dust*; it gives a good description of radiation in the high-frequency, geometric-optics approximation. It is easy to check that the form (function of u)$/r^2$ for the energy density is dictated by energy-momentum conservation. You may also verify that all the standard energy conditions are satisfied by $T_{\alpha\beta}$ if $dm/du < 0$, that is, if m decreases with increasing retarded time. We conclude that the metric of Eq. (4.94), called the *outgoing Vaidya metric*, makes a physically reasonable solution to the Einstein field equations.

We wish to compute the ADM and Bondi–Sachs masses for the Vaidya spacetime. The first step is to select a spacelike hypersurface Σ bounded by a closed two-surface S; this hypersurface must asymptotically coincide with a surface $\bar{t} = $ constant of Minkowski spacetime. A suitable choice is to let Σ be a surface of constant $t \equiv u + r$, for which the unit normal

$$n_\alpha = -(2 - f)^{-1/2} \partial_\alpha(u + r)$$

is everywhere timelike. From Eq. (4.94) we obtain that the induced metric on Σ is

$$h_{ab} \, dy^a \, dy^b = (2 - f) \, dr^2 + r^2 \, d\Omega^2.$$

For S we choose the two-sphere $r = R$, where R is a constant much larger than the maximum value of $2m(u)$; eventually we will take the limit $R \to \infty$. Recall that spatial infinity corresponds to keeping t fixed while taking the limit (which means that $u \to -\infty$), whereas null infinity corresponds to keeping u fixed while taking the limit. The unit normal on S is $r_a = (2 - f)^{1/2} \partial_a r$, and the induced metric is $\sigma_{AB} \, d\theta^A d\theta^B = R^2 \, d\Omega^2$.

First we calculate

$$M(S) \equiv -\frac{1}{8\pi} \oint_S (k - k_0) \sqrt{\sigma} \, d^2\theta$$

for the bounded two-surface S; the two different limits to infinity will be taken next. The extrinsic curvature of S embedded in Σ is calculated as

$$k = r^a{}_{|a} = \frac{2}{R} \left[1 + \frac{2m(u)}{R} \right]^{-1/2} = \frac{2}{R} \left[1 - \frac{m(u)}{R} + O(R^{-2}) \right],$$

and the extrinsic curvature of S embedded in flat space is $k_0 = 2/R$. Subtracting, we have that $k - k_0 = -2m(u)/R^2 + O(R^{-3})$ and integrating over S yields $M(S) = m(u) + O(R^{-1})$.

We may now take the limit $R \to \infty$. As was mentioned previously, the ADM mass is obtained by keeping $t = u + R$ fixed while taking the limit. This gives

$$M_{\text{ADM}}(t) = m(-\infty), \qquad (4.96)$$

and we see that the ADM mass is a constant, equal to the initial value of the mass function. We may therefore say that M_{ADM} represents *all* the mass initially present in the spacetime. (This interpretation is quite general and not limited to this specific example.) For the Bondi–Sachs mass we must keep u fixed while taking the limit. This gives

$$M_{\text{BS}}(u) = m(u), \qquad (4.97)$$

and we see that the Bondi–Sachs mass is identified with the mass function of the Vaidya spacetime. It decreases in response to the outflow of radiation described by the stress-energy tensor of Eq. (4.95). Notice that the field equation

$$\frac{dm}{du} = -4\pi r^2 T_{uu} = -4\pi r^2 \left(-T^r_{\ u}\right) \equiv -4\pi r^2 F$$

is compatible with the general mass-loss formula of Eq. (4.93).

It may appear paradoxical that the ADM mass of a dynamical spacetime should be a constant. This, however, is what should be expected of a *radiating* spacetime (Fig. 4.5). The ADM mass represents all the mass present on a spacelike hypersurface of constant t. This hypersurface intersects the central body whose mass does decrease as a consequence of radiation loss. But this does not mean that the ADM mass should decrease, because the hypersurface intersects also the radiation, and the ADM mass accounts for both forms of energy. The net result is a conserved quantity. On the other hand, the Bondi–Sachs mass represents all the mass present

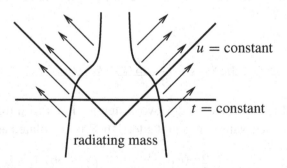

Figure 4.5   A radiating spacetime.

on a null hypersurface of constant u. Because this hypersurface *fails* to intersect any of the radiation that was emitted prior to the retarded time u, the net result is a quantity that decreases with increasing retarded time.

4.3.6 Transfer of mass and angular momentum

We shall now derive expressions for the transfer of mass and angular momentum across a hypersurface Σ in a stationary, axially symmetric spacetime.

To begin, consider the vector fields

$$\varepsilon^\alpha = -T^\alpha_{\ \beta}\,\xi^\beta_{(t)}, \qquad \ell^\alpha = T^\alpha_{\ \beta}\,\xi^\beta_{(\phi)}, \tag{4.98}$$

where $T^{\alpha\beta}$ is a *test* stress-energy tensor that does not influence the spacetime geometry. According to the definition of the stress-energy tensor, ε^α can be interpreted as an energy flux vector, while ℓ^α is interpreted as an angular-momentum flux vector.

To see this clearly, consider the simple case of dust, a perfect fluid with stress-energy tensor $T^{\alpha\beta} = \rho u^\alpha u^\beta$, where ρ is the rest-mass density and u^α the four-velocity. Energy-momentum conservation implies that u^α satisfies the geodesic equation, and that $j^\alpha = \rho u^\alpha$ is a conserved vector: $j^\alpha_{\ ;\alpha} = 0$. This vector can be interpreted as the dust's momentum density, or equivalently, as a rest-mass flux vector. Then $\varepsilon^\alpha = \tilde{E} j^\alpha$ and $\ell^\alpha = \tilde{L} j^\alpha$, where $\tilde{E} \equiv -u_\alpha \xi^\alpha_{(t)}$ is the conserved energy per unit rest mass and $\tilde{L} \equiv u_\alpha \xi^\alpha_{(\phi)}$ the conserved angular momentum per unit rest mass. (As we have indicated, both $\tilde{E}$ and $\tilde{L}$ are constants of the motion.) These relations show quite clearly that ε^α represents a flux of energy density, while ℓ^α is a flux of angular-momentum density.

The vectors ε^α and ℓ^α are divergence-free. For example,

$$\varepsilon^\alpha_{\ ;\alpha} = -T^{\alpha\beta}_{\ \ ;\alpha}\,\xi_{(t)\beta} + T^{\alpha\beta}\,\xi_{(t)\beta;\alpha} = 0;$$

the first term vanishes by virtue of energy-momentum conservation, and the second vanishes because $\xi_{(t)\beta;\alpha}$ is an antisymmetric tensor field. This implies that the integral of ε^α or ℓ^α over a hypersurface $\partial\mathscr{V}$ enclosing a four-dimensional region $\mathscr{V}$ is identically zero. For example,

$$\oint_{\partial\mathscr{V}} \varepsilon^\alpha \, d\Sigma_\alpha = 0.$$

This equation states that the *total transfer* of energy across a closed hypersurface $\partial\mathscr{V}$ is zero. This is clearly a statement of conservation of total energy – or total mass.

The boundary $\partial \mathcal{V}$ can be partitioned into any number of pieces. If one such piece is the hypersurface Σ, then the integral of ε^α over Σ represents the mass transferred across this piece of $\partial \mathcal{V}$. Thus,

$$\Delta M = -\int_\Sigma T^\alpha{}_\beta \, \xi^\beta_{(t)} \, d\Sigma_\alpha \qquad (4.99)$$

is the mass transferred across the hypersurface Σ, and similarly,

$$\Delta J = \int_\Sigma T^\alpha{}_\beta \, \xi^\beta_{(\phi)} \, d\Sigma_\alpha \qquad (4.100)$$

is the angular momentum transferred across Σ.

For illustration, let us return to our previous example, and let us choose Σ to be spacelike and orthogonal to the vector field u^α. Then $d\Sigma_\alpha = -u_\alpha \sqrt{h} \, d^3 y$ and we find that $\Delta M = \int_\Sigma \tilde{E} \rho \sqrt{h} \, d^3 y$ and $\Delta J = \int_\Sigma \tilde{L} \rho \sqrt{h} \, d^3 y$. The first equation states that the transfer of energy across Σ is the integral of $\tilde{E} \rho$, the energy density. The second equation comes with a very similar interpretation.

4.4 Bibliographical notes

During the preparation of this chapter I have relied on the following references: Arnowitt, Deser, and Misner (1962); Bondi, van der Burg, and Metzner (1962); Brown and York (1993), Brown, Lau, and York (1997); Carter (1979); Hawking and Horowitz (1996); Sachs (1962); Sudarsky and Wald (1992); and Wald (1984).

More specifically:

An overview of the Lagrangian and Hamiltonian formulations of general relativity is given in Appendix E of Wald. The Hamiltonian formulation was initiated by Arnowitt, Deser, and Misner, who also introduced the ADM mass. Early treatments of the Hamiltonian formulation often discarded the all-important boundary terms; careful treatments are given in Sudarsky and Wald, Brown and York, and Hawking and Horowitz. (Problem 7 below is based on this last paper.) The Hamiltonian definitions for mass and angular momentum are taken from Brown and York; the discussion of Section 4.2.4 is also based on their paper. Sections 4.3.3 and 4.3.5 are based on Carter's Sections 6.6.1 and 6.6.2, respectively. The Bondi–Sachs mass was introduced by Bondi and his collaborators in an effort to put the notion of gravitational-wave energy on a firm footing. The definition given in Section 4.3.4 is due to Brown, Lau, and York. I would like to point out that the first occurrence of the $(k - k_0)$ formula for the ADM mass can be found in a 1988 paper by Katz, Lynden-Bell, and Israel.

Suggestions for further reading:

The numerical integration of the Einstein field equations is currently one of the most active areas of research in gravitational physics. While the starting point of most numerical methods is the $3 + 1$ decomposition of the field equations presented in Section 4.2.8, the story by no means ends there. For its comprehensive review of various methods and its summary of the field's achievements, the 2001 article by Lehner is a very useful reference.

We have seen that the gravitational action, and the gravitational Hamiltonian, must include a subtraction term in order to be well defined for asymptotically-flat spacetimes. The subtraction term, however, is not unique, and alternative proposals were put forth by Mann (1999 and 2000) and Lau (1999). Generalizations to spacetimes that are not asymptotically flat have been considered, mostly in the context of spacetimes with a negative cosmological constant; see the 1999 paper by Kraus, Larsen, and Siebelink.

Who was the first to discover the field equations of general relativity: Einstein or Hilbert? The story used to be that after formulating a variational principle for general relativity, Hilbert published the field equations first, just a few days before Einstein did (on November 25, 1915). Recent historical investigation reveals, however, that Hilbert did not, in fact, anticipate Einstein. The revised story is told by Corry, Renn, and Stachel (1997).

4.5 Problems

1. The Lagrangian density for the free electromagnetic field is

$$\mathscr{L} = -\frac{1}{16\pi} F^{\alpha\beta} F_{\alpha\beta},$$

where $F_{\alpha\beta} = A_{\beta;\alpha} - A_{\alpha;\beta}$ is the Faraday tensor, expressed in terms of the vector potential A_α.

(a) Derive the Maxwell field equations in vacuum, $F^{\alpha\beta}{}_{;\beta} = 0$, on the basis of this Lagrangian density.

(b) Show that the stress-energy tensor for the electromagnetic field is given by

$$T_{\alpha\beta} = \frac{1}{4\pi} \left(F_{\alpha\mu} F_\beta{}^\mu - \frac{1}{4} g_{\alpha\beta} F^{\mu\nu} F_{\mu\nu} \right).$$

2. The Lagrangian density for a point particle of mass m moving on a world line $z^\alpha(\lambda)$ is given by

$$\mathscr{L} = -m \int \sqrt{-g_{\alpha\beta} \dot{z}^\alpha \dot{z}^\beta}\, \delta_4(x, z)\, d\lambda,$$

where $\delta_4(x, x')$ is a four-dimensional, scalarized δ-function satisfying

$$\int_{\mathcal{V}} \delta_4(x, x')\sqrt{-g}\, d^4x = 1$$

if x' is within the domain of integration; we also have $\dot{z}^\alpha = dz^\alpha/d\lambda$, and the parameterization of the world line is arbitrary.

(a) Derive an expression for the stress-energy tensor of a point particle. To simplify this expression, set $d\lambda = d\tau$ (with τ denoting proper time on the world line) at the end of the calculation.

(b) Prove that when it is applied to a point particle, the statement $T^{\alpha\beta}{}_{;\beta} = 0$ gives rise to the geodesic equation for $u^\alpha = dz^\alpha/d\tau$.

(c) Explain whether the result of part (b) constitutes a valid proof of the statement that the Einstein field equations predict the motion of a massive body to be geodesic.

3. Calculate the gravitational action S_G for a region $\mathcal{V}$ of Schwarzschild spacetime. Take $\mathcal{V}$ to be bounded by the hypersurfaces Σ_{t_1}, Σ_{t_2}, Σ_R, and Σ_ρ, where Σ_{t_1} (Σ_{t_2}) is the spacelike hypersurface described by $t = t_1$ $(t = t_2)$, and where Σ_R (Σ_ρ) is the three-cylinder at $r = R$ $(r = \rho)$. Here, $2M < \rho < R$. At the end of the calculation, take the limits $R \to \infty$ and $\rho \to 2M$.

4. Derive Eq. (4.79), the evolution equation for the extrinsic curvature. You may use $\dot{p}^{ab} = -\mathcal{P}^{ab}$ as a starting point, or proceed from scratch with the definition $\dot{K}_{ab} = \pounds_t(n_{\alpha;\beta}e_a^\alpha e_b^\beta)$. [Either way, the calculation is tedious! You may want to consult York (1979).]

5. Recall that in Section 3.6.5 we introduced a mass function $m(r)$ that determines the three-metric of a spherically symmetric hypersurface. Prove that

$$-\frac{1}{8\pi} \oint_{S(r)} (k - k_0)\sqrt{\sigma}\, d^2\theta = r\left(1 - \sqrt{1 - 2m/r}\right),$$

where $S(r)$ is a two-surface of constant r. Use this to show that $m(\infty)$ is the ADM mass of this hypersurface.

6. In this problem we explore some consequences of Eq. (4.89), which gives an expression for the ADM mass of a stationary spacetime.

(a) Prove that the right-hand side of Eq. (4.89) is independent of the choice of hypersurface Σ.

(b) Show that if $T^{\alpha\beta}$ is the stress-energy tensor of a static perfect fluid, then Eq. (4.89) reduces to

$$M = \int_\Sigma (\rho + 3p)\, e^\Phi \sqrt{h}\, d^3y,$$

where ρ is the mass density, p the pressure, and $e^{2\Phi} \equiv -g_{\alpha\beta}\xi^\alpha_{(t)}\xi^\beta_{(t)}$. [Hints: A perfect fluid is static if its four-velocity u^α is parallel to $\xi^\alpha_{(t)}$. You may assume that the spacetime also is static.]

(c) Specialize to spherical symmetry, and write the spacetime metric as

$$ds^2 = -e^{2\Phi}\,dt^2 + (1 - 2m/r)^{-1}\,dr^2 + r^2\,d\Omega^2,$$

in which Φ and m are functions of r. Refer back to the result of Problem 5 and deduce the identity

$$\int_\Sigma \rho(1 - 2m/r)^{1/2}\,dV = \int_\Sigma (\rho + 3p)\,e^\Phi\,dV,$$

where $dV = \sqrt{h}\,d^3y$ is the invariant volume element on the hypersurface Σ.

(d) Specialize now to a weak-field situation, for which the metric can be expressed as

$$ds^2 = -(1 + 2\Phi)\,dt^2 + (1 - 2\Phi)(dx^2 + dy^2 + dz^2);$$

the Newtonian potential Φ is a function of $\bar{r} = \sqrt{x^2 + y^2 + z^2}$. Working consistently in the weak-field approximation, show that the identity derived in part (c) reduces to

$$\frac{1}{8\pi}\int_\Sigma |\nabla\Phi|^2\,dV = 3\int_\Sigma p\,dV,$$

in which dV and all vectorial operations refer to the three-dimensional flat space of ordinary vector calculus. The left-hand side represents (minus) the total gravitational potential energy of the system. For a monoatomic ideal gas in thermodynamic equilibrium, the right-hand side represents twice the total kinetic energy of the system. This equation is therefore a formulation of the virial theorem of Newtonian gravitational physics. The identity of part (c) can then be interpreted as a general-relativistic version of the virial theorem.

7. The ADM mass is usually defined by

$$M = \frac{1}{16\pi}\oint_{S\to\infty}(D^b\gamma_{ab} - D_a\gamma)r^a\sqrt{\sigma}\,d^2\theta,$$

which is a very different expression from the one appearing in Section 4.3.1. Here, S is the two-surface that encloses the spacelike hypersurface Σ. If h_{ab} is the metric on Σ in arbitrary coordinates y^a, then $\gamma_{ab} \equiv h_{ab} - h^0_{ab}$, where h^0_{ab} is the metric of flat space in the same coordinates. We also have $\gamma \equiv \gamma^a_a$, and D_a is the covariant derivative associated with the flat metric h^0_{ab}, which is used to raise and lower all indices. Finally, r^a is the unit normal of the surface S, and $\sqrt{\sigma}\,d^2\theta$ is the surface element on S.

The purpose of this problem is to prove that this definition is equivalent to the one given in the text,

$$M = -\frac{1}{8\pi}\oint_{S\to\infty}(k-k_0)\sqrt{\sigma}\,d^2\theta,$$

where k (k_0) is the extrinsic curvature of S embedded in Σ (flat space). You may proceed along the following lines:

Because both expressions are invariant under a coordinate transformation, we may use, in a neighbourhood of S, the coordinates (ℓ, θ^A), where ℓ is proper distance off S in the direction orthogonal to S, and θ^A are coordinates on S which are Lie transported off S along curves orthogonal to S. In these coordinates the metric on Σ is given by

$$h_{ab}\,dy^a\,dy^b = d\ell^2 + \tilde{\sigma}_{AB}(\ell)\,d\theta^A\,d\theta^B,$$

where $\tilde{\sigma}_{AB}(\ell)$ (which also depends on θ^A) is such that $\tilde{\sigma}_{AB}(0) = \sigma_{AB}$, the induced metric on S. Similarly,

$$h^0_{ab}\,dy^a\,dy^b = d\ell^2 + \tilde{\sigma}^0_{AB}(\ell)\,d\theta^A\,d\theta^B.$$

Because the induced metrics must agree on S, we also have $\tilde{\sigma}^0_{AB}(0) = \sigma_{AB}$. This implies that $\gamma_{ab} = 0$ on S.

Using this information, show that both expressions for M reduce to the same form,

$$M = -\frac{1}{16\pi}\oint_{S\to\infty}h^{0ab}\gamma_{ab,\ell}\sqrt{\sigma}\,d^2\theta.$$

This is sufficient to prove that the two expressions are indeed equivalent.

8. In this problem we study the transport of energy and angular momentum by a scalar field ψ in flat spacetime. The metric is $ds^2 = -dt^2 + dr^2 + r^2\,d\Omega^2$ and the scalar field satisfies the wave equation $g^{\alpha\beta}\psi_{;\alpha\beta} = -4\pi\rho$, where ρ is a specified source. It can be shown that in the wave zone (where r is much larger than a typical wavelength of the radiation), the field is given by

$$\psi(t, r, \theta, \phi) = \frac{1}{r}\sum_{\ell=0}^{\infty}\sum_{m=-\ell}^{\ell}a_{\ell m}(u)\,Y_{\ell m}(\theta, \phi) + O(r^{-2}),$$

where $Y_{\ell m}(\theta, \phi)$ are spherical harmonics; the amplitudes $a_{\ell m}$ are constructed from ρ, and they are functions of retarded time $u \equiv t - r$. The scalar field comes with a stress-energy tensor

$$T_{\alpha\beta} = \psi_{,\alpha}\psi_{,\beta} - \frac{1}{2}\left(\psi^{,\mu}\psi_{,\mu}\right)g_{\alpha\beta},$$

and we are interested in the transfer of energy and angular momentum across a null hypersurface Σ defined by $v = \text{constant}$, where $v \equiv t + r$ is advanced time.

(a) Show that $d\Sigma_\alpha = -r^2 k_\alpha \, du \, d\Omega$, where $k_\alpha = -\frac{1}{2}\partial_\alpha v$ and $d\Omega = \sin\theta \, d\theta \, d\phi$, is a surface element on Σ.

(b) Prove that for any test field producing a stress-energy tensor $T^{\alpha\beta}$, the amount of energy crossing Σ per unit retarded time is

$$\frac{dE}{du} = \oint_S r^2 T_{\alpha\beta} k^\alpha t^\beta \, d\Omega,$$

where $t^\alpha = \partial x^\alpha / \partial t$ and S is a two-sphere of constant u and v. Prove also that the amount of angular momentum flowing across Σ is given by

$$\frac{dJ}{du} = -\oint_S r^2 T_{\alpha\beta} k^\alpha \phi^\beta \, d\Omega,$$

where $\phi^\alpha = \partial x^\alpha / \partial \phi$.

(c) Show that for scalar radiation, the preceding expressions reduce to

$$\frac{dE}{du} = \sum_{\ell=0}^{\infty} \sum_{m=-\ell}^{\ell} |\dot{a}_{\ell m}(u)|^2$$

and

$$\frac{dJ}{du} = \sum_{\ell=0}^{\infty} \sum_{m=-\ell}^{\ell} im \, \dot{a}_{\ell m}(u) a_{\ell m}^*(u)$$

in the limit $v \to \infty$. Here, an overdot indicates differentiation with respect to u and an asterisk denotes complex conjugation.

(d) Suppose that the source producing the scalar radiation is in *rigid rotation* around the z axis, in the sense that the t and ϕ dependence of ρ resides entirely in the combination $\phi - \Omega t$, where Ω is a constant angular velocity. Prove that in this situation the field satisfies

$$\psi_{,\alpha} \xi^\alpha = 0,$$

where $\xi^\alpha \equiv t^\alpha + \Omega\phi^\alpha$. Prove also that in the limit $v \to \infty$, the transfers of energy and angular momentum are related by

$$\frac{\mathrm{d}E}{\mathrm{d}u} = \Omega\,\frac{\mathrm{d}J}{\mathrm{d}u}.$$

This relation applies to any type of radiation emitted by a source in rigid rotation. It is valid also in curved spacetimes, provided that the spacetime is stationary, axially symmetric, and asymptotically flat.

5

Black holes

The final chapter of this book is devoted to one of the most successful applications of general relativity, the mathematical theory of black holes. In the first part of the chapter we explore three exact solutions to the Einstein field equations that describe black holes; these are the Schwarzschild (Section 5.1), Reissner–Nordström (Section 5.2), and Kerr (Section 5.3) solutions. In Section 5.4 we move away from the specifics of those solutions and consider properties of black holes that can be formulated quite generally, without relying on the details of a particular metric. In the final section of this chapter, Section 5.5, we present the four fundamental laws of black-hole mechanics.

The most important feature of a black-hole spacetime is the event horizon, a null hypersurface which acts as a causal boundary between two regions of the spacetime, the interior and exterior of the black hole. Many physical quantities associated with the black hole, such as its mass, angular momentum, and surface area, are defined by integration over the event horizon. The integration techniques introduced in Chapter 3 will be put to direct use here, as well as the notions of mass and angular momentum encountered in Chapter 4. And since the event horizon is generated by a congruence of null geodesics, the methods developed in Chapter 2 will also be part of our discussion. So here it all comes together in one final glorious moment!

5.1 Schwarzschild black hole

5.1.1 Birkhoff's theorem

The Schwarzschild metric,

$$ds^2 = -\left(1 - \frac{2M}{r}\right) dt^2 + \left(1 - \frac{2M}{r}\right)^{-1} dr^2 + r^2 d\Omega^2, \tag{5.1}$$

is the unique solution to the Einstein field equations that describes the vacuum spacetime outside a spherically symmetric body of mass M. While this object could have a time-dependent mass distribution, the external spacetime is necessarily static and its metric is given by Eq. (5.1). This statement, known as *Birkhoff's theorem*, implies that a spherical mass distribution cannot emit gravitational waves.

The proof of the theorem goes as follows. The metric of a spherically symmetric spacetime can always be cast in the form

$$ds^2 = -e^{2\psi} f \, dt^2 + f^{-1} \, dr^2 + r^2 \, d\Omega^2, \qquad (5.2)$$

involving the two arbitrary functions $\psi(t, r)$ and $f(t, r)$. (This statement is not quite true: It could happen that t and r fail to be good coordinates in some region of the spacetime, and this would invalidate this form of the metric in that region. We will encounter such cases shortly, but they can be ignored for the time being.) It is convenient to also introduce a mass function $m(t, r)$ defined by

$$f = 1 - \frac{2m}{r}. \qquad (5.3)$$

For the metric of Eq. (5.2), the Einstein field equations are

$$\frac{\partial m}{\partial r} = 4\pi r^2 (-T^t_{\ t}), \qquad \frac{\partial m}{\partial t} = -4\pi r^2 (-T^r_{\ t}),$$

$$\frac{\partial \psi}{\partial r} = 4\pi r f^{-1} (-T^t_{\ t} + T^r_{\ r}). \qquad (5.4)$$

The first two equations motivate the name 'mass' for the function $m(t, r)$, as $-T^t_{\ t}$ represents the density of mass-energy and $-T^r_{\ t}$ its outward flux; they imply that in vacuum, $m(t, r) = M$, a constant. The third gives $\psi' = 0$, and $\psi(t, r)$ can be set equal to zero without loss of generality. The Schwarzschild solution is thereby recovered.

5.1.2 Kruskal coordinates

The difficulties of the Schwarzschild metric at $r = 2M$ are well known. While the spacetime is perfectly well behaved there, the coordinates (t, r) become singular at $r = 2M$ – they are no longer in a one-to-one correspondence with spacetime events. This problem can be circumvented by introducing another coordinate system. The following construction originates from the independent work of Kruskal (1960) and Szekeres (1960).

Consider a swarm of massless particles moving radially in the Schwarzschild spacetime – t and r vary, but not θ and ϕ. It is easy to check that ingoing particles move along curves $v = $ constant, while outgoing particles move along curves

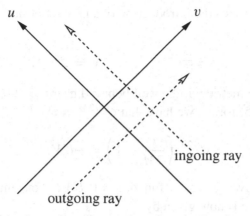

Figure 5.1   Spacetime diagram based on the (u, v) coordinates.

u = constant, where

$$u = t - r^*, \qquad v = t + r^*,$$
$$r^* = \int \frac{\mathrm{d}r}{1 - 2M/r} = r + 2M \ln\left|\frac{r}{2M} - 1\right|. \tag{5.5}$$

In a spacetime diagram using v (advanced time) and u (retarded time) as oblique coordinates (both oriented at 45 degrees), the massless particles propagate at 45 degrees, just as in flat spacetime (Fig. 5.1). The null coordinates (u, v) are therefore well suited to the description of (radial) null geodesics. In these coordinates the Schwarzschild metric takes the form

$$\mathrm{d}s^2 = -(1 - 2M/r)\,\mathrm{d}u\,\mathrm{d}v + r^2\,\mathrm{d}\Omega^2. \tag{5.6}$$

Here, r appears no longer as a coordinate, but as the function of u and v defined implicitly by $r^*(r) = \frac{1}{2}(v - u)$. In these coordinates the surface $r = 2M$ appears at $v - u = -\infty$, and it is still the locus of a coordinate singularity.

To see how this coordinate singularity might be removed, we focus our attention on a small neighbourhood of the surface $r = 2M$, in which the relation $r^*(r)$ can be approximated by $r^* \simeq 2M \ln |r/2M - 1|$. This implies that $r/2M \simeq 1 \pm e^{r^*/2M} = 1 \pm e^{(v-u)/4M}$, and $f \simeq \pm e^{(v-u)/4M}$. Here and below, the upper sign refers to the part of the neighbourhood corresponding to $r > 2M$, while the lower sign refers to $r < 2M$. The metric (5.6) becomes

$$\mathrm{d}s^2 \simeq \mp(e^{-u/4M}\,\mathrm{d}u)(e^{v/4M}\,\mathrm{d}v) + r^2\,\mathrm{d}\Omega^2.$$

This expression motivates the introduction of a new set of null coordinates, U and V, defined by

$$U = \mp e^{-u/4M}, \qquad V = e^{v/4M}. \tag{5.7}$$

In terms of these the metric will be well behaved near $r = 2M$. Going back to the exact expression (5.5) for r^*, we have that $e^{r^*/2M} = e^{(v-u)/4M} = \mp UV$, or

$$e^{r/2M}\left(\frac{r}{2M} - 1\right) = -UV, \tag{5.8}$$

which implicitly gives r as a function of UV. You may check that the Schwarzschild metric is now given by

$$ds^2 = -\frac{32M^3}{r}e^{-r/2M}\,dU\,dV + r^2\,d\Omega^2. \tag{5.9}$$

This is manifestly regular at $r = 2M$. The coordinates U and V are called *null Kruskal coordinates*. In a Kruskal diagram (a map of the U-V plane; see Fig. 5.2), outgoing light rays move along curves $U =$ constant, while ingoing light rays move along curves $V =$ constant.

In the Kruskal coordinates, a surface of constant r is described by an equation of the form $UV =$ constant, which corresponds to a two-branch hyperbola in the U-V plane. For example, $r = 2M$ becomes $UV = 0$, while $r = 0$ becomes $UV = 1$. There are *two copies* of each surface $r =$ constant in a Kruskal diagram. For example, $r = 2M$ can be either $U = 0$ or $V = 0$. The Kruskal coordinates therefore reveal the existence of a much larger manifold than the portion covered by the original Schwarzschild coordinates. In a Kruskal diagram, this portion is labeled **I**. The Kruskal coordinates do not only allow the continuation of the metric through $r = 2M$ into region **II**, they also allow continuation into regions **III** and

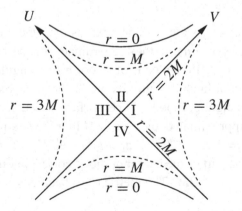

Figure 5.2 Kruskal diagram.

IV. These additional regions, however, exist only in the maximal extension of the Schwarzschild spacetime. If the black hole is the result of gravitational collapse, then the Kruskal diagram must be cut off at a timelike boundary representing the surface of the collapsing body. Regions **III** and **IV** then effectively disappear below the surface of the collapsing star.

5.1.3 Eddington–Finkelstein coordinates

Because of the implicit nature of the relation between r and UV, the Kruskal coordinates can be awkward to use in some computations. In fact, it is rarely necessary to employ coordinates that cover all four regions of the Kruskal diagram, although it is often desirable to have coordinates that are well behaved at $r = 2M$. In such situations, choosing v and r as coordinates, or u and r, does the trick. These coordinate systems are called *ingoing* and *outgoing Eddington–Finkelstein* coordinates, respectively.

It is easy to check that in the ingoing coordinates, the Schwarzschild metric takes the form

$$ds^2 = -(1 - 2M/r)\,dv^2 + 2\,dv\,dr + r^2\,d\Omega^2, \tag{5.10}$$

while in the outgoing coordinates,

$$ds^2 = -(1 - 2M/r)\,du^2 - 2\,du\,dr + r^2\,d\Omega^2. \tag{5.11}$$

It may also be verified that the (v, r) coordinates cover regions **I** and **II** of the Kruskal diagram, while u and r cover regions **IV** and **I**.

The Eddington–Finkelstein coordinates can also be used to construct spacetime diagrams (Fig. 5.3), but these do not have the property that both ingoing and outgoing null geodesics propagate at 45 degrees. For example, it follows

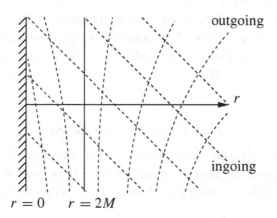

Figure 5.3   Spacetime diagram based on the (v, r) coordinates.

from Eq. (5.10) that ingoing light rays move with $dv = 0$, that is, along coordinate lines that can be oriented at 45 degrees, but the outgoing rays move with $dv/dr = 2/(1 - 2M/r)$, that is, with a varying slope.

5.1.4 Painlevé–Gullstrand coordinates

Another useful set of coordinates for the Schwarzschild spacetime are the *Painlevé–Gullstrand* coordinates first considered in Section 3.13, Problem 1. Here, as with the Eddington–Finkelstein coordinates, the spatial coordinates (r, θ, ϕ) are the same as in the original form of the metric, Eq. (5.1), but the time coordinate is different: T is proper time as measured by a free-falling observer starting from rest at infinity and moving radially inward.

The four-velocity of such an observer is given by $u^\alpha \, \partial_\alpha = f^{-1} \partial_t - \sqrt{1 - f} \, \partial_r$, where $f = 1 - 2M/r$. From this we deduce that $u_\alpha = -\partial_\alpha T$, where the time function T is obtained by integrating $dT = dt + f^{-1} \sqrt{1 - f} \, dr \equiv d\tau$, where τ is proper time. (Integration is elementary, and the result appears in Section 3.13, Problem 1.) After inserting this expression for dt into Eq. (5.1), we obtain the Painlevé–Gullstrand form of the Schwarzschild metric:

$$ds^2 = -dT^2 + \left(dr + \sqrt{2M/r} \, dT\right)^2 + r^2 \, d\Omega. \tag{5.12}$$

The coordinates (T, r, θ, ϕ) give rise to a metric that is regular at $r = 2M$, in correspondence with the fact that our free-falling observer does not consider this surface to be in any way special. Because this observer originates in region **I** of the spacetime (at $r = \infty$) and ends up in region **II** (at $r = 0$), the new coordinates cover only these two regions of the Kruskal diagram. By reversing the motion – letting dr become $-dr$ in Eq. (5.12) – an alternative coordinate system is produced that covers regions **VI** and **I** instead.

From Eq. (5.12) we infer a rather striking property of the Painlevé–Gullstrand coordinates: The hypersurfaces $T = $ constant are all intrinsically flat. This can be seen directly from the fact that the induced metric on any such hypersurface is given by $ds^2 = dr^2 + r^2 \, d\Omega$.

5.1.5 Penrose–Carter diagram

The double-null Kruskal coordinates make the causal structure of the Schwarzschild spacetime very clear, and this is their main advantage. Another useful set of double-null coordinates is obtained by applying the transformation

$$\tilde{U} = \arctan U, \qquad \tilde{V} = \arctan V. \tag{5.13}$$

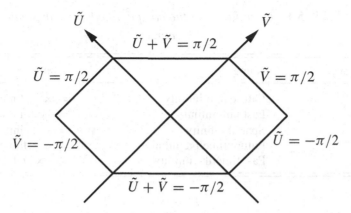

Figure 5.4   Compactified coordinates for the Schwarzschild spacetime.

This rescaling of the null coordinates does not affect the appearance of radial light rays, which still propagate at 45 degrees in a spacetime diagram based on the new coordinates (Fig. 5.4). However, while the range of the initial coordinates was infinite (for example, $-\infty < U < \infty$), it is finite for the new coordinates (for example, $-\pi/2 < \tilde{U} < \pi/2$). The entire spacetime is therefore mapped onto a finite domain of the $\tilde{U}$-$\tilde{V}$ plane. This compactification of the manifold introduces bad coordinate singularities at the boundaries of the new coordinate system, but these are of no concern when the purpose is simply to construct a compact map of the entire spacetime.

In the new coordinates the surfaces $r = 2M$ are located at $\tilde{U} = 0$ and $\tilde{V} = 0$, and the singularities at $r = 0$, or $UV = 1$, are now at $\tilde{U} + \tilde{V} = \pm\pi/2$. The spacetime is also bounded by the surfaces $\tilde{U} = \pm\pi/2$ and $\tilde{V} = \pm\pi/2$. The four points $(\tilde{U}, \tilde{V}) = (\pm\pi/2, \pm\pi/2)$ are singularities of the coordinate transformation: In the actual spacetime the surfaces $U = 0$, $U = \infty$, and $UV = 1$ never meet.

It is useful to assign names to the various boundaries of the compactified spacetime (Fig. 5.5). The surfaces $\tilde{U} = \pi/2$ and $\tilde{V} = \pi/2$ are called *future null infinity* and are labelled $\mathscr{I}^{+}$ (pronounced 'scri plus'). The diagram makes it clear that $\mathscr{I}^{+}$ contains the future endpoints of all outgoing null geodesics (those along which r increases). Similarly, the surfaces $\tilde{U} = -\pi/2$ and $\tilde{V} = -\pi/2$ are called *past null infinity* and are labelled $\mathscr{I}^{-}$. These contain the past endpoints of all ingoing null geodesics (those along which r decreases). The points at which $\mathscr{I}^{+}$ and $\mathscr{I}^{-}$ meet are called *spacelike infinity* and are labelled i^{0}. These contain the endpoints of all spacelike geodesics. The points $(\tilde{U}, \tilde{V}) = (0, \pi/2)$ and $(\tilde{U}, \tilde{V}) = (\pi/2, 0)$ are called *future timelike infinity* and are labelled i^{+}. These contain the future endpoints of all timelike geodesics that do not terminate at $r = 0$. Finally, the points $(\tilde{U}, \tilde{V}) = (0, -\pi/2)$ and $(\tilde{U}, \tilde{V}) = (-\pi/2, 0)$ are called *past timelike infinity* and

Table 5.1   *Boundaries of the compactified Schwarzschild spacetime.*

Label	Name	Definition
$\mathscr{I}^+$	Future null infinity	$v = \infty$, u finite
$\mathscr{I}^-$	Past null infinity	$u = -\infty$, v finite
i^0	Spatial infinity	$r = \infty$, t finite
i^+	Future timelike infinity	$t = \infty$, r finite
i^-	Past timelike infinity	$t = -\infty$, r finite

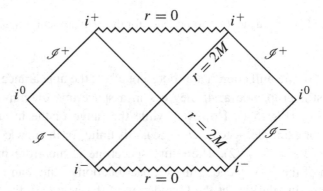

Figure 5.5   Penrose–Carter diagram of the Schwarzschild spacetime.

are labelled i^-. These contain the past endpoints of all timelike geodesics that do not originate at $r = 0$. Table 5.1 provides a summary of these definitions.

Compactified maps such as the one displayed in Fig. 5.5 are called *Penrose–Carter diagrams*. They display, at a glance, the complete causal structure of the spacetime under consideration. They make a very useful tool in general relativity.

5.1.6 Event horizon

On a Kruskal diagram (Fig. 5.2), all radial light rays move along curves $U =$ constant or $V =$ constant. The light cones are therefore oriented at 45 degrees, and timelike world lines, which lie *within* the light cones, move with a slope larger than unity. The one-way character of the surface $r = 2M$ separating regions **I** and **II** of the Schwarzschild spacetime is then clear: An observer crossing this surface can never retrace her steps, and cannot elude an encounter with the curvature singularity at $r = 0$. It is also clear that after crossing $r = 2M$, the observer can no longer send signals to the outside world, although she may continue to receive them. The surface $r = 2M$ therefore prevents any external observer from detecting what goes on inside. In this context it is called the black-hole's *event horizon*. The region within the event horizon (region **II**) is called the *black-hole* region of the Schwarzschild spacetime.

The surface $r = 2M$ that separates regions **I** and **II** must be distinguished from the surface $r = 2M$ that separates regions **IV** and **I**. It is clear that the latter is an event horizon to all observers living inside region **IV** (who cannot perceive what goes on in region **I**). It is also a one-way surface, because observers from the outside cannot cross it. To distinguish between the two surfaces $r = 2M$, it is usual to refer to the first as a *future horizon* and to the second as a *past horizon*. The region within the past horizon (region **IV**) is called the *white-hole* region of the Schwarzschild spacetime.

5.1.7 Apparent horizon

Another important property of the surface $r = 2M$ has to do with the behaviour of outgoing light rays in a neighbourhood of this surface. Here, the term *outgoing* will refer specifically to those rays which move on curves $U = $ constant. This is potentially confusing because the radial coordinate r does not necessarily increase along those rays; in fact, r increases only if $U < 0$ (outside the black hole) and it decreases if $U > 0$ (inside the black hole). While the term 'outgoing' should perhaps be reserved to designate rays along which r always increases, this terminology is nevertheless widely used. Similarly, we will use the term *ingoing* to designate light rays which move on curves $V = $ constant. If $V > 0$, then r decreases along the ingoing rays; if $V < 0$, r increases.

We will show that the expansion of a congruence of outgoing light rays (as defined above) changes sign at $r = 2M$. (This should be obvious just from the fact that r increases along the geodesics that are outside $r = 2M$, but decreases along geodesics that are inside.) Outgoing light rays have

$$k_\alpha = -\partial_\alpha U \tag{5.14}$$

as their (affinely parameterized) tangent vector, and their expansion is calculated as $\theta = k^\alpha{}_{;\alpha} = |g|^{-1/2}(|g|^{1/2}k^\alpha)_{,\alpha}$. In Kruskal coordinates $k^V = |g_{UV}|^{-1}$ is the only nonvanishing component of k^α, and $|g|^{1/2} = |g_{UV}|r^2 \sin^2\theta$. This gives $\theta = 2r_{,V}/r|g_{UV}|$, and using Eq. (5.8) and (5.9) we obtain

$$\theta = k^\alpha{}_{;\alpha} = -\frac{U}{2Mr}. \tag{5.15}$$

As was previously claimed, the expansion is positive for $U < 0$ (in the past of $r = 2M$) and negative for $U > 0$ (in the future of $r = 2M$). The expansion

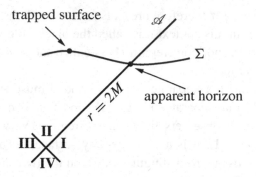

Figure 5.6   Trapped surfaces and apparent horizon of a spacelike hypersurface.

therefore changes sign at $r = 2M$, and in this context, this surface is called an *apparent horizon*. (A similar calculation would reveal that for ingoing light rays, the expansion is negative everywhere in regions **I** and **II**.)

To give a proper definition to the term 'apparent horizon,' we must first introduce the notion of a *trapped surface* (Fig. 5.6). Let Σ be a spacelike hypersurface. A trapped surface on Σ is a closed, two-dimensional surface S such that for both congruences (ingoing and outgoing) of future-directed null geodesics orthogonal to S, the expansion θ is negative everywhere on S. (It should be clear that each two-sphere $U, V =$ constant in region **II** of the Kruskal diagram is a trapped surface.) Let $\mathscr{T}$ be the portion of Σ that contains trapped surfaces; this is known as the *trapped region* of Σ. The boundary of the trapped region, $\partial \mathscr{T}$, is what is defined to be the *apparent horizon* of the spacelike hypersurface Σ. (In Schwarzschild spacetime this would be any two-sphere at $r = 2M$.) Notice that the apparent horizon is a marginally trapped surface: For one congruence of null geodesics orthogonal to $\partial \mathscr{T}$, $\theta = 0$. Notice also that the apparent horizon designates a specific two-surface S on a given hypersurface Σ. The apparent horizon can generally be extended toward the future (and past) of Σ, because hypersurfaces to the future (and past) of Σ also contain apparent horizons. The union of all these apparent horizons forms a three-dimensional surface $\mathscr{A}$ called the *trapping horizon* of the spacetime. (In Schwarzschild spacetime this would be the entire hypersurface $r = 2M$.) In the following we will not distinguish between the two-dimensional apparent horizon and the three-dimensional trapping horizon; we will refer to both as the apparent horizon. (This sloppiness of language is not uncommon.)

5.1.8 Distinction between event and apparent horizons: Vaidya spacetime

The event and apparent horizons of the Schwarzschild spacetime coincide, and it may not be clear why the two concepts need to be distinguished. This coincidence, however, is a consequence of the fact that the spacetime is stationary; for more general black-hole spacetimes the event and apparent horizons are distinct hypersurfaces. To illustrate this we introduce a simple, non-stationary black-hole spacetime.

We express the Schwarzschild metric in terms of ingoing Eddington–Finkelstein coordinates,

$$ds^2 = -f\,dv^2 + 2\,dv\,dr + r^2\,d\Omega^2, \tag{5.16}$$

and we allow the mass function to depend on advanced time v:

$$f = 1 - \frac{2m(v)}{r}. \tag{5.17}$$

This gives the *ingoing Vaidya metric*, a solution to the Einstein field equations with stress-energy tensor

$$T_{\alpha\beta} = \frac{dm/dv}{4\pi r^2}\,l_\alpha l_\beta, \tag{5.18}$$

where $l_\alpha = -\partial_\alpha v$ is tangent to ingoing null geodesics. This stress-energy tensor describes *null dust*, a pressureless fluid with energy density $(dm/dv)/(4\pi r^2)$ and four-velocity l^α. (A similar, *outgoing* Vaidya solution was considered in Section 4.3.5. A notable difference between these solutions is that here, the mass function must *increase* for $T_{\alpha\beta}$ to satisfy the standard energy conditions.)

Consider the following situation. A black hole, initially of mass m_1, is irradiated (with ingoing null dust) during a finite interval of advanced time (between v_1 and v_2) so that its mass increases to m_2. Such a spacetime is described by the Vaidya metric, with a mass function given by

$$m(v) = \begin{cases} m_1 & v \le v_1 \\ m_{12}(v) & v_1 < v < v_2 \\ m_2 & v \ge v_2 \end{cases},$$

where $m_{12}(v)$ increases smoothly from m_1 to m_2. We would like to determine the physical significance of the surfaces $r = 2m_1$, $r = 2m_{12}(v)$, and $r = 2m_2$, and find the precise location of the event horizon.

It should be clear that $r = 2m_1$ and $r = 2m_2$ describe the apparent horizon when $v \le v_1$ and $v \ge v_2$, respectively. More generally, we will show that the apparent horizon of the Vaidya spacetime is always located at $r = 2m(v)$.

The null vector field $k_\alpha\,dx^\alpha = -f\,dv + 2\,dr$ is tangent to a congruence of outgoing null geodesics. It does not, however, satisfy the geodesic equation in affine-parameter form: As a brief calculation reveals, $k_{\alpha;\beta}k^\beta = \kappa\,k_\alpha$ where $\kappa = 2m(v)/r^2$. To calculate the expansion of the outgoing null geodesics, we need to introduce an affine parameter λ^* and a rescaled tangent vector k_*^α. (The calculation can also be handled via the results of Section 2.6, Problem 8.) As was shown in Section 1.3, the desired relation between these vectors is $k_*^\alpha = e^{-\Gamma} k^\alpha$, where $d\Gamma/d\lambda = \kappa(\lambda)$ with λ denoting the original parameter. We have

$$\theta = k_{*;\alpha}^\alpha$$
$$= e^{-\Gamma}\left(k_{;\alpha}^\alpha - \Gamma_{,\alpha}k^\alpha\right)$$
$$= e^{-\Gamma}\left(k_{;\alpha}^\alpha - \frac{d\Gamma}{d\lambda}\right)$$
$$= e^{-\Gamma}\left(k_{;\alpha}^\alpha - \kappa\right).$$

Here, the factor $k_{;\alpha}^\alpha - \kappa$ is the congruence's expansion when measured in terms of the initial parameter λ – it is equal to $(\delta A)^{-1}d(\delta A)/d\lambda$, where δA is the congruence's cross-sectional area. The factor $e^{-\Gamma}$ converts it to $(\delta A)^{-1}d(\delta A)/d\lambda^*$, and this operation does not affect the sign of θ. A simple computation gives $k_{;\alpha}^\alpha = 2(r-m)/r^2$ and we arrive at

$$e^\Gamma \theta = \frac{2}{r^2}\left[r - 2m(v)\right].$$

So $\theta = 0$ on the surface $r = 2m(v)$, and we conclude that the apparent horizon begins at $r = 2m_1$ for $v \leq v_1$, follows $r = 2m_{12}(v)$ in the interval $v_1 < v < v_2$, and remains at $r = 2m_2$ for $v \geq v_2$.

We may now show that while the apparent horizon is a null hypersurface before $v = v_1$ and after $v = v_2$, it is *spacelike* in the interval $v_1 \leq v \leq v_2$. This follows at once from the fact that if $\Phi \equiv r - 2m(v) = 0$ describes the apparent horizon, then

$$g^{\alpha\beta}\Phi_{,\alpha}\Phi_{,\beta} = -4\frac{dm}{dv}$$

is negative (so that the normal $\Phi_{,\alpha}$ is timelike) if $dm/dv > 0$ (so that the energy conditions are satisfied). We therefore see that the apparent horizon is null when the spacetime is stationary, but that it is spacelike otherwise.

Where is the event horizon? Clearly it must coincide with the surface $r = 2m_2$ in the future of $v = v_2$. But what is its extension to the past of $v = v_2$? Because the event horizon is defined as a *causal boundary* in spacetime, it must be a null hypersurface generated by null geodesics (more will be said on this in Section 5.4). The event horizon can therefore *not* coincide with the apparent horizon in the past of $v = v_2$. Instead, its location is determined by finding the outgoing null geodesics of the Vaidya spacetime that connect smoothly with the generators of the surface $r = 2m_2$. (See Fig. 5.7; a particular example is worked out in Section 5.7, Problem 2.)

It is clear that the generators of the event horizon have to be expanding in the past of $v = v_2$ if they are to be stationary (in the sense that $\theta = 0$) in the future. Indeed, supposing that the null energy condition is satisfied (which will be true

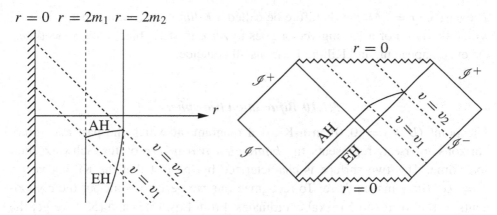

Figure 5.7 Black hole irradiated with ingoing null dust.

if $dm/dv > 0$), the focusing theorem (Section 2.4) implies that the congruence formed by the null generators of the event horizon will be focused by the infalling null dust; a zero expansion in the future of $v = v_2$ guarantees a positive expansion in the past. The event horizon is therefore generated by those null geodesics that undergo just the right amount of focusing, so that after encountering the last of the infalling matter, their expansion goes to zero.

The event horizon coincides with the apparent horizon only in the future of $v = v_2$. In the past, because the apparent horizon has a spacelike segment while the event horizon is everywhere null, the apparent horizon lies *within* the event horizon, that is, inside the black hole (Fig. 5.7). As we shall see in Section 5.4, this observation is quite general.

It is a remarkable property of the event horizon that the entire future history of the spacetime must be known before its position can be determined: The black hole's final state must be known before the horizon's null generators can be identified. This *teleological* property is not shared by the apparent horizon, whose location at any given time (as represented by a spacelike hypersurface) depends only on the properties of the spacetime at that time.

5.1.9 Killing horizon

The vector $t^\alpha = \partial x^\alpha/\partial t$ is a Killing vector of the Schwarzschild spacetime. While this vector is timelike outside the black hole, it is *null* on the event horizon, and it is *spacelike* inside:

$$g_{\alpha\beta}t^\alpha t^\beta = 1 - \frac{2M}{r}.$$

The surface $r = 2M$ can therefore be called a *Killing horizon*, a hypersurface on which the norm of a Killing vector goes to zero. In static black-hole spacetimes, the event, apparent, and Killing horizons all coincide.

5.1.10 Bifurcation two-sphere

The point $(U, V) = (0, 0)$ in a Kruskal diagram, at which the past and future horizons intersect, represents the *bifurcation two-sphere* of the Schwarzschild spacetime. This two-surface is characterized by the fact that the Killing vector $t^\alpha = \partial x^\alpha / \partial t$ vanishes there. To recognize this we need to work out the components of this vector in Kruskal coordinates. From Eqs. (5.5) and (5.7) we get the relation $e^{t/2M} = -V/U$, and after using Eq. (5.8) we obtain

$$U^2 = e^{(r-t)/2M} \left(\frac{r}{2M} - 1 \right), \qquad V^2 = e^{(r+t)/2M} \left(\frac{r}{2M} - 1 \right).$$

Taking partial derivatives with respect to t, we arrive at

$$t^U = -\frac{U}{4M}, \qquad t^V = \frac{V}{4M}. \tag{5.19}$$

It follows immediately that $t^\alpha = 0$ at the bifurcation two-sphere. It should be noted that the bifurcation two-sphere exists only in the maximally extended Schwarzschild spacetime. If the black hole is the result of gravitational collapse, then the bifurcation two-sphere is not part of the actual spacetime.

According to our previous calculation, t^V is the only nonvanishing component of the Killing vector on the future horizon. This implies that $t_\alpha \propto -\partial_\alpha U$ at $U = 0$, and we have the important result that t^α is tangent to the null generators of the event horizon. This was to be expected from the fact that the event horizon of the Schwarzschild spacetime is also a Killing horizon.

5.2 Reissner–Nordström black hole

5.2.1 Derivation of the Reissner–Nordström solution

The Reissner–Nordström (RN) metric describes a static, spherically symmetric black hole of mass M possessing an electric charge Q. We begin our discussion with a derivation of this solution to the Einstein–Maxwell equations.

We assume that the electromagnetic-field tensor $F^{\alpha\beta}$ has no components along the θ and ϕ directions; this ensures that the field is purely electric when measured by stationary observers. Under this assumption the only nonvanishing component is F^{tr}. Maxwell's equations in vacuum are $0 = F^{\alpha\beta}{}_{;\beta} = |g|^{-1/2}(|g|^{1/2}F^{\alpha\beta})_{,\beta}$.

Using the metric of Eq. (5.2), this implies $(e^{\psi} r^2 F^{tr})' = 0$, or

$$F^{tr} = e^{-\psi} \frac{Q}{r^2},$$

where Q is a constant of integration, to be interpreted as the black-hole charge. The stress-energy tensor for the electromagnetic field is

$$T^{\alpha}{}_{\beta} = \frac{1}{4\pi} \left(F^{\alpha\mu} F_{\beta\mu} - \frac{1}{4} \delta^{\alpha}{}_{\beta} F^{\mu\nu} F_{\mu\nu} \right),$$

and a few steps of algebra yield

$$T^{\alpha}{}_{\beta} = \frac{Q^2}{8\pi r^4} \, \mathrm{diag}(-1, -1, 1, 1). \tag{5.20}$$

The Einstein field equations (5.4) imply $m' = Q^2/2r^2$, or $m(r) = M - Q^2/2r$. The fact that $T^t{}_t = T^r{}_r$ implies $\psi' = 0$, so that ψ can be set to zero without loss of generality. The RN solution is therefore

$$ds^2 = -\left(1 - \frac{2M}{r} + \frac{Q^2}{r^2}\right) dt^2 + \left(1 - \frac{2M}{r} + \frac{Q^2}{r^2}\right)^{-1} dr^2 + r^2 d\Omega^2, \tag{5.21}$$

with an electromagnetic-field tensor whose only nonvanishing component is

$$F^{tr} = \frac{Q}{r^2}. \tag{5.22}$$

Here, M is total (ADM) mass of the spacetime and Q is the black hole's electric charge.

To see that Q is indeed the charge, consider a nonsingular charge distribution on a spacelike hypersurface Σ, described by a current density j^{α}. An appropriate definition for total charge is $Q = \int_{\Sigma} j^{\alpha} \, d\Sigma_{\alpha}$, or $Q = (4\pi)^{-1} \int_{\Sigma} F^{\alpha\beta}{}_{;\beta} \, d\Sigma_{\alpha}$ after using Maxwell's equations. Using Stokes' theorem (Section 3.3.3), we rewrite this as an integral over a closed two-surface S bounding the charge distribution. This yields Gauss' law,

$$Q = \frac{1}{8\pi} \oint_S F^{\alpha\beta} \, dS_{\alpha\beta}. \tag{5.23}$$

The advantage of this expression for the total charge is that it is applicable even when the charge distribution is singular, which is the case in the present application. Also, this definition of total charge is in the same spirit as the previously encountered definitions for total mass and angular momentum (Section 4.3). Substituting Eq. (5.22) and evaluating for a two-sphere of constant t and r confirms that the Q appearing in Eq. (5.22) is indeed the black hole's electric charge.

5.2.2 Kruskal coordinates

The function $f(r) = 1 - 2M/r + Q^2/r^2$ has zeroes at $r = r_\pm$, where

$$r_\pm = M \pm \sqrt{M^2 - Q^2}. \tag{5.24}$$

The roots are both real, and the RN spacetime truly contains a black hole, when $|Q| \leq M$. The special case of a black hole with $|Q| = M$ is referred to as an *extreme* RN black hole. If $|Q| > M$, then the RN solution describes a *naked singularity* at $r = 0$.

The coordinates (t, r) are singular at the outer horizon ($r = r_+$), and new coordinates must be introduced to extend the metric across this surface. This can be done with Kruskal coordinates. As we shall see, however, these coordinates fail to be regular at the inner horizon ($r = r_-$), and another coordinate transformation will be required to extend the metric beyond this surface. Thus, Kruskal coordinates are specific to a given horizon, and a single coordinate patch is not sufficient to cover the entire RN manifold (Fig. 5.8). We will see that the outer horizon in an event horizon for the RN spacetime, and the inner horizon is an apparent horizon.

Let us first take care of the extension across the outer horizon. We express the RN metric in the form

$$ds^2 = -f\, dt^2 + f^{-1}\, dr^2 + r^2\, d\Omega^2,$$

where $f = 1 - 2M/r + Q^2/r^2$. Near $r = r_+$ this function can be approximated by

$$f(r) \simeq 2\kappa_+(r - r_+),$$

where $\kappa_+ \equiv \frac{1}{2}f'(r_+)$. It follows that near $r = r_+$,

$$r^* \equiv \int \frac{dr}{f} \simeq \frac{1}{2\kappa_+} \ln\bigl|\kappa_+(r - r_+)\bigr|.$$

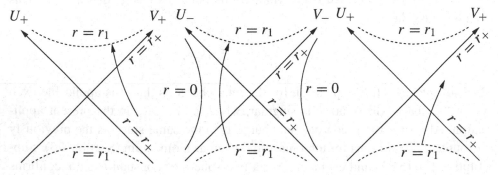

Figure 5.8   Kruskal patches for the Reissner–Nordström spacetime.

Introducing the null coordinates $u = t - r^*$ and $v = t + r^*$, the surface $r = r_+$ appears at $v - u = -\infty$ and we define the Kruskal coordinates U_+ and V_+ by

$$U_+ = \mp e^{-\kappa_+ u}, \qquad V_+ = e^{\kappa_+ v}. \tag{5.25}$$

Here, the upper sign refers to $r > r_+$ and the lower sign refers to $r < r_+$. It is easy to check that $f \simeq -2U_+ V_+$ near $r = r_+$, so that the metric becomes

$$ds^2 \simeq -\frac{2}{\kappa_+^2} \, dU_+ dV_+ + r_+^2 \, d\Omega^2.$$

This shows that when it is expressed in the coordinates (U_+, V_+), the metric is well behaved at the outer horizon. On the other hand, an exact integration for $r^*(r)$ would reveal that $r^* \to +\infty$ at the inner horizon, which is then located at $v - u = \infty$, or $U_+ V_+ = \infty$. The Kruskal coordinates are *singular* at the inner horizon.

The coordinates (U_+, V_+) should be used only in the interval $r_1 < r < \infty$, where $r_1 > r_-$ is some cutoff radius. Inside $r = r_1$ another coordinate system must be introduced. One such system is (t, r), in which the metric takes the standard form of Eq. (5.21). It is important to understand that this new coordinate patch, which covers the portion of the RN spacetime corresponding to the interval $r_- < r < r_+$, is *distinct* from the original patch that covers the external region $r > r_+$. And indeed, because f is now negative, the new t must be interpreted as a spacelike coordinate (because $g_{tt} > 0$) while r must be interpreted as a timelike coordinate (because $g_{rr} < 0$).

There still remains the issue of extending the spacetime beyond $r = r_-$, where the new (t, r) coordinates fail. We want to construct a new set of Kruskal coordinates, U_- and V_-, adapted to the inner horizon. Retracing the same steps as before, we have that near $r = r_-$ the function f can be approximated by

$$f(r) \simeq -2\kappa_-(r - r_-),$$

where $\kappa_- = \frac{1}{2}|f'(r_-)|$. It follows that

$$r^* \simeq -\frac{1}{2\kappa_-} \, \ln|\kappa_-(r - r_-)|.$$

With $u = t - r^*$ and $v = t + r^*$, the surface $r = r_-$ appears at $v - u = +\infty$ and we define the new Kruskal coordinates by

$$U_- = \mp e^{\kappa_- u}, \qquad V_- = -e^{-\kappa_- v}. \tag{5.26}$$

Here, the upper sign refers to $r > r_-$ and the lower sign refers to $r < r_-$. Then $f \simeq -2U_- V_-$ and the metric becomes

$$ds^2 \simeq -\frac{2}{\kappa_-^2} \, dU_- dV_- + r_-^2 \, d\Omega^2.$$

This is manifestly regular across $r = r_-$. The new Kruskal coordinates, however, are singular at $r = r_+$.

What happens now on the other side of the inner horizon? The most notice-able feature is that the singularity at $r = 0$ appears as a *timelike* surface – this is markedly different from what happens inside a Schwarzschild black hole, where the singularity is spacelike. Because $f > 0$ when $r < r_-$, r re-acquires its inter-pretation as a spacelike coordinate; any surface $r = $ constant $< r_-$ is therefore a timelike hypersurface, and this includes the singularity. Because it is timelike, the singularity can be *avoided* by observers moving within the black hole. This is a striking new phenomenon, and we should examine it very carefully.

Consider the motion of a typical observer inside a RN black hole (Fig. 5.8). Before crossing the inner horizon (but after going across the outer horizon), r is a timelike coordinate and the motion necessarily proceeds with r decreasing. After crossing $r = r_-$, however, r becomes spacelike and both types of motion (r de-creasing or increasing) become possible. Our observer may therefore decide to re-verse course, and if she does, she will avoid $r = 0$ altogether. Her motion inside the inner horizon will then proceed with r increasing and she will cross, once more, the surface $r = r_-$. This, however, is *another copy* of the inner horizon, distinct from the one encountered previously. (Recall that there are two copies of each surface $r = $ constant in a Kruskal diagram.) After entering this new $r > r_-$ region, our ob-server notices that r has once again become timelike, and she finds that reversing course is no longer possible: Her motion must proceed with r increasing and this brings her in the vicinity of another surface $r = r_+$. Because there is no reason for spacetime to just stop there, yet another Kruskal patch (U_+, V_+) must be intro-duced to extend the RN metric beyond this horizon. The new Kruskal coordinates take over where the old patch (U_-, V_-) leaves off, at the spacelike hypersurface $r = r_1$.

The ultimate conclusion to these considerations is that our observer eventually emerges out of the black hole, through another copy of the outer horizon, into a new asymptotically-flat universe. Her trip may not end there: Our observer could now decide to enter the RN black hole that resides in this new universe, and this en-tire cycle would repeat! It therefore appears that the RN metric describes more than just a single black hole. Indeed, it describes an infinite lattice of asymptotically-flat universes connected by black-hole tunnels.

Such a fantastic spacetime structure is best represented with a Penrose–Carter diagram (Fig. 5.9). The diagram shows that the region bounded by the surfaces $r = r_+$ and $r = r_-$ contains trapped surfaces: Both ingoing and outgoing light rays originating from this region converge toward the singularity. The outer and inner horizons are therefore apparent horizons, but only the outer horizon is an event horizon.

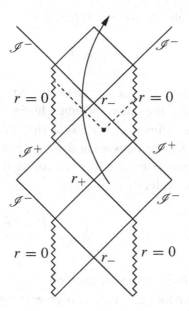

Figure 5.9 Penrose–Carter diagram of the Reissner–Nordström spacetime.

5.2.3 Radial observers in Reissner–Nordström spacetime

The discovery of black-hole tunnels is so bizarre that it should be backed up by a solid calculation. Here we consider the geodesic motion of a free-falling observer in the RN spacetime. It is assumed that the motion proceeds entirely in the radial direction and that initially, it is directed inward.

We will first work with the (v, r) coordinates, in which the metric takes the form

$$ds^2 = -f\,dv^2 + 2\,dv\,dr + r^2\,d\Omega^2, \qquad (5.27)$$

where $f = 1 - 2M/r + Q^2/r^2$. The observer's four-velocity is $u^\alpha\,\partial_\alpha = \dot{v}\,\partial_v + \dot{r}\,\partial_r$, where an overdot denotes differentiation with respect to proper time τ. The quantity $\tilde{E} = -u_\alpha t^\alpha = -u_v$, the observer's energy per unit mass, is a constant of the motion. In terms of $\dot{v}$ and $\dot{r}$ this is given by $\tilde{E} = f\dot{v} - \dot{r}$. On the other hand, the normalization condition $u^\alpha u_\alpha = -1$ gives $f\dot{v}^2 - 2\dot{v}\dot{r} = 1$, and these equations imply

$$\dot{r}_{\text{in}} = -\left(\tilde{E}^2 - f\right)^{1/2}, \qquad \dot{v}_{\text{in}} = \frac{\tilde{E} - \left(\tilde{E}^2 - f\right)^{1/2}}{f}, \qquad (5.28)$$

where the sign in front of the square root was chosen appropriately for an ingoing observer.

The equation for $\dot{r}$ can also be written in the form

$$\dot{r}^2 + f = \tilde{E}^2, \tag{5.29}$$

which comes with a nice interpretation as an energy equation (Fig. 5.10). Its message is clear: After crossing the outer and inner horizons the observer reaches a turning point ($\dot{r} = 0$) at a radius $r_{\min} < r_-$ such that $f(r_{\min}) = \tilde{E}^2$. The motion, which initially was inward, turns outward and the observer eventually emerges out of the black hole, into a new external universe. During the outward portion of the motion, the observer's four-velocity is given by

$$\dot{r}_{\text{out}} = +\left(\tilde{E}^2 - f\right)^{1/2}, \qquad \dot{v}_{\text{out}} = \frac{\tilde{E} + \left(\tilde{E}^2 - f\right)^{1/2}}{f}, \tag{5.30}$$

with the opposite sign in front of the square root.

Let us examine the behaviour of $\dot{v}$ as the observer traverses a horizon. When the motion is inward we have that $\dot{v} \simeq (2\tilde{E})^{-1}$ in the limit $f \to 0$. This means that v stays finite during the first crossings of the outer and inner horizons. When the motion is outward, $\dot{v} \simeq 2\tilde{E}/f$ in the limit $f \to 0$, and this means that the coordinates (v, r) become singular during the second crossing of the inner horizon. The observer's motion cannot be followed beyond this point, unless new coordinates are introduced.

Let us therefore switch to the coordinates (u, r), in which the RN metric takes the form

$$ds^2 = -f\,du^2 - 2\,du\,dr + r^2\,d\Omega^2. \tag{5.31}$$

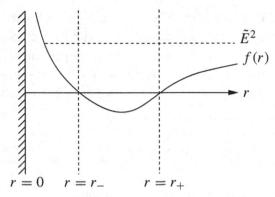

Figure 5.10 Effective potential for radial motion in Reissner–Nordström spacetime.

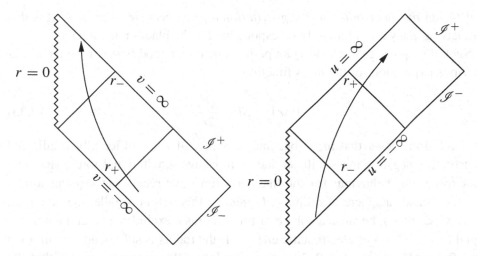

Figure 5.11 Eddington–Finkelstein patches for the Reissner–Nordström spacetime.

In these coordinates, and during the outward portion of the motion (after the bounce at $r = r_{\text{min}}$), the four-velocity is given by

$$\dot{r}_{\text{out}} = +\left(\tilde{E}^2 - f\right)^{1/2}, \qquad \dot{u}_{\text{out}} = \frac{\tilde{E} - \left(\tilde{E}^2 - f\right)^{1/2}}{f}. \tag{5.32}$$

We have that $\dot{u} \simeq (2\tilde{E})^{-1}$ when $f \to 0$, which shows that u stays finite during the second crossings of the inner and outer horizons.

We see that a large portion of the RN spacetime is covered by the two coordinate patches employed here (Fig. 5.11). This includes two asymptotically-flat regions connected by a black-hole tunnel that contains two copies of the outer horizon, and two copies of the inner horizon. The complete spacetime is obtained by tessellation, using the patches (v, r) and (u, r) as tiles; this gives rise to the diagram of Fig. 5.9. Because the complete spacetime contains an infinite number of black-hole tunnels, an infinite number of coordinate patches is required for its description.

The presence of black-hole tunnels in the RN spacetime is now well established. These tunnels, of course, have a lot to do with the occurrence of a turning point in the motion of our free-falling observer. This is a rather striking feature of the RN spacetime. While turning points are a familiar feature of Newtonian mechanics, in this context they are always associated with the presence of an angular-momentum term in the effective potential: The centrifugal force is repulsive and it prevents an observer from reaching the centre at $r = 0$. This, however, cannot explain what is happening here, because the motion was restricted from the start to be radial – there is no angular momentum present to produce a repulsive force. The gravitational field alone must be responsible for the repulsion, and we are forced to conclude that *inside the inner horizon, the gravitational force becomes repulsive!* It is this repulsive gravity that ultimately is responsible for the black-hole tunnels.

Such a surprising conclusion can perhaps be understood better if we recall our previous expression for the mass function:

$$m(r) = M - \frac{Q^2}{2r}. \tag{5.33}$$

This relation shows that $m(r)$ becomes *negative* if r is sufficiently small, and clearly, this negative mass will produce a repulsive gravitational force. How can we explain this behaviour for the mass function? We recall that $m(r)$ measures the mass inside a sphere of radius r. In general this will be smaller than the total mass $M \equiv m(\infty)$, because a sphere of finite radius r excludes a certain amount – equal to $Q^2/(2r)$ – of electrostatic energy. If the radius is sufficiently small, then $Q^2/(2r) > M$ and $m(r) < 0$. You may check that this always occurs within the inner horizon.

The conclusion that the RN spacetime contains black-hole tunnels is firm. Should we then feel confident that a trip inside a charged black hole will lead us to a new universe? This answer is no. The reason is that the existence of such tunnels depends very sensitively on the assumed symmetries of the RN spacetime, namely, staticity and spherical symmetry. These symmetries would not be exact in a realistic black hole, and slight perturbations have a dramatic effect on the hole's internal structure. The tunnels are unstable, and they do not appear in realistic situations. (More will be said on this in Section 5.7, Problem 3.)

5.2.4 Surface gravity

In Section 5.2.2, a quantity $\kappa_+ \equiv \frac{1}{2} f'(r_+)$ was introduced during the construction of Kruskal coordinates adapted to the outer horizon. We shall name this quantity the *surface gravity* of the black hole, and henceforth denote it simply by κ. As we shall see in Section 5.5, the surface gravity provides an important characterization of black holes, and it plays a key role in the laws of black-hole mechanics. For the RN black hole the surface gravity is given explicitly by

$$\kappa = \frac{r_+ - r_-}{2r_+^2} = \frac{\sqrt{M^2 - Q^2}}{r_+^2}, \tag{5.34}$$

where we have used Eq. (5.24). Notice that $\kappa = 0$ for an extreme RN black hole and that

$$\kappa = \frac{1}{4M} \tag{5.35}$$

for a Schwarzschild black hole.

The name 'surface gravity' deserves a justification. Consider, in a static and spherically symmetric spacetime with metric

$$ds^2 = -f \, dt^2 + f^{-1} \, dr^2 + r^2 \, d\Omega^2, \qquad (5.36)$$

a particle of unit mass held stationary at a radius r. (Here, f is not necessarily required to have the RN form, but this will be the case of interest.) The four-velocity of the stationary particle is $u^\alpha = f^{-1/2} t^\alpha$ and its acceleration is $a^\alpha = u^\alpha_{;\beta} u^\beta$. The only nonvanishing component is $a^r = \frac{1}{2} f'$ and the magnitude of the acceleration vector is

$$a(r) \equiv \left(g_{\alpha\beta} a^\alpha a^\beta \right)^{1/2} = \frac{1}{2} f^{-1/2} f'(r). \qquad (5.37)$$

This is the force required to hold the particle at r if the force is applied locally, at the particle's position. This, not surprisingly, diverges in the limit $r \to r_+$. But suppose instead that the particle is held in place by an observer at infinity, by means of an infinitely long, massless string. What is $a_\infty(r)$, the force applied by this observer?

To answer this we consider the following thought experiment. Let the observer at infinity raise the string by a small proper distance δs, thereby doing an amount $\delta W_\infty = a_\infty \delta s$ of work. At the particle's position the displacement is also δs, but the work done is $\delta W = a \, \delta s$. (You may justify this statement by working in a local Lorentz frame at r.) Suppose now that the work δW is converted into radiation that is then collected at infinity. The received energy is redshifted by a factor $f^{1/2}$, so that $\delta E_\infty = f^{1/2} a \, \delta s$. But energy conservation demands that the energy extracted be equal to the energy put in, so that $\delta E_\infty = \delta W_\infty$. This implies

$$a_\infty(r) = f^{1/2} a(r) = \frac{1}{2} f'(r). \qquad (5.38)$$

This is the force applied by the observer at infinity. This quantity is well behaved in the limit $r \to r_+$, and it is appropriate to call $a_\infty(r_+)$ the surface gravity of the black hole. Thus,

$$\kappa \equiv a_\infty(r_+) = \frac{1}{2} f'(r_+). \qquad (5.39)$$

The surface gravity is therefore the force required of an observer at infinity to hold a particle (of unit mass) stationary at the event horizon.

The surface gravity can also be defined in terms of the Killing vector t^α. We have seen in Section 5.1.9 that the event horizon of a static spacetime is also a Killing horizon, so that t^α is tangent to the horizon's null generators. Because t^α is orthogonal to itself on the horizon, it is also *normal* to the horizon. But $\Phi \equiv -t^\alpha t_\alpha = 0$ on the horizon, and since the normal vector is proportional to

$\Phi_{,\alpha}$, there must exist a scalar κ such that

$$(-t^\mu t_\mu)_{;\alpha} = 2\kappa t_\alpha \tag{5.40}$$

on the horizon. A brief calculation confirms that this κ is the surface gravity: Using the coordinates (v, r) we have that $t^\alpha \partial_\alpha = \partial_v$ and $t_\alpha \, dx^\alpha = dr$ on the horizon; with $\Phi = -g_{vv} = f$ we obtain $\Phi_{,\alpha} = f' \partial_\alpha r$, which is just Eq. (5.40) with $\kappa = \frac{1}{2} f'(r_+)$. This calculation reveals also that the horizon's null generators are parameterized by v.

Because t^α is tangent to the horizon's null generators, it must satisfy the geodesic equation at $r = r_+$. This comes as an immediate consequence of Eq. (5.40) and Killing's equation: On the horizon,

$$t^\alpha_{;\beta} t^\beta = \kappa t^\alpha, \tag{5.41}$$

and we see that v is not an affine parameter on the generators. An affine parameter λ can be obtained by integrating the equation $d\lambda/dv = e^{\kappa v}$ (Section 1.3). This gives $\lambda = V/\kappa$, where $V \equiv e^{\kappa v}$ is one of the Kruskal coordinates adapted to the event horizon – it was denoted V_+ in Section 5.2.2. It follows that on the horizon, the null vector

$$k^\alpha = V^{-1} t^\alpha \tag{5.42}$$

satisfies the geodesic equation in affine-parameter form.

It is possible to derive an explicit formula for κ. Because the congruence of null generators is necessarily hypersurface orthogonal, Frobenius' theorem (Section 2.4.3) guarantees that the relation

$$t_{[\alpha;\beta} t_{\gamma]} = 0$$

is satisfied on the event horizon. Using Killing's equation, this implies

$$t_{\alpha;\beta} t_\gamma + t_{\gamma;\alpha} t_\beta + t_{\beta;\gamma} t_\alpha = 0,$$

and contracting with $t^{\alpha;\beta}$ yields

$$\begin{aligned}
t^{\alpha;\beta} t_{\alpha;\beta} t_\gamma &= -t_{\gamma;\alpha} t^\alpha_{;\beta} t^\beta + t_{\beta;\gamma} t^\beta_{;\alpha} t^\alpha \\
&= -\kappa \, t_{\gamma;\alpha} t^\alpha + \kappa \, t_{\beta;\gamma} t^\beta \\
&= -2\kappa^2 t_\gamma.
\end{aligned}$$

We have obtained

$$\kappa^2 = -\frac{1}{2} t^{\alpha;\beta} t_{\alpha;\beta}, \tag{5.43}$$

in which it is understood that the right-hand side is evaluated at $r = r_+$. Equations (5.40), (5.41), and (5.43) can all be regarded as fundamental definitions of the surface gravity; they are of course all equivalent.

5.3 Kerr black hole

5.3.1 The Kerr metric

A solution to the Einstein field equations describing a *rotating* black hole was discovered by Roy Kerr in 1963. (There is also a solution to the Einstein–Maxwell equations that describes a charged, rotating black hole. It is known as the Kerr–Newman solution, and it is described in Section 5.7, Problem 8.) As we shall see, the Kerr metric can be written in a number of different ways. In the standard Boyer–Lindquist coordinates it is given by

$$
\begin{aligned}
ds^2 &= -\left(1 - \frac{2Mr}{\rho^2}\right) dt^2 - \frac{4Mar \sin^2 \theta}{\rho^2} dt\, d\phi + \frac{\Sigma}{\rho^2} \sin^2 \theta\, d\phi^2 \\
&\quad + \frac{\rho^2}{\Delta} dr^2 + \rho^2\, d\theta^2 \\
&= -\frac{\rho^2 \Delta}{\Sigma} dt^2 + \frac{\Sigma}{\rho^2} \sin^2 \theta\, (d\phi - \omega\, dt)^2 + \frac{\rho^2}{\Delta} dr^2 + \rho^2\, d\theta^2,
\end{aligned}
\tag{5.44}
$$

where

$$
\begin{aligned}
\rho^2 &= r^2 + a^2 \cos^2 \theta, \quad \Delta = r^2 - 2Mr + a^2, \\
\Sigma &= (r^2 + a^2)^2 - a^2 \Delta \sin^2 \theta, \quad \omega \equiv -\frac{g_{t\phi}}{g_{\phi\phi}} = \frac{2Mar}{\Sigma}.
\end{aligned}
\tag{5.45}
$$

The Kerr metric is stationary and axially symmetric; it therefore admits the Killing vectors $t^\alpha = \partial x^\alpha / \partial t$ and $\phi^\alpha = \partial x^\alpha / \partial \phi$. It is also asymptotically flat. The Komar formulae (Section 4.3) confirm that M is the spacetime's ADM mass, and show that $J \equiv aM$ is the angular momentum (so that a is the ratio of angular momentum to mass).

The components of the inverse metric are

$$
g^{tt} = -\frac{\Sigma}{\rho^2 \Delta}, \quad g^{t\phi} = -\frac{2Mar}{\rho^2 \Delta}, \quad g^{\phi\phi} = \frac{\Delta - a^2 \sin^2 \theta}{\rho^2 \Delta \sin^2 \theta},
$$

$$
g^{rr} = \frac{\Delta}{\rho^2}, \quad g^{\theta\theta} = \frac{1}{\rho^2}.
\tag{5.46}
$$

The metric and its inverse have singularities at $\Delta = 0$ and $\rho^2 = 0$. To distinguish between coordinate and curvature singularities, we examine the squared Riemann tensor of the Kerr spacetime:

$$R^{\alpha\beta\gamma\delta}R_{\alpha\beta\gamma\delta} = \frac{48M^2(r^2 - a^2\cos^2\theta)(\rho^4 - 16a^2r^2\cos^2\theta)}{\rho^{12}}. \tag{5.47}$$

This reveals that the singularity of the metric at $\Delta = 0$ is just a coordinate singularity, but that the Kerr spacetime is truly singular at $\rho^2 = 0$. The nature of the curvature singularity will be clarified in Section 5.3.8.

Various properties of the Kerr spacetime will be examined in the following subsections. To facilitate this discussion we will introduce three families of observers: zero-angular-momentum observers (ZAMOs), static observers, and stationary observers.

5.3.2 Dragging of inertial frames: ZAMOs

ZAMOs are observers with zero angular momentum: if u^α is the four-velocity, then $\tilde{L} \equiv u_\alpha \phi^\alpha = 0$. This implies that $g_{\phi t}\dot{t} + g_{\phi\phi}\dot{\phi} = 0$, where an overdot indicates differentiation with respect to proper time τ. Using Eqs. (5.44) this translates to

$$\Omega \equiv \frac{\mathrm{d}\phi}{\mathrm{d}t} = \omega, \tag{5.48}$$

and we see that ZAMOs possess an angular velocity equal to $\omega = -g_{t\phi}/g_{\phi\phi}$. This angular velocity increases as the observer approaches the black hole, and it goes in the same direction as the hole's own rotation – the ZAMOs rotate with the black hole. This striking property of the Kerr black hole, which in fact is shared by all rotating bodies, is called the *dragging of inertial frames* (see Section 3.10). At large distances from the black hole, $\omega \simeq 2J/r^3$, and the dragging disappears completely at infinity.

5.3.3 Static limit: static observers

We now consider *static* observers in the Kerr spacetime. Such observers have a four-velocity proportional to the Killing vector t^α:

$$u^\alpha = \gamma t^\alpha, \tag{5.49}$$

where the factor $\gamma \equiv (-g_{\alpha\beta}t^\alpha t^\beta)^{-1/2}$ ensures that the four-velocity is properly normalized. Because these observers must be held in place by an external agent (a rocket engine, for example), the motion is not geodesic.

Static observers cannot exist everywhere in the Kerr spacetime. This can be seen from the fact that t^α is not everywhere timelike, but becomes null when $\gamma^{-2} = -g_{tt} = 0$; when this occurs, Eq. (5.49) breaks down. The *static limit* is

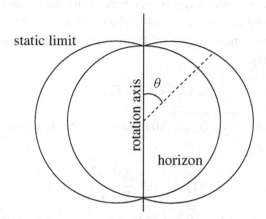

Figure 5.12   Static limit and event horizon of the Kerr spacetime.

therefore described by $g_{tt} = 0$ or, after using Eqs. (5.44) and (5.45), $r^2 - 2Mr + a^2 \cos^2 \theta = 0$. Solving for r reveals that the static limit is located at $r = r_{\rm sl}(\theta)$, where

$$r_{\rm sl}(\theta) = M + \sqrt{M^2 - a^2 \cos^2 \theta}. \tag{5.50}$$

Thus, observers cannot remain static when $r \le r_{\rm sl}(\theta)$, even if an arbitrarily large force is applied. Instead, the dragging of inertial frames compels them to rotate with the black hole. As we shall see, the static limit does *not* coincide with the hole's event horizon. The finite region between the horizon and the static limit is called the *ergosphere* of the Kerr spacetime (Fig. 5.12).

5.3.4 Event horizon: stationary observers

We now consider observers moving in the ϕ direction with an arbitrary, but uniform, angular velocity $d\phi/dt = \Omega$. Because such observers do not perceive any time variation in the black hole's gravitational field, they are called *stationary* observers. They move with a four-velocity

$$u^\alpha = \gamma(t^\alpha + \Omega \phi^\alpha), \tag{5.51}$$

where $t^\alpha + \Omega \phi^\alpha$ is a Killing vector for the Kerr spacetime, and γ a new normalization factor given by

$$\begin{aligned}
\gamma^{-2} &= -g_{\alpha\beta}(t^\alpha + \Omega \phi^\alpha)(t^\beta + \Omega \phi^\beta) \\
&= -g_{tt} - 2\Omega g_{t\phi} - \Omega^2 g_{\phi\phi} \\
&= -g_{\phi\phi}(\Omega^2 - 2\omega\Omega + g_{tt}/g_{\phi\phi}),
\end{aligned}$$

where $\omega = -g_{t\phi}/g_{tt}$.

Stationary observers cannot exist everywhere in the Kerr spacetime: The vector $t^\alpha + \Omega\phi^\alpha$ must be timelike, and this fails to be true when γ^{-2} is nonpositive. It is easy to check that the condition $\gamma^{-2} > 0$ gives rise to the following requirement on the angular velocity:

$$\Omega_- < \Omega < \Omega_+, \qquad (5.52)$$

where $\Omega_\pm = \omega \pm \sqrt{\omega^2 - g_{tt}/g_{\phi\phi}}$. After some algebra, using Eqs. (5.44) and (5.45), this reduces to

$$\Omega_\pm = \omega \pm \frac{\Delta^{1/2}\rho^2}{\Sigma \sin\theta}. \qquad (5.53)$$

A stationary observer with $\Omega = 0$ is a static observer, and we know that static observers exist only outside the static limit. It must therefore be true that Ω_- changes sign at $r = r_{sl}(\theta)$. This is confirmed by a few lines of algebra, using Eqs. (5.50) and (5.53). As r decreases further from $r_{sl}(\theta)$, Ω_- increases while Ω_+ decreases. Eventually we arrive at the situation $\Omega_- = \Omega_+$, which implies $\Omega = \omega$; at this point the stationary observer is forced to move around the black hole with an angular velocity equal to ω. This occurs when $\Delta = 0$, or $r^2 - 2Mr + a^2 = 0$. The largest solution is $r = r_+$, where

$$r_+ = M + \sqrt{M^2 - a^2}. \qquad (5.54)$$

Notice that the roots of $\Delta = 0$ are real if and only if $a \leq M$, or $J \leq M^2$: There is an upper limit on the angular momentum of a black hole. Kerr black holes with $a = M$ are said to be *extremal*. For $a > M$ the Kerr metric describes a naked singularity.

The vector $t^\alpha + \Omega\phi^\alpha$ becomes null at $r = r_+$ and stationary observers cannot exist inside this surface, which we identify with the black hole's *event horizon* (Fig. 5.12). The quantity

$$\Omega_H \equiv \omega(r_+) = \frac{a}{r_+^2 + a^2} \qquad (5.55)$$

is then interpreted as the angular velocity of the black hole. Stationary observers just outside the horizon have an angular velocity equal to Ω_H – they are in a state of *corotation* with the black hole.

To confirm that $r = r_+$ is truly the event horizon, we use the property that in a stationary spacetime, the event horizon is also an apparent horizon – a surface of zero expansion for a congruence of outgoing null geodesics orthogonal to the surface. The event horizon must therefore be a null, stationary surface. Now, the normal to any stationary surface must be proportional to $\partial_\alpha r$, and such a surface will be null if $g^{\alpha\beta}(\partial_\alpha r)(\partial_\beta r) = g^{rr} = 0$. Using Eq. (5.46) gives

$$\Delta \equiv r^2 - 2Mr + a^2 = 0. \qquad (5.56)$$

The largest solution, $r = r_+$, designates the event horizon. The other root,

$$r_- = M - \sqrt{M^2 - a^2}, \qquad (5.57)$$

describes the black hole's inner apparent horizon, which is analogous to the inner horizon of the Reissner–Nordström black hole.

We have found that the vector

$$\xi^\alpha \equiv t^\alpha + \Omega_H \phi^\alpha \qquad (5.58)$$

is null at the event horizon. It is tangent to the horizon's null generators, which wrap around the horizon with an angular velocity Ω_H. Because it is a linear combination of two Killing vectors, ξ^α is also a Killing vector, and the event horizon of the Kerr spacetime is a Killing horizon. Notice an important difference between stationary and static black holes: For a static black hole, t^α becomes null at the event horizon; for a stationary black hole, t^α is null at the static limit and ξ^α becomes null at the event horizon.

5.3.5 The Penrose process

The fact that t^α is spacelike in the ergosphere – the region $r_+ < r < r_{sl}(\theta)$ of the Kerr spacetime – implies that the (conserved) energy $E = -p_\alpha t^\alpha$ of a particle with four-momentum p^α can be of either sign. Particles with negative energy can therefore exist in the ergosphere, but they would never be able to escape from this region. (Note that $E < 0$ refers to the energy that *would* be measured at infinity *if* the particle could be brought there. Any local measurement of the particle's energy inside the static limit would return a positive value.)

It is easy to elaborate a scenario in which negative-energy particles created in the ergosphere are used to extract positive energy from a Kerr black hole. Imagine that a particle of energy $E_1 > 0$ comes from infinity and enters the ergosphere. There, it decays into two new particles, one with energy $-E_2 < 0$, the other with energy $E_3 = E_1 + E_2 > E_1$. While the negative-energy particle remains inside the static limit, the positive-energy particle escapes to infinity where its energy is extracted. Because E_3 is larger than the energy of the initial particle, the black hole must have given off some of its own energy. This is the Penrose process, by which some of the energy of a rotating black hole can be extracted.

The Penrose process is self-limiting: Only a fraction of the hole's total energy can be tapped. Suppose that in order to exploit the Penrose process, a rotating black hole is made to absorb a particle of energy $E = -p_\alpha t^\alpha < 0$ and angular momentum $L = p_\alpha \phi^\alpha$. Because the Killing vector $\xi^\alpha = t^\alpha + \Omega_H \phi^\alpha$ is timelike just outside the event horizon, the combination $E - \Omega_H L \equiv -p_\alpha \xi^\alpha$ must be positive; otherwise the particle would not be able to penetrate the horizon. Thus $L < E/\Omega_H$ and L must be negative if $E < 0$. The black hole will therefore lose angular momentum during the Penrose process. Eventually the hole's angular momentum will go to zero, the ergosphere will disappear, and the Penrose process will stop. We might say that only the hole's *rotational energy* can be extracted by the Penrose process.

Note that in a process by which a black hole absorbs a particle of energy E (of either sign) and angular momentum L, its parameters change by amounts $\delta M = E$ and $\delta J = L$. Since $E - \Omega_H L$ must be positive, we have

$$\delta M - \Omega_H \, \delta J > 0.$$

As we shall see, this inequality is a direct consequence of the first and second laws of black-hole mechanics.

5.3.6 Principal null congruences

The Boyer–Lindquist coordinates, like the Schwarzschild coordinates, are singular at the event horizon: While a trip down to the event horizon requires a finite proper time, the interval of coordinate time t is infinite. Moreover, because the angular velocity $d\phi/dt$ tends to a finite limit at the horizon, ϕ also increases by an infinite amount. We therefore need another coordinate system to extend the Kerr metric beyond the event horizon. It is advantageous to tailor these new coordinates to the behaviour of null geodesics. The two congruences considered here (which are known as the *principal null congruences* of the Kerr spacetime) are especially simple to deal with; we will use them to construct new coordinates for the Kerr metric.

It is a remarkable feature of the Kerr metric that the equations for geodesic motion can be expressed in a decoupled, first-order form. These equations involve three constants of the motion: the energy parameter $\tilde{E}$, the angular-momentum parameter $\tilde{L}$, and the 'Carter constant' $\mathcal{Q}$. (This last constant appears because of the existence of a Killing tensor. This is explained in Section 5.7, Problem 4, which also provides a derivation of the geodesic equations.) For null geodesics the equations are

$$\rho^2 \dot{t} = -a(a\tilde{E} \sin^2\theta - \tilde{L}) + (r^2 + a^2)P/\Delta,$$
$$\rho^2 \dot{r} = \pm\sqrt{R},$$
$$\rho^2 \dot{\theta} = \pm\sqrt{\Theta},$$
$$\rho^2 \dot{\phi} = -(a\tilde{E} - \tilde{L}/\sin^2\theta) + aP/\Delta,$$

in which an overdot indicates differentiation with respect to an affine parameter λ, and

$$P = \tilde{E}(r^2 + a^2) - a\tilde{L},$$

$$R = P^2 - \Delta\left[(\tilde{L} - a\tilde{E})^2 + \mathscr{Q}\right],$$

$$\Theta = \mathscr{Q} + \cos^2\theta(a^2\tilde{E}^2 - \tilde{L}^2/\sin^2\theta).$$

We simplify these equations by making the following choices:

$$\tilde{L} = a\tilde{E}\sin^2\theta, \qquad \mathscr{Q} = -(\tilde{L} - a\tilde{E})^2 = -(a\tilde{E}\cos^2\theta)^2.$$

It is easy to check that these imply $\Theta = 0$, so that our geodesics move with a constant value of θ. We also have $P = \tilde{E}\rho^2$ and $R = (\tilde{E}\rho^2)^2$, which give

$$\dot{t} = \tilde{E}(r^2 + a^2)/\Delta, \quad \dot{r} = \pm\tilde{E}, \quad \dot{\theta} = 0, \quad \dot{\phi} = a\tilde{E}/\Delta.$$

The constant $\tilde{E}$ can be absorbed into the affine parameter λ. We obtain an *ingoing* congruence by choosing the negative sign for $\dot{r}$, and we shall use l^α to denote its tangent vector field:

$$l^\alpha\partial_\alpha = \frac{r^2 + a^2}{\Delta}\partial_t - \partial_r + \frac{a}{\Delta}\partial_\phi. \tag{5.59}$$

Choosing instead the positive sign gives an *outgoing* congruence, with

$$k^\alpha\partial_\alpha = \frac{r^2 + a^2}{\Delta}\partial_t + \partial_r + \frac{a}{\Delta}\partial_\phi \tag{5.60}$$

as its tangent vector field.

To give the simplest description of the *ingoing* congruence, we introduce new coordinates v and ψ defined by

$$v = t + r^*, \qquad \psi = \phi + r^\sharp, \tag{5.61}$$

where

$$r^* = \int \frac{r^2 + a^2}{\Delta}\,dr$$

$$= r + \frac{Mr_+}{\sqrt{M^2 - a^2}}\ln\left|\frac{r}{r_+} - 1\right| - \frac{Mr_-}{\sqrt{M^2 - a^2}}\ln\left|\frac{r}{r_-} - 1\right| \tag{5.62}$$

and

$$r^\sharp = \int \frac{a}{\Delta}\,dr = \frac{a}{2\sqrt{M^2 - a^2}}\ln\left|\frac{r - r_+}{r - r_-}\right|. \tag{5.63}$$

It is easy to check that in these coordinates, $l^r = -1$ is the only nonvanishing component of the tangent vector. This means that v and ψ (as well as θ) are constant on each of the ingoing null geodesics, and that $-r$ is the affine parameter.

The simplest description of the *outgoing* congruence is provided by the coordinates (u, r, θ, χ), where

$$u = t - r^*, \qquad \chi = \phi - r^\sharp. \tag{5.64}$$

In these coordinates $k^r = +1$ is the only nonvanishing component of the tangent vector. This shows that u and χ (as well as θ) are constant along the outgoing null geodesics, and that r is the affine parameter.

The Kerr metric can be expressed in either one of these new coordinate systems. While the coordinates (v, r, θ, ψ) are well behaved on the future horizon but singular on the past horizon, the coordinates (u, r, θ, χ) are well behaved on the past horizon but singular on the future horizon. For example, a straightforward computation reveals that after a transformation to the ingoing coordinates, the Kerr metric becomes

$$ds^2 = -\left(1 - \frac{2Mr}{\rho^2}\right) dv^2 + 2\, dv\, dr - 2a \sin^2 \theta\, dr\, d\psi$$

$$- \frac{4Mar \sin^2 \theta}{\rho^2}\, dv\, d\psi + \frac{\Sigma}{\rho^2} \sin^2 \theta\, d\psi^2 + \rho^2\, d\theta^2. \tag{5.65}$$

These coordinates produce an extension of the Kerr metric across the future horizon. Several coordinate patches, both ingoing and outgoing, are required to cover the entire Kerr spacetime, whose causal structure is very similar to that of the Reissner–Nordström spacetime. We shall return to this topic in Section 5.3.9.

5.3.7 Kerr–Schild coordinates

Another useful set of coordinates for the Kerr metric is (t', x, y, z), the pseudo-Lorentzian *Kerr–Schild* coordinates in terms of which the metric takes a particularly interesting form. These are constructed as follows.

We start with Eq. (5.65) and separate out the terms that are proportional to M. After some algebra we obtain

$$ds^2 = -dv^2 + 2\, dv\, dr - 2a \sin^2 \theta\, dr\, d\psi + (r^2 + a^2) \sin^2 \theta\, d\psi^2 + \rho^2\, d\theta^2$$

$$+ \frac{2Mr}{\rho^2} (dv - a \sin^2 \theta\, d\psi)^2.$$

The terms that do not involve M have a simple interpretation: They give the metric of flat spacetime in an unfamiliar coordinate system. The rest of the line element can be written neatly in terms of l_α: Recalling that $l^r = -1$ is the only nonvanishing component of l^α, we find that

$$-l_\alpha\, dx^\alpha = dv - a \sin^2 \theta\, d\psi$$

and the line element becomes

$$ds^2 = (ds^2)_{\text{flat}} + \frac{2Mr}{\rho^2} \left(l_\alpha \, dx^\alpha\right)^2. \tag{5.66}$$

The Kerr metric can therefore be expressed as

$$g_{\alpha\beta} = \eta_{\alpha\beta} + \frac{2Mr}{\rho^2} l_\alpha l_\beta, \tag{5.67}$$

where $\eta_{\alpha\beta}$ is the metric of flat spacetime in the coordinates (v, r, θ, ψ).

Equation (5.67) gives us the Kerr metric in a rather attractive form. Any metric that can be written as $g_{\alpha\beta} = \eta_{\alpha\beta} + H l_\alpha l_\beta$, where H is a scalar function and l_α a null vector field, is known as a *Kerr–Schild* metric. It is by adopting such an expression that Kerr discovered his solution in 1963. (Some general aspects of the Kerr–Schild decomposition are worked out in Section 5.7, Problem 5.)

The next order of business is to find the coordinate transformation that brings $\eta_{\alpha\beta}$ to the standard Minkowski form. The answer is

$$x + iy = (r + ia) \sin\theta \, e^{i\psi}, \quad z = r \cos\theta, \quad t' = v - r. \tag{5.68}$$

Going through the necessary algebra does indeed reveal that in these coordinates,

$$(ds^2)_{\text{flat}} = -dt'^2 + dx^2 + dy^2 + dz^2. \tag{5.69}$$

It is easy to work out the components of l_α in this coordinate system. Because the null geodesics move with constant values of v, θ, and ψ, we have that $\dot{x} + i\dot{y} = -\sin\theta \, e^{i\psi}$, $\dot{z} = -\cos\theta$, and $\dot{t'} = 1$, where we have used $\dot{r} = -1$. Lowering the indices is a trivial matter (see Section 5.7, Problem 5), and expressing the right-hand sides in terms of the new coordinates gives

$$-l_\alpha \, dx^\alpha = dt' + \frac{rx + ay}{r^2 + a^2} dx + \frac{ry - ax}{r^2 + a^2} dy + \frac{z}{r} dz. \tag{5.70}$$

The quantity r must now be expressed in terms of x, y, and z. Starting with $x^2 + y^2 = (r^2 + a^2) \sin^2\theta$, it is easy to show that

$$r^4 - (x^2 + y^2 + z^2 - a^2)r^2 - a^2 z^2 = 0, \tag{5.71}$$

which may be solved for $r(x, y, z)$. Equations (5.66), (5.69)–(5.71) give the explicit form of the Kerr metric in the Kerr–Schild coordinates.

5.3.8 The nature of the singularity

We have seen that the Kerr spacetime possesses a curvature singularity at

$$\rho^2 \equiv r^2 + a^2 \cos^2\theta = 0.$$

According to this equation, the singularity occurs only in the equatorial plane ($\theta = \pi/2$), at $r = 0$. The Kerr–Schild coordinates can help us make sense of this statement. The relations $x^2 + y^2 = (r^2 + a^2)\sin^2\theta$ and $z = r\cos\theta$ indicate that the 'point' $r = 0$ corresponds in fact to the entire disk $x^2 + y^2 \le a^2$ in the plane $z = 0$. The points interior to the disk correspond to angles such that $\sin^2\theta < 1$. The boundary,

$$x^2 + y^2 = a^2,$$

corresponds to the equatorial plane, and this is where the Kerr metric is singular. The curvature singularity of the Kerr spacetime is therefore located on a ring of (coordinate) radius a in the x-y plane. This singularity can be avoided: Observers at $r = 0$ can stay away from the equatorial plane, and they never have to encounter the singularity; such observers end up *going through* the ring.

5.3.9 Maximal extension of the Kerr spacetime

We have already constructed coordinate systems that allow the continuation of the Kerr metric across the event horizon. We now complete the discussion and show how the spacetime can also be extended beyond the inner horizon. For simplicity we shall work with the two-dimensional section of the Kerr spacetime obtained by setting $\theta = 0$. This is the rotation axis, and because the Kerr metric is not spherically symmetric, this does represent a loss of generality.

Going back to Eq. (5.44) and the original Boyer–Lindquist coordinates, we find that when $\theta = 0$, the Kerr metric reduces to

$$ds^2 = -\left(1 - \frac{2Mr}{r^2 + a^2}\right)dt^2 + \frac{r^2 + a^2}{\Delta}dr^2$$

$$= -\frac{\Delta}{r^2 + a^2}\left(dt - \frac{r^2 + a^2}{\Delta}dr\right)\left(dt + \frac{r^2 + a^2}{\Delta}dr\right),$$

or

$$ds^2 = -f\, du\, dv, \qquad (5.72)$$

where $u = t - r^*$ and $v = t + r^*$ are the coordinates of Section 5.3.6. Here,

$$f = \frac{\Delta}{r^2 + a^2} = \frac{(r - r_+)(r - r_-)}{r^2 + a^2}, \qquad (5.73)$$

and $r_\pm = M \pm \sqrt{M^2 - a^2}$ denote the positions of the outer and inner horizons, respectively. The metric of Eq. (5.72) is extremely simple, and the construction of Kruskal coordinates for the $\theta = 0$ section of the Kerr spacetime proceeds just as for the Reissner–Nordström (RN) black hole (Section 5.2.2).

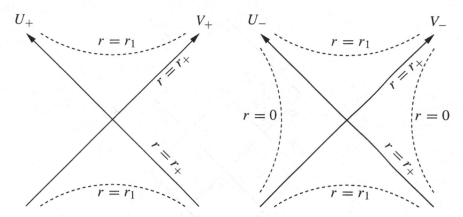

Figure 5.13   Kruskal patches for the Kerr spacetime.

We first consider the continuation of the metric across the event horizon. Near $r = r_+$ Eq. (5.73) can be approximated by

$$f \simeq 2\kappa_+(r - r_+),$$

where $\kappa_+ = \frac{1}{2}f'(r_+)$. It follows that

$$r^* = \int \frac{dr}{f} \simeq \frac{1}{2\kappa_+} \ln|\kappa_+(r - r_+)|$$

and $f \simeq \pm 2\,e^{2\kappa_+ r^*} = \pm 2\,e^{\kappa_+(v-u)}$; the upper sign refers to $r > r_+$ and the lower sign to $r < r_+$. Introducing the new coordinates

$$U_+ = \mp e^{-\kappa_+ u}, \qquad V_+ = e^{\kappa_+ v}, \tag{5.74}$$

we find that near $r = r_+$ the Kerr metric admits the manifestly regular form $ds^2 \simeq -2\kappa_+^{-2}\,dU_+ dV_+$.

Just as for the RN spacetime, the coordinates U_+ and V_+ are singular at the inner horizon, and another coordinate patch is required to extend the Kerr metric beyond this horizon (Fig. 5.13). The procedure is now familiar. Near $r = r_-$ we approximate Eq. (5.73) by $f \simeq -2\kappa_-(r - r_-)$, where $\kappa_- = \frac{1}{2}|f'(r_-)|$, so that $f \simeq \mp 2\,e^{-2\kappa_- r^*} = \mp 2\,e^{\kappa_-(u-v)}$. The appropriate coordinate transformation is now

$$U_- = \mp e^{\kappa_- u}, \qquad V_+ = -e^{-\kappa_- v}, \tag{5.75}$$

and the metric becomes $ds^2 \simeq -2\kappa_-^{-2}\,dU_- dV_-$.

Just as for the RN spacetime, another copy of the outer horizon presents itself in the future of the inner horizon, and another Kruskal patch is required to extend the spacetime beyond this new horizon. This continues ad nauseam, and we see that the maximally extended Kerr spacetime contains an infinite succession

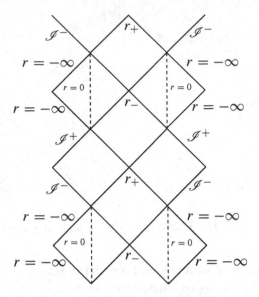

Figure 5.14   Penrose–Carter diagram of the Kerr spacetime.

of asymptotically-flat universes connected by black-hole tunnels. There is more, however. It is easy to check that in a spacetime diagram based on the (U_-, V_-) coordinates, the surface $r = 0$ is represented by $U_- V_- = -1$. This is a timelike surface, and on the rotation axis this surface is *nonsingular*. The Kerr spacetime can therefore be extended beyond $r = 0$, into a region in which r adopts negative values. This new region has no analogue in the RN spacetime; it contains no horizons and it becomes flat in the limit $r \to -\infty$. Observers in this region interpret the Kerr metric as describing the gravitational field of a (naked) ring singularity. You should be able to convince yourself that this singularity has a *negative* mass.

The maximally extended Kerr spacetime can be represented by a Penrose–Carter diagram (Fig. 5.14). The resulting causal structure is extremely complex. It should be kept in mind, however, that the interior of a Kerr black hole is subject to the same instability as that of a RN black hole (see Section 5.2.3 and Section 5.7, Problem 3). The tunnels to other universes, and the regions of negative r, are not present inside physically realistic black holes.

5.3.10 Surface gravity

As was pointed out in Section 5.3.4, the vector

$$\xi^\alpha = t^\alpha + \Omega_H \phi^\alpha, \tag{5.76}$$

where Ω_H is given by Eq. (5.55), is null at the event horizon and is in fact tangent to the horizon's null generators. From the same arguments as those presented in Section 5.2.4, the black hole's surface gravity κ can be defined by

$$\left(-\xi^\beta \xi_\beta\right)_{;\alpha} = 2\kappa \xi_\alpha, \tag{5.77}$$

or by

$$\xi^\alpha_{;\beta} \xi^\beta = \kappa \xi^\alpha, \tag{5.78}$$

or finally, by

$$\kappa^2 = -\frac{1}{2} \xi^{\alpha;\beta} \xi_{\alpha;\beta}. \tag{5.79}$$

These definitions are all equivalent.

Let us use Eq. (5.77) to calculate the surface gravity. The norm of ξ^α is given by

$$\xi^\beta \xi_\beta = \frac{\Sigma \sin^2 \theta}{\rho^2} (\Omega_H - \omega)^2 - \frac{\rho^2 \Delta}{\Sigma},$$

and differentiation yields

$$\left(-\xi^\beta \xi_\beta\right)_{;\alpha} = \frac{\rho^2}{\Sigma} \Delta_{,\alpha}$$

on the horizon, at which $\omega = \Omega_H$ and $\Delta = 0$. We have that $\Delta_{,\alpha} = 2(r_+ - M)\,\partial_\alpha r$ and $\xi_\alpha = (1 - a\Omega_H \sin^2 \theta)\,\partial_\alpha r$ on the horizon, and a few lines of algebra reveal that the surface gravity is

$$\kappa = \frac{r_+ - M}{r_+^2 + a^2} = \frac{\sqrt{M^2 - a^2}}{r_+^2 + a^2}. \tag{5.80}$$

Notice that this is the same quantity that was denoted κ_+ in Section 5.3.9. Notice also that $\kappa = 0$ for an extreme Kerr black hole. And finally, notice that in the general case κ does not depend on θ – the surface gravity is uniform on the event horizon. We shall return to this remarkable property in Section 5.5.1.

5.3.11 Bifurcation two-sphere

In the coordinates (v, r, θ, ψ) which are regular on the event horizon, $\xi^\alpha \partial_\alpha = \partial_v + \Omega_H \partial_\psi$. This shows that the horizon's null generators are parameterized by the advanced-time coordinate v, but as Eq. (5.78) reveals, v is not affine. An affine parameter λ is obtained by integrating

$$\frac{d\lambda}{dv} = e^{\kappa v},$$

so that $\kappa \lambda = e^{\kappa v} \equiv V$. It follows that on the horizon, the vector $k^\alpha = V^{-1}\xi^\alpha$ satisfies the geodesic equation in affine-parameter form: $k^\alpha_{\;;\beta}k^\beta = 0$. [This vector is not equal to the k^α introduced in Section 5.3.6, but it is proportional to it. It is easy to check that these vectors are indeed related by $k^\alpha_{\text{new}} = \frac{1}{2}\Delta k^\alpha_{\text{old}}/(r^2 + a^2)$, where the right-hand side is to be evaluated on the horizon.] If $\kappa \neq 0$ and the event horizon is geodesically complete (in the sense that the null generators can be extended arbitrarily far into the past), the relation

$$\xi^\alpha = V k^\alpha \tag{5.81}$$

implies that $\xi^\alpha = 0$ at $V = 0$. This defines a closed two-surface called the *bifurcation two-sphere* of the Kerr spacetime. The conditions are sometimes violated: The event horizon of a black hole formed by gravitational collapse is not geodesically complete, because the horizon was necessarily formed in the finite past; and as we have seen, the surface gravity of an extreme Kerr black hole (for which $M = a$) vanishes. In either one of these situations the bifurcation two-sphere does not exist.

5.3.12 Smarr's formula

There exists a simple algebraic relation between the black-hole mass M, its angular momentum $J \equiv Ma$, and its surface area A. This is defined by

$$A = \oint_{\mathscr{H}} \sqrt{\sigma}\, d^2\theta, \tag{5.82}$$

where $\mathscr{H}$ is a two-dimensional cross section of the event horizon, described by $v = \text{constant}$, $r = r_+$, $0 \leq \theta \leq \pi$, and $0 \leq \psi < 2\pi$. From Eq. (5.65) we find that the induced metric is given by

$$\sigma_{AB}\, d\theta^A\, d\theta^B = \rho^2\, d\theta^2 + \frac{\Sigma}{\rho^2}\sin^2\theta\, d\psi^2,$$

so that $\sqrt{\sigma}\, d^2\theta = \sqrt{\Sigma}\sin\theta\, d\theta\, d\psi = (r_+^2 + a^2)\sin\theta\, d\theta\, d\psi$. Integration yields

$$A = 4\pi\left(r_+^2 + a^2\right). \tag{5.83}$$

The algebraic relation, which was discovered by Larry Smarr in 1973, reads

$$M = 2\,\Omega_H J + \frac{\kappa A}{4\pi}, \tag{5.84}$$

where Ω_H is the hole's angular velocity and κ its surface gravity. Smarr's formula is established by straightforward algebra: Substituting Eqs. (5.55), (5.80), and (5.83) into the right-hand side of Eq. (5.84) reveals that it is indeed equal to

M. We will generalize Smarr's formula, and present an alternative derivation, in Section 5.5.2.

5.3.13 Variation law

It is clear that the surface area of a black hole is a function of its mass and angular momentum: $A = A(M, J)$. Suppose that a black hole of mass M and angular momentum J is perturbed so that its parameters become $M + \delta M$ and $J + \delta J$. (For example, the black hole might absorb a particle, as was considered in Section 5.3.5.) How does the area change? There exists a simple formula relating δA to the changes in mass and angular momentum. It is

$$\frac{\kappa}{8\pi} \delta A = \delta M - \Omega_H \, \delta J. \tag{5.85}$$

To derive this we start with Eq. (5.83), which immediately implies

$$\frac{\delta A}{8\pi} = r_+ \, \delta r_+ + a \, \delta a.$$

But the horizon radius r_+ depends on M and a; the defining relation is $r_+^2 - 2Mr_+ + a^2 = 0$ and this gives us

$$(r_+ - M) \, \delta r_+ = r_+ \, \delta M - a \, \delta a.$$

This result can be substituted into the preceding expression for δA. The final step is to relate a to the hole's angular momentum J; we have that $a = J/M$ and this implies $M \, \delta a = \delta J - a \, \delta M$. Collecting these results, we arrive at Eq. (5.85) after involving Eqs. (5.55) and (5.80).

In Section 5.3.5 we found that the right-hand side of Eq. (5.85) must be positive. What we have, therefore, is the statement that the surface area of a Kerr black hole *always increases* during a process by which it absorbs a particle. This is a restricted version of the second law of black-hole mechanics, to which we shall return in Section 5.5.4.

5.4 General properties of black holes

The Kerr family of solutions to the Einstein field equations plays an extremely important role in the description of black holes, but this does not mean that *all* black holes are Kerr black holes. For example, a black hole accreting matter is not stationary, and a stationary hole is not a Kerr black hole if it is tidally distorted by nearby masses. In this section we consider those properties of black holes that are quite general, and not specific to any particular solution to the Einstein field equations.

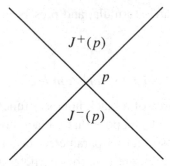

Figure 5.15   Causal future and past of an event p.

5.4.1 General black holes

A spacetime containing a black hole possesses two distinct regions, the interior and exterior of the black hole; they are distinguished by the property that all external observers are causally disconnected from events occurring inside. Physically speaking, this corresponds to the fact that once she has entered a black hole, an observer can no longer send signals to the outside world.

These fundamental notions can be cast in mathematical terms. Consider an event p and the set of all events that can be reached from p by future-directed curves, either timelike or null (Fig. 5.15). This set is denoted $J^+(p)$ and is called the *causal future* of p. A similar definition can be given for its *causal past*, $J^-(p)$. These definitions can be extended to whole sets of events: If S is such a set, then $J^+(S)$ is the union of the causal futures of all the events p contained in S; a similar definition can be given for $J^-(S)$.

Loosely speaking, a spacetime contains a black hole if there exist outgoing null geodesics that never reach future null infinity, denoted $\mathscr{I}^+$. These originate from the black-hole interior, a region characterized by the very fact that all future-directed curves starting from it fail to reach $\mathscr{I}^+$. Thus, events lying within the black-hole interior cannot be in the causal past of $\mathscr{I}^+$. The *black-hole region B* of the spacetime manifold $\mathscr{M}$ is therefore the set of all events p that do not belong to the causal past of future null infinity:

$$B = \mathscr{M} - J^-(\mathscr{I}^+). \qquad (5.86)$$

The *event horizon H* is then defined to be the boundary of the black-hole region:

$$H = \partial B = \partial\big(J^-(\mathscr{I}^+)\big). \qquad (5.87)$$

The two-dimensional surface obtained by intersecting the event horizon with a spacelike hypersurface Σ is denoted $\mathscr{H}(\Sigma)$; it is called a *cross section* of the horizon.

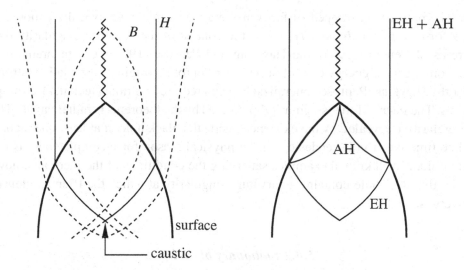

Figure 5.16   Event and apparent horizons of a black-hole spacetime.

Because the event horizon is a causal boundary, it must be a null hypersurface. Penrose (1968) was able to establish that *the event horizon is a null hypersurface generated by null geodesics that have no future end points*. This means that: (i) when followed into the past, a generator may, but does not have to, leave the horizon; (ii) once a generator has entered the horizon, it cannot leave; (iii) two generators can never intersect, except possibly when they both enter the horizon; and finally, (iv) through every point on the event horizon, except for those at which new generators enter, there passes one and only one generator. It should be clear that the entry points into the event horizon are caustics of the congruence of null generators (Fig. 5.16).

The black-hole region typically contains *trapped surfaces*, closed two-surfaces S with the property that for both ingoing and outgoing congruences of null geodesics orthogonal to S, the expansion is negative everywhere on S. (Exceptions are the extreme cases of Kerr, Kerr–Newman, or Reissner–Nordström black holes, which do not have trapped surfaces.) The three-dimensional boundary of the region of spacetime that contains trapped surfaces is the *trapping horizon*, and its two-dimensional intersection with a spacelike hypersurface Σ is called an *apparent horizon*. The apparent horizon is therefore a marginally trapped surface – a closed two-surface on which *one* of the congruences has zero expansion. The apparent horizon of a stationary black hole typically coincides with the event horizon. In dynamical situations, however, the apparent horizon always lies *within* the black-hole region (Fig. 5.16), unless the null energy condition is violated. (Refer back to Section 5.1.8 for a specific example.)

   The presence of trapped surfaces inside a black hole unequivocally announces
the formation of a *singularity*; this is the content of the beautiful singularity theorems of Penrose (1965) and Hawking and Penrose (1970). The theorems rely
on some form of energy condition (null for Penrose's original formulation, strong
for the Hawking–Penrose generalization) and require additional technical assumptions. The nature of the 'singularity' predicted by the theorems is rather vague: The
singularity is revealed by the existence inside the black hole of at least one incomplete timelike or null geodesic, but the physical reason for incompleteness is not
identified. In all known examples satisfying the conditions of the theorems, however, the black hole contains a curvature singularity at which the Riemann tensor
diverges.

5.4.2 Stationary black holes

It was established by Hawking in 1972 that if a black hole is stationary, then it
must be *either* static *or* axially symmetric. This means that the stationary spacetime
of a rotating hole is necessarily axially symmetric and that it admits two Killing
vectors, t^α and ϕ^α. Hawking was also able to show that a linear combination of
these vectors,

$$\xi^\alpha = t^\alpha + \Omega_H \phi^\alpha, \tag{5.88}$$

is null at the event horizon. Here, Ω_H is the hole's angular velocity, which vanishes
if the spacetime is nonrotating (and therefore static). Thus, the event horizon is
a Killing horizon and ξ^α is tangent to the horizon's null generators. These are
parameterized by advanced time v, so that a displacement along a generator is
described by $dx^\alpha = \xi^\alpha\, dv$. The hole's surface gravity κ is then defined by the
relation

$$\xi^\alpha_{\;;\beta} \xi^\beta = \kappa \xi^\alpha, \tag{5.89}$$

which holds on the horizon. We will prove in Section 5.5.2 that κ is constant along
the horizon's null generators. (Indeed, it is uniform over the entire horizon.) This
means that we can replace v by an affine parameter $\lambda = V/\kappa$, where $V = e^{\kappa v}$
(Section 5.3.10). Then

$$k^\alpha \equiv V^{-1} \xi^\alpha \tag{5.90}$$

satisfies the geodesic equation in affine-parameter form. It follows that if $\kappa \neq 0$
and the horizon is geodesically complete (in the sense that its generators never
leave the horizon when followed into the past), then there exists a two-surface,
called the *bifurcation two-sphere*, on which $\xi^\alpha = 0$.

The properties of stationary black holes listed here were all encountered before, during our presentation of the Kerr solution. It should be appreciated, however, that Eqs. (5.88)–(5.90) hold by virtue of the sole fact that the black hole is stationary; these results do not depend on the specific details of a particular metric.

The observation that a stationary black hole must be axially symmetric if it is rotating might seem puzzling. After all, it should be possible to place a nonsymmetrical distribution of matter outside the hole, and let it tidally distort the event horizon in a nonsymmetrical manner. This, presumably, would produce a black hole that is still stationary and rotating, but not axially symmetric. Hawking and Hartle (1972) have shown that this, in fact, is false! The reason is that such a distribution of matter would impart a torque on the black hole, which would force it to spin down to a nonrotating (and static) configuration. Thus, such a situation would not leave the black hole stationary.

Additional properties of stationary black holes can be inferred from Raychaudhuri's equation (Section 2.4.4),

$$\frac{\mathrm{d}\theta}{\mathrm{d}\lambda} = -\frac{1}{2}\theta^2 - \sigma^{\alpha\beta}\sigma_{\alpha\beta} - R_{\alpha\beta}k^\alpha k^\beta, \tag{5.91}$$

in which we have put $\omega_{\alpha\beta} = 0$ to reflect the fact that the congruence of null generators is necessarily hypersurface orthogonal. The event horizon will be stationary if θ and $\mathrm{d}\theta/\mathrm{d}\lambda$ are both zero. Using the Einstein field equations and the null energy condition, Eq. (5.91) implies that the stress-energy tensor must satisfy

$$T_{\alpha\beta}\xi^\alpha\xi^\beta = 0 \tag{5.92}$$

on the horizon. This means that matter cannot be flowing across the event horizon; if it were, the generators would get focused and the black hole would not be stationary. Raychaudhuri's equation also implies

$$\sigma_{\alpha\beta} = 0; \tag{5.93}$$

the null generators of the event horizon have a vanishing shear tensor.

5.4.3 Stationary black holes in vacuum

In the absence of any matter in their exterior, stationary black holes admit an extremely simple description.

If the black hole is *static*, then it must be spherically symmetric and it can only be described by the Schwarzschild solution. This beautiful uniqueness theorem, the first of its kind, was established by Werner Israel in 1967. It implies that in the absence of angular momentum, complete gravitational collapse must result in a Schwarzschild black hole. This might seem puzzling, because the statement is true irrespective of the initial shape of the progenitor, which might have been strongly nonspherical. The mechanism by which a nonspherical star shakes off its higher multipole moments during gravitational collapse was elucidated by Richard Price in 1972: These multipole moments are simply radiated away, either out to infinity or into the black hole. After the radiation has faded away the hole settles down to its final, spherical state.

If the black hole is *axially symmetric*, then it must be a Kerr black hole. This extension of Israel's uniqueness theorem was established by Brandon Carter (1971) and D.C. Robinson (1975).

The black-hole uniqueness theorems can be generalized to include situations in which the black hole carries an electric charge. If the black hole is static, then it must be a Reissner–Nordström black hole (Israel, 1968). If it is axially symmetric, then it must be a Kerr–Newman black hole (Mazur, 1982; Bunting, unpublished).

We see that a black hole in isolation can be characterized, uniquely and completely, by just three parameters: its mass, angular momentum, and charge. No other parameter is required, and this remarkable property is at the origin of John Wheeler's famous phrase, 'a black hole has no hair.' Chandrasekhar (1987) was well justified to write:

Black holes are macroscopic objects with masses varying from a few solar masses to millions of solar masses. To the extent that they may be considered as stationary and isolated, to that extent, they are all, every single one of them, described *exactly* by the Kerr solution. This is the only instance we have of an exact description of a macroscopic object. Macroscopic objects, as we see them all around us, are governed by a variety of forces, derived from a variety of approximations to a variety of physical theories. In contrast, the only elements in the construction of black holes are our basic concepts of space and time. They are, thus, almost by definition, the most perfect macroscopic objects there are in the universe. And since the general theory of relativity provides a single unique two-parameter family of solutions for their descriptions, they are the simplest objects as well.

5.5 The laws of black-hole mechanics

In 1973, Jim Bardeen, Brandon Carter, and Stephen Hawking formulated a set of four laws governing the behaviour of black holes. These laws of black-hole mechanics bear a striking resemblance to the four laws of thermodynamics. While this analogy was at first perceived to be purely formal and coincidental, it soon became clear that black holes do indeed behave as thermodynamic systems. The crucial step in this realization was Hawking's remarkable discovery of 1974 that quantum processes allow a black hole to emit a thermal flux of particles. It is thus possible for a black hole to be in thermal equilibrium with other thermodynamic systems. The laws of black-hole mechanics, therefore, are nothing but a description of the thermodynamics of black holes.

5.5.1 Preliminaries

We begin our discussion of the four laws by collecting a few important results from preceding chapters; these will form the bulk of the mathematical framework required for the derivations.

Let $y^a = (v, \theta^A)$ be coordinates on the event horizon. The advanced-time coordinate v is a non-affine parameter on the horizon's null generators, and θ^A labels the generators. The vectors

$$\xi^\alpha = \left(\frac{\partial x^\alpha}{\partial v}\right)_{\theta^A}, \qquad e_A^\alpha = \left(\frac{\partial x^\alpha}{\partial \theta^A}\right)_v \qquad (5.94)$$

are tangent to the horizon; they satisfy $\xi_\alpha e_A^\alpha = 0 = \pounds_\xi e_A^\alpha$ and $\xi^\alpha = t^\alpha + \Omega_H \phi^\alpha$ is a Killing vector. We complete the basis by introducing an auxiliary null vector N^α, normalized by $N_\alpha \xi^\alpha = -1$. This basis gives us the completeness relations (Section 3.1)

$$g^{\alpha\beta} = -\xi^\alpha N^\beta - N^\alpha \xi^\beta + \sigma^{AB} e_A^\alpha e_B^\beta,$$

where σ^{AB} is the inverse of $\sigma_{AB} = g_{\alpha\beta} e_A^\alpha e_B^\beta$, the metric on the two-dimensional space transverse to the generators. The determinant of the two-metric will be denoted σ.

The vectorial surface element on the event horizon can be expressed as (Section 3.2)

$$d\Sigma_\alpha = -\xi_\alpha \, dS \, dv, \qquad (5.95)$$

where $dS = \sqrt{\sigma} \, d^2\theta$. The two-dimensional surface element on a cross section $v =$ constant is

$$dS_{\alpha\beta} = 2\xi_{[\alpha} N_{\beta]} \, dS. \qquad (5.96)$$

We shall denote such a cross section by $\mathscr{H}(v)$.

Finally, we will need Raychaudhuri's equation for the congruence of null generators, expressed in a form that does not require the parameter to be affine. This was worked out in Section 2.6, Problem 8 and the answer is

$$\frac{d\theta}{dv} = \kappa\,\theta - \frac{1}{2}\theta^2 - \sigma^{\alpha\beta}\sigma_{\alpha\beta} - 8\pi T_{\alpha\beta}\xi^\alpha\xi^\beta; \qquad (5.97)$$

the last term would normally involve the Ricci tensor, but we have used the Einstein field equations to write it in terms of the stress-energy tensor. We recall that θ is the fractional rate of change of the congruence's cross-sectional area: $\theta = (dS)^{-1} d(dS)/dv$.

5.5.2 Zeroth law

The zeroth law of black-hole mechanics states that *the surface gravity of a station-ary black hole is uniform over the entire event horizon*. We saw in Section 5.3.10 that this statement is indeed true for the specific case of a Kerr black hole, but the scope of the zeroth law is much wider: The black hole need not be isolated and its metric need not be the Kerr metric.

To prove that κ is uniform on the event horizon, we need to establish that (i) κ is constant along the horizon's null generators, and (ii) κ does not vary from generator to generator. We will prove both statements in turn, starting with

$$\kappa^2 = -\frac{1}{2}\,\xi^{\alpha;\beta}\xi_{\alpha;\beta} \tag{5.98}$$

as our definition for the surface gravity. (We saw in Section 5.2.4 that this relation is equivalent to $\xi^{\alpha}_{\ ;\beta}\xi^{\beta} = \kappa\xi^{\alpha}$.) We shall need the identity

$$\xi_{\alpha;\mu\nu} = R_{\alpha\mu\nu\beta}\xi^{\beta}, \tag{5.99}$$

which is satisfied by any Killing vector ξ^{α}. (This was derived in Section 1.13, Problem 9.)

We differentiate Eq. (5.98) in the directions tangent to the horizon. (Because κ is defined only on the event horizon, its normal derivative does not exist.) Using Eq. (5.99) we obtain

$$2\kappa\kappa_{,\alpha} = -\xi^{\mu;\nu}R_{\mu\nu\alpha\beta}\xi^{\beta}. \tag{5.100}$$

The fact that κ is constant on each generator follows immediately from this:

$$\kappa_{,\alpha}\xi^{\alpha} = 0. \tag{5.101}$$

We must now examine how κ changes in the transverse directions. Equation (5.100) implies

$$2\kappa\kappa_{,\alpha}\,e^{\alpha}_{A} = -\xi^{\mu;\nu}R_{\mu\nu\alpha\beta}\,e^{\alpha}_{A}\xi^{\beta},$$

and we would like to show that the right-hand side is zero. Let us first assume that the event horizon is geodesically complete, so that it contains a bifurcation two-sphere, at which $\xi^{\alpha} = 0$. Then the last equation implies that $\kappa_{,\alpha}e^{\alpha}_{A} = 0$ at the bifurcation two-sphere. Because $\kappa_{,\alpha}e^{\alpha}_{A}$ is constant on the null generators (Section 5.7, Problem 6), we have that $\kappa_{,\alpha}e^{\alpha}_{A} = 0$ on all cross sections $v = $ constant of the event horizon. This shows that the value of κ does not change from generator to generator, and we conclude that κ is uniform over the entire event horizon.

It is easy to see that the property $\kappa_{,\alpha}e^{\alpha}_{A} = 0$ must be independent of the exis-tence of a bifurcation two-sphere. Consider two stationary black holes, identical in every respect in the future of $v = 0$ (say), but different in the past, so that only one of them possesses a bifurcation two-sphere. (We imagine that the first black hole has existed forever, and that the second black hole was formed prior to $v = 0$ by gravitational collapse; the second black hole is stationary only for $v > 0$.) Our proof that $\kappa_{,\alpha} e_A^\alpha = 0$ on all cross sections $v = $ constant of the event horizon applies to the first black hole. But since the spacetimes are identical for $v > 0$, the property $\kappa_{,\alpha} e_A^\alpha = 0$ must apply also to the second black hole. Thus, the zeroth law is established for all stationary black holes, whether or not they are geodesically complete.

It is clear that the relation $\xi^{\mu;\nu} R_{\mu\nu\alpha\beta} e_A^\alpha \xi^\beta = 0$ must hold everywhere on a stationary event horizon, but it is surprisingly difficult to prove this. In their original discussion Bardeen, Carter, and Hawking establish this identity by using the Einstein field equations and the dominant energy condition (Section 5.7, Problem 7). This restriction was lifted in a 1996 paper by Rácz and Wald.

5.5.3 Generalized Smarr formula

Before moving on to the first law, we generalize Smarr's formula (Section 5.3.12) that relates the black-hole mass M to its angular momentum J, angular velocity Ω_H, surface gravity κ, and surface area A. In the present context the black hole is stationary and axially symmetric, but it is not assumed to be a Kerr black hole.

Our starting point is the Komar expressions for total mass and angular momentum (Section 4.3.3):

$$ M = -\frac{1}{8\pi} \oint_S \nabla^\alpha t^\beta \, dS_{\alpha\beta}, \qquad J = \frac{1}{16\pi} \oint_S \nabla^\alpha \phi^\beta \, dS_{\alpha\beta}, $$

where the integrations are over a closed two-surface at infinity. We consider a spacelike hypersurface Σ extending from the event horizon to spatial infinity (Fig. 5.17). Its inner boundary is $\mathcal{H}$, a two-dimensional cross section of the event horizon, and its outer boundary is S. Using Gauss' theorem, as was done in Section 4.3.3 (but without the inner boundary), we find that M and J can be expressed as

$$ M = M_H + 2 \int_\Sigma \left(T_{\alpha\beta} - \frac{1}{2} T g_{\alpha\beta} \right) n^\alpha t^\beta \sqrt{h} \, d^3 y \tag{5.102} $$

and

$$ J = J_H - \int_\Sigma \left(T_{\alpha\beta} - \frac{1}{2} T g_{\alpha\beta} \right) n^\alpha \phi^\beta \sqrt{h} \, d^3 y, \tag{5.103} $$

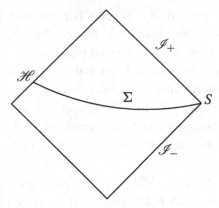

Figure 5.17   A spacelike hypersurface in a black-hole spacetime.

where M_H and J_H are the black-hole mass and angular momentum, respectively. They are given by surface integrals over $\mathcal{H}$:

$$M_H = -\frac{1}{8\pi} \oint_{\mathcal{H}} \nabla^\alpha t^\beta \, \mathrm{d}S_{\alpha\beta} \tag{5.104}$$

and

$$J_H = \frac{1}{16\pi} \oint_{\mathcal{H}} \nabla^\alpha \phi^\beta \, \mathrm{d}S_{\alpha\beta}, \tag{5.105}$$

where $\mathrm{d}S_{\alpha\beta}$ is the surface element of Eq. (5.96). The interpretation of Eqs. (5.102) and (5.103) is clear: The total mass M (angular momentum J) is given by a contribution M_H (J_H) from the black hole, plus a contribution from the matter distribution outside the hole. If the black hole is in vacuum, then $M = M_H$ and $J = J_H$.

Smarr's formula emerges after a few simple steps. Using Eqs. (5.96), (5.104) and (5.105) we have

$$M_H - 2\,\Omega_H J_H = -\frac{1}{8\pi} \oint_{\mathcal{H}} \nabla^\alpha \left(t^\beta + \Omega_H \phi^\beta\right) \mathrm{d}S_{\alpha\beta}$$

$$= -\frac{1}{8\pi} \oint_{\mathcal{H}} \nabla^\alpha \xi^\beta \, \mathrm{d}S_{\alpha\beta}$$

$$= -\frac{1}{4\pi} \oint_{\mathcal{H}} \xi^{\beta;\alpha} \xi_\alpha N_\beta \, \mathrm{d}S$$

$$= -\frac{1}{4\pi} \oint_{\mathcal{H}} \kappa \xi^\beta N_\beta \, \mathrm{d}S$$

$$= \frac{\kappa}{4\pi} \oint_{\mathcal{H}} \mathrm{d}S,$$

where we have used the relation $\xi^\alpha N_\alpha = -1$ and the fact that κ is constant over $\mathcal{H}$. The last integration gives the horizon's surface area and we arrive at

$$M_H = 2\,\Omega_H J_H + \frac{\kappa A}{4\pi}, \tag{5.106}$$

the generalized Smarr formula.

5.5.4 First law

We consider a quasi-static process during which a stationary black hole of mass M, angular momentum J, and surface area A is taken to a new stationary black hole with parameters $M + \delta M$, $J + \delta J$, and $A + \delta A$. The first law of black-hole mechanics states that *the changes in mass, angular momentum, and surface area are related by*

$$\delta M = \frac{\kappa}{8\pi}\,\delta A + \Omega_H\,\delta J. \tag{5.107}$$

If the initial and final black holes are in vacuum, then they are Kerr black holes by virtue of the uniqueness theorems, and a derivation of Eq. (5.107) was already presented in Section 5.3.13. That derivation, however, relied heavily on the details of the Kerr metric. We will now present a derivation that is largely independent of those details. In particular we will not assume that the black hole is in vacuum.

We suppose that a black hole, initially in a stationary state, is perturbed by a small quantity of matter described by the (infinitesimal) stress-energy tensor $T_{\alpha\beta}$. As a result the mass and angular momentum of the black hole increase by amounts (Section 4.3.4)

$$\delta M = -\int_H T^\alpha_{\ \beta}\,t^\beta\,d\Sigma_\alpha \tag{5.108}$$

and

$$\delta J = \int_H T^\alpha_{\ \beta}\,\phi^\beta\,d\Sigma_\alpha, \tag{5.109}$$

where the integrations are over the entire event horizon. We will be working to first order in the perturbation $T_{\alpha\beta}$, keeping t^α, ϕ^α, and $d\Sigma_\alpha$ at their unperturbed values. We assume that at the end of the process, the black hole is returned to another stationary state.

Substituting the surface element of Eq. (5.95) into Eqs. (5.108) and (5.109) we find

$$\delta M - \Omega_H \delta J = \int_H T_{\alpha\beta}(t^\beta + \Omega_H \phi^\beta)\xi^\alpha\,dS\,dv$$

$$= \int dv \oint_{\mathcal{H}(v)} T_{\alpha\beta}\xi^\alpha\xi^\beta\,dS.$$

To work out the integral we turn to Eq. (5.97). Because θ and $\sigma_{\alpha\beta}$ are quantities of the first order in $T_{\alpha\beta}$, it is appropriate to neglect the quadratic terms and Raychaudhuri's equation simplifies to

$$\frac{d\theta}{dv} = \kappa\,\theta - 8\pi\,T_{\alpha\beta}\xi^\alpha\xi^\beta.$$

Then

$$\delta M - \Omega_H \delta J = -\frac{1}{8\pi}\int dv \oint_{\mathcal{H}(v)} \left(\frac{d\theta}{dv} - \kappa\,\theta\right) dS$$

$$= -\frac{1}{8\pi}\oint_{\mathcal{H}(v)} \theta\,dS\,\Big|_{-\infty}^{\infty} + \frac{\kappa}{8\pi}\int dv \oint_{\mathcal{H}(v)} \theta\,dS.$$

Because the black hole is stationary both before and after the perturbation, $\theta(v = \pm\infty) = 0$ and the boundary terms vanish. Using the fact that θ is the fractional rate of change of the congruence's cross-sectional area, we obtain

$$\delta M - \Omega_H \delta J = \frac{\kappa}{8\pi}\int dv \oint_{\mathcal{H}(v)} \left(\frac{1}{dS}\frac{d}{dv}\,dS\right) dS$$

$$= \frac{\kappa}{8\pi}\oint_{\mathcal{H}(v)} dS\,\Big|_{-\infty}^{\infty}$$

$$= \frac{\kappa}{8\pi}\,\delta A,$$

where δA is the change in the black hole's surface area. This is Eq. (5.107), the statement of the first law of black-hole mechanics.

5.5.5 Second law

The second law of black-hole mechanics states that if the null energy condition is satisfied, then *the surface area of a black hole can never decrease*: $\delta A \geq 0$. This *area theorem* was established by Stephen Hawking in 1971.

Glossing over various technical details, the area theorem follows directly from the focusing theorem (Section 2.4.5) and Penrose's observation that the event horizon is generated by null geodesics with no future end points. This statement means that the generators of the event horizon can never run into caustics. (A generator can enter the horizon at a caustic point, but once in H it will never meet another caustic.) The focusing theorem then implies that θ, the expansion of the congruence of null generators, must be positive, or zero, everywhere on the event horizon. To see this, suppose that $\theta < 0$ for some of the generators. The focusing theorem then guarantees that these generators will converge into a caustic, at which $\theta = -\infty$. We have a contradiction and we must conclude that $\theta \geq 0$ everywhere on the event horizon. This implies that the horizon's surface area will not decrease, which is just the statement of the area theorem. (The fact that new generators can enter the event horizon contributes even further to the growth of its area.)

5.5.6 Third law

The third law of black-hole mechanics states that if the stress-energy tensor is bounded and satisfies the weak energy condition, then *the surface gravity of a black hole cannot be reduced to zero within a finite advanced time*. A precise formulation of this law was given by Werner Israel in 1986.

We have seen that a black hole of zero surface gravity is an extreme black hole. (Recall that a Kerr black hole is extremal if $a = M$; for a Reissner–Nordström black hole, the condition is $|Q| = M$.) An equivalent statement of the third law is therefore that under the stated conditions on the stress-energy tensor, it is impossible for a black hole to become extremal within a finite advanced time.

The proof of the third law is rather involved and we will not attempt to go through it here. Instead of presenting a proof, we will illustrate the fact that the third law is essentially a consequence of the weak energy condition.

For the purpose of this discussion we need a black-hole spacetime which is sufficiently dynamical that it has the potential of becoming extremal at a finite advanced time v. A simple choice is the charged generalization of the ingoing Vaidya spacetime, whose metric is given by

$$ds^2 = -f \, dv^2 + 2 \, dv \, dr + r^2 \, d\Omega^2, \tag{5.110}$$

with

$$f = 1 - \frac{2m(v)}{r} + \frac{q^2(v)}{r^2}. \tag{5.111}$$

This metric describes a black hole whose mass m and charge q change with time because of irradiation by charged null dust, a fictitious form of matter. This interpretation is confirmed by inspection of the stress-energy tensor,

$$T^{\alpha\beta} = T^{\alpha\beta}_{\text{dust}} + T^{\alpha\beta}_{\text{em}}, \tag{5.112}$$

where

$$T^{\alpha\beta}_{\text{dust}} = \rho \, l^\alpha l^\beta, \qquad \rho = \frac{1}{4\pi r^2} \frac{\partial}{\partial v} \left(m - \frac{q^2}{2r} \right) \tag{5.113}$$

is the contribution from the null dust ($l_\alpha = -\partial_\alpha v$ is a null vector), and

$$T_{\text{em}}{}^\alpha{}_\beta = P \, \text{diag}(-1, -1, 1, 1), \qquad P = \frac{q^2}{8\pi r^4} \tag{5.114}$$

is the contribution from the electromagnetic field. The spacetime of Eqs. (5.110) and (5.111) will produce a violation of the third law if $m(v_0) = q(v_0)$ for some advanced time $v_0 < \infty$.

An essential aspect of this discussion is the weak energy condition (Section 2.1), which states that the energy density measured by an observer with four-velocity u^α will always be positive:

$$T_{\alpha\beta} u^\alpha u^\beta > 0.$$

Here, $T^{\alpha\beta}$ is the stress-energy tensor of Eq. (5.112). If our observer is restricted to move in the radial direction only, then $T_{\alpha\beta} u^\alpha u^\beta = \rho(dv/d\tau)^2 + P$. Because $dv/d\tau$ can be arbitrarily large, the weak energy condition requires $\rho > 0$. In particular ρ must be positive at the apparent horizon, $r = r_+(v)$, where $r_+ = m + (m^2 - q^2)^{1/2}$. This gives us the following condition:

$$4\pi r_+{}^3 \rho(r_+) = m\dot{m} - q\dot{q} + (m^2 - q^2)^{1/2}\, \dot{m} > 0, \tag{5.115}$$

where an overdot indicates differentiation with respect to v.

Let us imagine a situation in which the black hole becomes extremal at a finite advanced time v_0. This means that $\Delta(v_0) = 0$, where $\Delta(v) \equiv m(v) - q(v)$. Because the black hole was not extremal before $v = v_0$, we have that $\Delta(v) > 0$ for $v < v_0$ and $\Delta(v)$ must be *decreasing* as v approaches v_0. However, Eq. (5.115) implies

$$m(v_0)\, \dot{\Delta}(v_0) > 0,$$

according to which $\Delta(v)$ must be *increasing*. We have a contradiction and we conclude that the weak energy condition prevents the black hole from ever becoming extremal at a finite advanced time.

5.5.7 Black-hole thermodynamics

The four laws of black-hole mechanics bear a striking resemblance to the laws of thermodynamics, with κ playing the role of temperature, A that of entropy, and M that of internal energy. Hawking's discovery that quantum processes give rise to a thermal flux of particles from black holes implies they do indeed behave as thermodynamic systems. Black holes have a well-defined temperature, which as a matter of fact is proportional to the hole's surface gravity:

$$T = \frac{\hbar}{2\pi} \kappa. \tag{5.116}$$

The zeroth law is therefore a special case of the corresponding law of thermodynamics, which states that a system in thermal equilibrium has a uniform temperature. The first law, when recognized as a special case of the corresponding law of thermodynamics, implies that the black-hole entropy must be given by

$$S = \frac{1}{4\hbar} A. \tag{5.117}$$

The second law is therefore also a special case of the corresponding law of thermodynamics, which states that the entropy of an isolated system can never decrease. In this regard it should be noted that Hawking radiation actually causes the black-hole area to *decrease*, in violation of the area theorem. (The radiation's stress-energy tensor does not satisfy the null energy condition.) However, the process of black-hole evaporation does not violate the *generalized* second law, which states that the *total* entropy, the sum of radiation and black-hole entropies, does not decrease.

The fact that black holes behave as thermodynamic systems reveals a deep connection between such disparate fields as gravitation, quantum mechanics, and thermodynamics. This connection is still poorly understood today.

5.6 Bibliographical notes

During the preparation of this chapter I have relied on the following references: Bardeen, Carter, and Hawking (1973); Carter (1979); Chandrasekhar (1983); Hayward (1994); Israel (1986a); Israel (1986b); Misner, Thorne, and Wheeler (1973); Sullivan and Israel (1980); Wald (1984); and Wald (1992).

More specifically:

The term 'trapping horizon,' used in Sections 5.1.7 and 5.4.1, was introduced by Sean Hayward in his 1994 paper. The various definitions for the surface gravity (Sections 5.2.4, 5.3.10, 5.4.2, and 5.5.2) are taken from Section 12.5 of Wald (1984). The discussion of the Kerr black hole is based on Sections 33.1–5 of Misner, Thorne, and Wheeler, and Sections 57 and 58 of Chandrasekhar. The definitions for black-hole region and event horizon are taken from Section 12.1 of Wald (1984); trapped surfaces and apparent horizons are defined in Wald's Section 9.5 and 12.2, respectively. Penrose's theorem on the structure of the event horizon (Section 5.4.1) is very nicely discussed in Section 34.4 of Misner, Thorne, and Wheeler. Section 9.5 of Wald (1984) provides a thorough discussion of the singularity theorems. The general properties of stationary black holes (Section 5.4.2) are discussed in Section 12.3 of Wald (1984) and Section 6.3.1 of Carter. An overview of the uniqueness theorems of black-hole spacetimes (Section 5.4.3) can be found in Section 12.3 of Wald (1984) and Section 6.7 of Carter. In Section 5.5 the derivations of the zeroth and first laws are taken from Wald's 1992 Erice lectures. The generalized Smarr formula is derived in Section 6.6.1 of Carter. The discussion of the second law is adapted from Section 6.1.2 of Carter. The final form of the third law was given in Israel (1986b); my discussion is based on Sullivan and Israel. Finally, in the problems below, the material on the Majumdar–Papapetrou solution is taken from Section 113 of Chandrasekhar, the description of null-dust collapse is adapted from Israel (1986a), and the alternative derivation of the zeroth law is based on Bardeen, Carter, and Hawking.

Suggestions for further reading:

There is a lot more than can be said on black holes. The book by Frolov and Novikov is a genuine encyclopedia that is well worth consulting. The book by Thorne, Price, and Macdonald focuses on astrophysical aspects of black-hole physics, and presents them in an interesting package known as the 'membrane paradigm.' The book by Birrell and Davies gives a complete account of the quantum production of particles by black holes (Hawking evaporation). The uniqueness theorems of black-hole spacetimes are presented by Heusler in his 2003 book.

Since their original formulation by Bardeen, Carter, and Hawking, the laws of black-hole mechanics have been reformulated in terms of the apparent horizon (instead of the event horizon). See the recent papers by Ashtekar, Beetle, Krishnan, and Lewandowski (2001 and 2002) for an account of this interesting development.

The search for a statistical understanding of black-hole entropy continues. For a survey of various possibilities you might consult the articles by Jacobson (1999), Sorkin (1998), and Peet (1998).

5.7 Problems

1. The metric of an extreme ($Q = \pm M$) Reissner–Nordström black hole is given by

$$ds^2 = -\left(1 - \frac{M}{r}\right)^2 dt^2 + \left(1 - \frac{M}{r}\right)^{-2} dr^2 + r^2 d\Omega^2.$$

   (a) Find an appropriate set of Kruskal coordinates for this spacetime.
   (b) Show that the region $r \leq M$ does not contain trapped surfaces.
   (c) Sketch a Penrose–Carter diagram for this spacetime.
   (d) Find a coordinate transformation that brings the metric to the form

$$ds^2 = -\left(1 + \frac{M}{\bar{r}}\right)^{-2} dt^2 + \left(1 + \frac{M}{\bar{r}}\right)^2 (dx^2 + dy^2 + dz^2),$$

   where $\bar{r}^2 = x^2 + y^2 + z^2$. Show that in these coordinates, the electromagnetic field tensor can be generated from the vector potential $A_\alpha dx^\alpha = \mp(1 + M/\bar{r})^{-1} dt$, in which the upper (lower) sign gives rise to a positive (negative) electric charge.

(e) Show that the metric

$$ds^2 = -\Phi^{-2}\,dt^2 + \Phi^2\left(dx^2 + dy^2 + dz^2\right)$$

and the vector potential $A_\alpha\,dx^\alpha = \mp\Phi^{-1}\,dt$ produce an exact solution to the Einstein–Maxwell equations provided that $\Phi(x)$ satisfies Laplace's equation $\nabla^2\Phi = 0$. Here, ∇^2 is the usual Laplacian operator of three-dimensional flat space and $x = (x, y, z)$. This metric is known as the *Majumdar–Papapetrou* solution. Prove that if the spacetime is asymptotically flat, then the total charge Q and the ADM mass M are related by $Q = \pm M$. Finally, find an expression for $\Phi(x)$ that corresponds to a collection of N black holes situated at arbitrary positions x_n $(n = 1, 2, \ldots, N)$.

2. A black hole is formed by the gravitational collapse of null dust. During the collapse the metric is given by an ingoing Vaidya solution with mass function $m(v) = v/16$. Spacetime is assumed to be flat before the collapse ($v < 0$), and after the collapse ($v > v_0$) the metric is given by a Schwarzschild solution with mass $m_0 = m(v_0) = v_0/16$. We want to study various properties of this spacetime.

(a) Show that in the interval $0 < v < v_0$, outgoing light rays are described by the parametric equations

$$r(\lambda) = c\,\lambda\,e^{-\lambda}, \qquad v(\lambda) = 4c\,(1 + \lambda)\,e^{-\lambda},$$

where c is a constant. Show that $v = 4r$ also describes an outgoing light ray. Plot a few of these curves in the (v, r) plane, using both positive and negative values of c. Plot also the position of the apparent horizon.

(b) Find the parametric equations that describe the event horizon.

(c) Prove that the curvature singularity at $r = 0$ is *naked*, in the sense that it is visible to observers at large distances. Prove also that at the moment it is visible, the singularity is *massless*. [It is generally true that the central singularity of a spherical collapse must be massless if it is naked. This was established by Lake (1992).]

3. In this problem we have a closer look at the instability of black-hole tunnels, a topic that was mentioned briefly in Sections 5.2.3 and 5.3.9. We will see that the instability is caused by the pathological behaviour of the ingoing branch ($v = \infty$) of the inner horizon ($r = r_-$). For reasons that will become clear, we shall call this the *Cauchy horizon* of the black-hole spacetime. For simplicity we shall restrict our attention to the Reissner–Nordström (RN) spacetime. [The physics of the Cauchy-horizon instability was this author's Ph.D. topic; see

Poisson and Israel (1990). The book by Burko and Ori (1997) presents a rather complete review of this fascinating part of black-hole physics.]

(a)   Consider an event P located anywhere in the future of the Cauchy horizon. Argue that the conditions at P are not uniquely determined by initial data placed on a spacelike hypersurface Σ located outside the black hole. Then argue that the Cauchy horizon is the boundary of the region of spacetime for which the evolution of this data is unique. (This region is called the *domain of dependence* of Σ, and we say that the *Cauchy problem* of general relativity is well posed in this region. The Cauchy horizon is the place at which the evolution ceases to be uniquely determined by the initial data; the Cauchy problem breaks down. In effect, the predictive power of the theory is lost at the Cauchy horizon.)

(b)   Consider a test null fluid with stress-energy tensor $T^{\alpha\beta} = \rho\, l^\alpha l^\beta$, where ρ is the energy density and $l_\alpha = -\partial_\alpha v$ the four-velocity. The fluid moves parallel to the Cauchy horizon, along ingoing null geodesics. Prove that ρ must be of the form

$$\rho = \frac{L(v)}{4\pi r^2},$$

where $L(v)$ is an arbitrary function of advanced time v. Show that if a finite quantity of energy is to enter the black hole, then $L \to 0$ as $v \to \infty$. (How fast must L vanish?) Typically, radiative fields outside black holes decay in time according to an inverse power law (Price 1972). We shall therefore take $L(v) \sim v^{-p}$ as $v \to \infty$, with p larger than, say, 2.

(c)   Consider now a free-falling observer inside the black hole. This observer moves in the outward radial direction, encounters the null dust, and measures its energy density to be $T_{\alpha\beta}\, u^\alpha u^\beta$, where u^α is the observer's four-velocity. Show that as the observer crosses the Cauchy horizon,

$$T_{\alpha\beta}\, u^\alpha u^\beta = \frac{\tilde{E}^2}{4\pi r_-^{\,2}}\, L(v)\, e^{2\kappa_- v},$$

where $\tilde{E} = -u_\alpha t^\alpha$ and κ_- was defined in Section 5.2.2. Conclude that the measured energy density diverges at the Cauchy horizon, even though the total amount of energy entering the hole is finite. This is the pathology of the Cauchy horizon, which ultimately is responsible for the instability of black-hole tunnels.

4.   The equations governing geodesic motion in the Kerr spacetime were given without justification in Section 5.3.6. Here we provide a derivation, which is

valid both for timelike and null geodesics. [The general form of the geodesic equations can be found in Section 33.5 of Misner, Thorne, and Wheeler (1973).]

(a) By definition, a Killing *tensor* field $\xi_{\alpha\beta}$ is one which satisfies the equation $\xi_{(\alpha\beta;\gamma)} = 0$. Show that if $\xi_{\alpha\beta}$ is a Killing tensor and u^α satisfies the geodesic equation ($u^\alpha{}_{;\beta}u^\beta = 0$), then $\xi_{\alpha\beta}u^\alpha u^\beta$ is a constant of the motion.

(b) Verify that

$$\xi_{\alpha\beta} = \Delta\, k_{(\alpha}l_{\beta)} + r^2 g_{\alpha\beta}$$

is a Killing tensor of the Kerr spacetime. Here, k^α and l^α are the null vectors defined in Section 5.3.6.

(c) Write the relations $\tilde{E} = -t^\alpha u_\alpha$ and $\tilde{L} = \phi^\alpha u_\alpha$ explicitly in terms of $u^\alpha = (\dot{t}, \dot{r}, \dot{\theta}, \dot{\phi})$. Then invert these relations to obtain the equations for $\dot{t}$ and $\dot{\phi}$. [Hint: Make sure to involve the inverse metric.]

(d) The Carter constant $\mathcal{Q}$ is defined by

$$\xi_{\alpha\beta}u^\alpha u^\beta = \mathcal{Q} + \left(\tilde{L} - a\tilde{E}\right)^2.$$

By working out the left-hand side, derive the equation for $\dot{r}$. [Hint: Express k_α and l_α in terms of the Killing vectors, and then $\xi_{\alpha\beta}u^\alpha u^\beta$ in terms of $\tilde{E}$, $\tilde{L}$, and $\dot{r}^2$.]

(e) Finally, use the normalization condition $g_{\alpha\beta}u^\alpha u^\beta = -\zeta$ (where $\zeta = 1$ for timelike geodesics and $\zeta = 0$ for null geodesics) to obtain the equation for $\dot{\theta}$.

5. Let l^α be a null, geodesic vector field in flat spacetime. With this vector and an arbitrary scalar function H we construct a new metric tensor $g_{\alpha\beta}$:

$$g_{\alpha\beta} = \eta_{\alpha\beta} + H l_\alpha l_\beta,$$

where $\eta_{\alpha\beta}$ is the Minkowski metric and $l_\alpha = \eta_{\alpha\beta}l^\beta$. Such a metric is called a Kerr–Schild metric.

(a) Show that l^α is null with respect to both metrics.

(b) Show that $g^{\alpha\beta} = \eta^{\alpha\beta} - H l^\alpha l^\beta$ is the inverse metric.

(c) Prove that $l_\alpha = g_{\alpha\beta}l^\beta$ and $l^\alpha = g^{\alpha\beta}l_\beta$. Thus, indices on the null vector can be lowered and raised with either metric.

(d) Calculate the Christoffel symbols for $g_{\alpha\beta}$. Show that they satisfy the relations

$$l_\mu \Gamma^\mu{}_{\alpha\beta} = -\frac{1}{2}\dot{H}l_\alpha l_\beta, \qquad l^\mu \Gamma^\alpha{}_{\mu\beta} = \frac{1}{2}\dot{H}l^\alpha l_\beta,$$

where $\dot{H} \equiv H_{,\mu}l^\mu$.

(e) Prove that $l^\alpha_{;\beta} l^\beta = 0$. Thus, l^α is a geodesic vector field in both metrics.

(f) Prove that the component $R_{\alpha\beta} l^\alpha l^\beta$ of the Ricci tensor vanishes for any choice of function H.

6. Complete the discussion of the zeroth law by proving that $\kappa_{,\alpha} e^\alpha_A$ is constant along the null generators of a stationary event horizon.

7. In this problem we provide an alternative derivation of the zeroth law of black-hole mechanics. This derivation is based on the original presentation by Bardeen, Carter, and Hawking (1973); it uses the Einstein field equations and the dominant energy condition.

(a) We have seen that the vector ξ^α is tangent to the null generators of the event horizon. It possesses the following properties: (i) ξ^α is null on the horizon; (ii) $\xi^\alpha_{;\beta}\xi^\beta = \kappa \xi^\alpha$ on the horizon; (iii) ξ^α is a Killing vector; and (vi) the congruence of null generators has zero expansion, shear, and rotation. Use these facts to infer

$$\xi_{\alpha;\beta} = \left(\kappa N_\alpha + c^A e_{A\alpha}\right)\xi_\beta - \xi_\alpha\left(\kappa N_\beta + c^B e_{B\beta}\right),$$

where $c^A \equiv \sigma^{AB}\xi_{\alpha;\beta}N^\alpha e^\beta_B$. This relation holds on the horizon only.

(b) Prove that the gradient of the surface gravity (in the directions tangent to the horizon) is given by

$$\kappa_{,\alpha} = -R_{\alpha\beta\gamma\delta}\xi^\beta N^\gamma \xi^\delta - \left(\sigma_{ABC}c^A c^B\right)\xi_\alpha.$$

This immediately implies that κ is constant on each generator: $\kappa_{,\alpha}\xi^\alpha = 0$.

(c) Show that the result of part (b) also implies

$$\kappa_{,\alpha}e^\alpha_A = -R_{\alpha\beta}e^\alpha_A \xi^\beta - \sigma^{BC}R_{\alpha\beta\gamma\delta}e^\alpha_A e^\beta_B e^\gamma_C \xi^\delta.$$

(d) The quantities $B_{AB} \equiv \xi_{\alpha;\beta}e^\alpha_A e^\beta_B$ and their tangential derivatives must all vanish on the horizon. Use this observation to derive

$$R_{\alpha\beta\gamma\delta}e^\alpha_A e^\beta_B e^\gamma_C \xi^\delta = 0.$$

This relations holds on the horizon only.

(e) Collecting the results of parts (c) and (d), use the Einstein field equations to write

$$\kappa_{,\alpha}e^\alpha_A = 8\pi j_\alpha e^\alpha_A,$$

where $j^\alpha = -T^\alpha_\beta \xi^\beta$ represents a flux of momentum across the horizon.

(f) The dominant energy condition states that j^α should be either timelike or null, and future directed. Use this, together with the stationary condition $T_{\alpha\beta}\xi^\alpha\xi^\beta = 0$, to prove that j^α must be parallel to ξ^α. Under these conditions, therefore,

$$\kappa_{,\alpha}e^\alpha_A = 0,$$

and the zeroth law is established.

8. The unique solution to the Einstein–Maxwell equations describing an isolated black hole of mass M, angular momentum $J \equiv aM$, and electric charge Q is known as the *Kerr–Newman solution*; it was discovered by Newman *et al.* in 1965. The Kerr–Newman metric can be expressed as

$$ds^2 = -\frac{\rho^2\Delta}{\Sigma}\,dt^2 + \frac{\Sigma}{\rho^2}\sin^2\theta\,(d\phi - \omega\,dt)^2 + \frac{\rho^2}{\Delta}\,dr^2 + \rho^2\,d\theta^2,$$

where $\rho^2 = r^2 + a^2\cos^2\theta$, $\Delta = r^2 - 2Mr + a^2 + Q^2$, $\Sigma = (r^2 + a^2)^2 - a^2\Delta\sin^2\theta$, and $\omega = a(r^2 + a^2 - \Delta)/\Sigma$. The metric comes with a vector potential

$$A_\alpha\,dx^\alpha = -\frac{Qr}{\rho^2}\left(dt - a\sin^2\theta\,d\phi\right).$$

When $Q = 0$, $A_\alpha = 0$ and this reduces to the Kerr solution.

(a) Find expressions for r_+, the radius of the event horizon, and Ω_H, the angular velocity of the black hole.

(b) Prove that the vector field

$$l^\alpha\,\partial_\alpha = \frac{r^2 + a^2}{\Delta}\,\partial_t - \partial_r + \frac{a}{\Delta}\,\partial_\phi$$

is tangent to a congruence of ingoing null geodesics. Prove also that

$$v \equiv t + \int \frac{r^2 + a^2}{\Delta}\,dr$$

and

$$\psi \equiv \phi + \int \frac{a}{\Delta}\,dr$$

are constant on each member of the congruence.

(c) Show that in the coordinates (v, r, θ, ψ), the Kerr–Newman metric takes the form

$$ds^2 = -\frac{\Delta - a^2\sin^2\theta}{\rho^2}\,dv^2 + 2\,dv\,dr - 2a\frac{r^2 + a^2 - \Delta}{\rho^2}\sin^2\theta\,dv\,d\psi$$

$$- 2a\sin^2\theta\,dr\,d\psi + \frac{\Sigma}{\rho^2}\sin^2\theta\,d\psi^2 + \rho^2\,d\theta^2.$$

Find an expression for A_α is this coordinate system.

(d)   Show that the vectors

$$\xi^\alpha \, \partial_\alpha = \partial_v + \Omega_H \, \partial_\psi, \quad e_\theta^\alpha \, \partial_\alpha = \partial_\theta, \quad e_\psi^\alpha \, \partial_\alpha = \partial_\psi,$$

and

$$N^\alpha \, \partial_\alpha = -\frac{a^2 \sin^2 \theta}{2(r_+^2 + a^2 \cos^2 \theta)} \, \partial_v - \frac{r_+^2 + a^2}{r_+^2 + a^2 \cos^2 \theta} \, \partial_r$$
$$- \frac{a}{2(r_+^2 + a^2)} \frac{2r_+^2 + a^2(1 + \cos^2 \theta)}{r_+^2 + a^2 \cos^2 \theta} \, \partial_\psi$$

form a good basis on the event horizon. In particular, prove that they give rise to the completeness relations $g^{\alpha\beta} = -\xi^\alpha N^\beta - N^\alpha \xi^\beta + \sigma^{AB} e_A^\alpha e_B^\beta$, where σ^{AB} is the inverse of $\sigma_{AB} \equiv g_{\alpha\beta} e_A^\alpha e_B^\beta$.

(e)   Prove that the surface gravity of a Kerr–Newman black hole is given by

$$\kappa = \frac{r_+ - M}{r_+^2 + a^2}.$$

Prove also that the hole's surface area is

$$A = 4\pi \left(r_+^2 + a^2 \right).$$

(f)   Compute the black-hole mass M_H and the black-hole angular momentum J_H of a Kerr–Newman black hole. (These quantities are defined in Section 5.5.3.) Make sure that your results are compatible with the following expressions:

$$M_H = \frac{r_+^2 + a^2}{2r_+} \left[1 - \frac{Q^2}{ar_+} \arctan(a/r_+) \right]$$

and

$$J_H = a \frac{r_+^2 + a^2}{2r_+} \left\{ 1 + \frac{Q^2}{2a^2} \left[1 - \frac{r_+^2 + a^2}{ar_+} \arctan(a/r_+) \right] \right\}.$$

Verify that these expressions satisfy the generalized Smarr formula.

(g)   Derive the following alternative version of Smarr's formula:

$$M = 2\Omega_H J + \frac{\kappa A}{4\pi} + \Phi_H Q,$$

where

$$\Phi_H \equiv -A_\alpha \xi^\alpha \Big|_{r=r_+} = \frac{r_+ Q}{r_+^2 + a^2}$$

is the electrostatic potential at the horizon.

(h) Consider a quasi-static process during which a stationary black hole of mass M, angular momentum J, and electric charge Q is taken to a new stationary black hole with parameters $M + \delta M$, $J + \delta J$, and $Q + \delta Q$. Prove that during such a transformation, the hole's surface area A will change by an amount δA given by

$$\delta M = \frac{\kappa}{8\pi} \delta A + \Omega_H \, \delta J + \Phi_H \, \delta Q.$$

This is the first law of black-hole mechanics for charged, rotating black holes.

9. Consider a quasi-static process during which the surface area of a black hole changes. (By quasi-static we mean that dA/dv is very small.) Derive the *Hawking–Hartle formula*,

$$\frac{dA}{dv} = \frac{8\pi}{\kappa} \oint_{\mathcal{H}(v)} \left(\frac{1}{8\pi} \sigma^{\alpha\beta} \sigma_{\alpha\beta} + T_{\alpha\beta} \xi^\alpha \xi^\beta \right) dS,$$

in which ξ^α is tangent to the null generators of the event horizon and $\sigma_{\alpha\beta}$ is their shear tensor. The second term within the integral represents the effect of accreting matter on the surface area. The first term represents the effect of gravitational radiation flowing across the horizon.

References

Arnowitt, R., Deser, S. and Misner, C. W. 1962. The dynamics of general relativity. In *Gravitation: An Introduction to Current Research,* ed. L. Witten, pp. 227–65. New York: Wiley. [Section 4.4]

Ashtekar, A., Beetle, C. and Lewandowski, J. 2002. Geometry of generic isolated horizons. *Class. Quantum Grav.*, **19**, 1195–225. [Section 5.6]

Ashtekar, A. and Krishnan, B. 2002. Dynamical horizons: Energy, angular momentum, fluxes and balance laws. *Phys. Rev. Lett.*, **89**, 261101 (4 pages). [Section 5.6]

Bardeen, J. M., Carter, B. and Hawking, S. W. 1973. The four laws of black hole mechanics. *Commun. Math. Phys.*, **31**, 161–70. [Section 5.6]

Barrabès, C. and Israel, W. 1991. Thin shells in general relativity and cosmology: The lightlike limit. *Phys. Rev.*, D **43**, 1129–42. [Section 3.12]

Barrabès, C. and Hogan, P. A. 1998. Lightlike signals in general relativity and cosmology. *Phys. Rev.*, D **58**, 044013 (9 pages). [Section 3.12]

Birkhoff, G. D. 1923. *Relativity and Modern Physics*. Cambridge: Harvard University Press. [Section 5.1.1]

Birrell, N. D. and Davies, P. C. W. 1982. *Quantum Fields in Curved Space*. Cambridge: Cambridge University Press. [Section 5.6]

Bondi, H., van der Burg, M. G. J. and Metzner, A. W. K. 1962. Gravitational waves in general relativity: VII. Waves from axi-symmetric isolated systems. *Proc. Roy. Soc. London*, **A269**, 21–52. [Section 4.4]

Brady, P. R., Louko, J. and Poisson, E. 1991. Stability of a shell around a black hole. *Phys. Rev.*, D **44**, 1891–4. [Section 3.13, Problem 7]

Brax, P. and van de Bruck, C. 2003. Cosmology and brane worlds: A review. *Class. Quantum Grav.*, **20**, R201–R232. [Section 3.12]

Brown, J. D. and York, J. W. 1993. Quasilocal energy and conserved charges derived from the gravitational action. *Phys. Rev.*, D **41**, 1407–19. [Section 4.4]

Brown, J. D., Lau, S. R. and York, J. W. 1997. Energy of isolated systems at retarded times as the null limit of quasilocal energy. *Phys. Rev.*, D **55**, 1977–84. [Section 4.4]

Burko, L. M. and Ori, A. 1997. Internal Structure of Black Holes and Spacetime Singularities. *An International Research Workshop*, Haifa, June 29–July 3, 1997. Bristol: Institute of Physics. [Section 5.7, Problem 3]

Carter, B. 1968. Global structure of the Kerr family of gravitational fields. *Phys. Rev.*, **174**, 1559–71. [Sections 5.1.5, 5.2.3, and 5.3.9]

1971. Axisymmetric black hole has only two degrees of freedom. *Phys. Rev. Lett.*, **26**, 331–3. [Section 5.4.3]

1979. The general theory of the mechanical, electromagnetic and thermodynamic properties of black holes. In *General Relativity: An Einstein Centenary Survey*, ed. S. W. Hawking and W. Israel, pp. 294–369. Cambridge: Cambridge University Press. [Sections 2.5, 4.4, and 5.6]

Chandrasekhar, S. 1983. *The Mathematical Theory of Black Holes*. New York: Oxford University Press. [Section 5.6]

1987. *Truth and Beauty: Aesthetics and Motivations in Science*. Chicago: Chicago University Press, p. 153. [Section 5.4.3]

Cook, G. 2000. Initial data for numerical relativity. *Living Rev. Relativity* **3**, 5. (Online article available at http://www.livingreviews.org/lrr-2000-5). [Section 3.12]

Corry, L., Renn, J. and Stachel, J. 1997. Belated decision in the Hilbert–Einstein priority dispute. *Science*, **278**, 1270–73. [Section 4.4]

Darmois, G. 1927. *Les équations de la gravitation einsteinienne*. Ch. V, Mémorial de Sciences Mathematiques, Fascicule XXV. Paris: Gauthier-Villars. [Section 3.7.6]

de la Cruz, V. and Israel, W. 1968. Spinning shell as a source of the Kerr metric. *Phys. Rev.*, **170**, 1187–92. [Section 3.12]

d'Inverno, R. 1992. *Introducing Einstein's Relativity*. Oxford: Clarendon Press. [Section 1.12]

Eddington, A. S. 1924. A comparison of Whitehead's and Einstein's formulas. *Nature*, **113**, 192. [Section 5.1.3]

Finkelstein, D. 1958. Past-future asymmetry of the gravitational field of a point particle. *Phys. Rev.* **110**, 965–7. [Section 5.1.3]

Friedmann, A. 1922. Über die Krümmung des Raumes. *Z. Phys.* **10**, 377–86. [Section 2.6, Problem 2; Sections 3.6.2 and 3.8]

Frolov, V. P. and Novikov, I. D. 1998. *Black Hole Physics: Basic Concepts and New Developments*. Dordrecht: Kluwer. [Section 5.6]

Gullstrand, A. 1922. Allegemeine Lösung des statischen Einkörper-problems in der Einsteinschen Gravitations Theorie. *Arkiv. Mat. Astron. Fys.*, **16**(8), 1–15. [Section 3.13, Problem 1; Section 5.1.4]

Hartle, J. B. 2003. *Gravity: An Introduction to Einstein's General Relativity*. San Francisco: Addison-Wesley. [Section 1.12]

Hawking, S. W. 1971. Gravitational radiation from colliding black holes. *Phys. Rev. Lett.*, **26**, 1344–6. [Section 5.5.5]

1972. Black holes in general relativity. *Commun. Math. Phys.*, **25**, 152–66. [Section 5.4.2]

1975. Particle creation by black holes. *Commun. Math. Phys.*, **43**, 199–220. [Section 5.5.7]

Hawking, S. W. and Ellis, G. F. R. 1973. *The Large Scale Structure of Space-time*. Cambridge: Cambridge University Press. [Section 1.12]

Hawking, S. W. and Hartle, J. B. 1972. Energy and angular momentum flow into a black hole. *Commun. Math. Phys.*, **27**, 283–90. [Section 5.5.7; Section 5.7, Problem 9]

Hawking, S. W. and Horowitz, G. T. 1996. The gravitational Hamiltonian, action, entropy, and surface terms. *Class. Quantum Grav.*, **13**, 1487–98. [Section 4.4]

Hawking, S. W. and Penrose, R. 1970. The singularities of gravitational collapse and cosmology. *Proc. Roy. Soc. Lond.*, **A314**, 529–48. [Section 5.4.1]

Hayward, S. A. 1994. General laws of black-hole dynamics. *Phys. Rev.*, D **49**, 6467–74. [Section 5.6]

Heusler, M. 2003. *Black Hole Uniqueness Theorems*. Cambridge: Cambridge University Press. [Section 5.6]

Israel, W. 1966. Singular hypersurfaces and thin shells in general relativity. *Nuovo cimento*, **44B**, 1–14; erratum **48B** (1967), 463. [Section 3.7.6]

1967. Event horizons in static vacuum spacetimes. *Phys. Rev.*, **164**, 1776–9. [Section 5.4.3]

1968. Event horizons in static electrovac spacetimes. *Commun. Math. Phys.*, **8**, 245–60. [Section 5.4.3]

1986a. The formation of black holes in nonspherical collapse and cosmic censorship. *Can. J. Phys.*, **64**, 120–27. [Section 5.6]

1986b. Third law of black-hole dynamics: A formulation and proof. *Phys. Rev. Lett.*, **57**, 397–9. [Section 5.5.6]

Jacobson, T. 1999. On the nature of black hole entropy. In *General Relativity and Relativistic Astrophysics: Eighth Canadian Conference, AIP Conference Proceedings 493*, ed. C. P. Burgess and R. C. Myers, pp. 85–97. American Institute of Physics. [Section 5.6]

Katz, J., Lynden-Bell, D. and Israel, W. 1988. Quasilocal energy in static gravitational fields. *Class. Quantum Grav.*, **5**, 971–87. [Section 4.4]

Kerr, R. P. 1963. Gravitational field of a spinning mass as an example of algebraically special metrics. *Phys. Rev. Lett.*, **11**, 237–8. [Section 5.3.1]

Kerr, R. P. and Schild, A. 1965. A new class of vacuum solutions of the Einstein field equations. In *Proceedings of the Galileo Galilei Centenary Meeting on General Relativity: Problems of Energy and Gravitational Waves*, ed. G. Barbera, pp. 222–3. Florence: Comitato nazionale per le manifestazione celebrative. [Section 5.3.7; Section 5.7, Problem 5]

Kraus, P., Larsen, F. and Siebelink, R. 1999. The gravitational action in asymptotically AdS and flat spacetimes. *Nucl. Phys.* **B563**, 259–78. [Section 4.4]

Kruskal, M. D. 1960. Maximal extension of Schwarzschild metric. *Phys. Rev.*, **119**, 1743–5. [Sections 5.1.2, 5.2.2, and 5.3.9]

Lake, K. 1992. Precursory singularities in spherical gravitational collapse. *Phys. Rev. Lett.*, **68**, 3129–32. [Section 5.7, Problem 2]

Lanczos, C. 1922. Bemerkung zur de Sitterschen Welt. *Phys. Zeits.*, **23**, 539–43. [Section 3.7.6]

1924. Flöchenhafte Verteilung der Materie in der Einsteinschen Gravitationstheorie. *Ann. Phys.* (Germany), **74**, 518–40. [Section 3.7.6]

Lau, S. R. 1999. Light-cone reference for total gravitational energy. *Phys. Rev.*, D **60**, 104034 (4 pages). [Section 4.4]

Lehner, L. 2001. Numerical relativity: A review. *Class. Quantum Grav.*, **18**, R25–R86. [Section 4.4]

Majumdar, S. D. 1947. A class of exact solutions of Einstein's field equations. *Phys. Rev.*, **72**, 390–98. [Section 5.7, Problem 1]

Manasse, F. K. and Misner, C. W. 1963. Fermi normal coordinates and some basic concepts in differential geometry. *J. Math. Phys.*, **4**, 735–45. [Section 1.12]

Mann, R. B. 1999. Misner string entropy. *Phys. Rev.*, D **60**, 104047 (5 pages). [Section 4.4]

2000. Entropy of rotating Misner string spacetimes. *Phys. Rev.*, D **61**, 084013 (6 pages). [Section 4.4]

Martel, K. and Poisson, E. 2001. Regular coordinate systems for Schwarzschild and other spherical spacetimes. *Am. J. Phys.*, **69**, 476–80. [Section 3.13, Problem 1]

Mazur, P. O. 1982. Proof of uniqueness of the Kerr–Newman black hole solution. *J. Phys.*, A **15**, 3173–80. [Section 5.4.3]

Misner, C. W. 1960. Wormhole initial conditions. *Phys. Rev.*, **118**, 1110–11. [Section 3.6.7]

Misner, C. W., Thorne, K. S. and Wheeler, J. A. 1973. *Gravitation*. New York: Freeman. [Preface; Sections 1.12, 3.12, and 5.6; Section 5.7, Problem 4]

Morris, M. S. and Thorne, K. S. 1988. Wormholes in spacetime and their use for interstellar travel: A tool for teaching general relativity. *Am. J. Phys.*, **56**, 395–412. [Section 2.5]

Musgrave, P. and Lake, K. 1997. Junctions and thin shells in general relativity using computer algebra: II. The null formalism. *Class. Quantum Grav.*, **14**, 1285–94. [Section 3.12]

Nakahara, M. 1990. *Geometry, Topology, and Physics*. Bristol: A. Hilger. [Section 1.12]

Newman, E. T., Couch, E., Chinnapared, K., Exton, A., Prakash, A. and Torrence, R. 1965. Metric of a rotating charged mass. *J. Math. Phys.*, **6**, 918–19. [Section 5.7, Problem 8]

Nordström, G. 1918. On the energy of the gravitational field in Einstein's theory. *Proc. Kon. Ned. Akad. Wet.*, **20**, 1238–45. [Section 5.2.1]

Oppenheimer, J. R. and Snyder, H. 1939. On continued gravitational contraction. *Phys. Rev.*, **56**, 455–9. [Section 3.8]

Painlevé, P. 1921. La mécanique classique et la théorie de la relativité. *C. R. Acad. Sci. (Paris)*, **173**, 677–80. [Section 3.13, Problem 1; Section 5.1.4]

Papapetrou, A. 1947. A static solution of the equations of the gravitational field for an arbitrary charge distribution. *Proc. Roy. Irish Acad.*, **51**, 191–205. [Section 5.7, Problem 1]

Peet, A. W. 1998. The Bekenstein formula and string theory (N-brane theory). *Class. Quant. Grav.*, **15**, 3291–338. [Section 5.6]

Penrose, R. 1965. Gravitational collapse and spacetime singularities. *Phys. Rev. Lett.*, **14**, 57–9. [Section 5.4.1]

    1968. Structure of spacetime. In *Battelle Rencontres 1967*, ed. C. M. DeWitt and J. A. Wheeler. New York: Benjamin. [Section 5.4.1]

Poisson, E. and Israel, W. 1990. The internal structure of black holes. *Phys. Rev.*, D **41**, 1796–809. [Section 5.7, Problem 3]

Price, R. H. 1972. Nonspherical perturbations of relativistic gravitational collapse: I. Scalar and gravitational perturbations. *Phys. Rev.*, D **5**, 2419–38. [Section 5.4.3]

    1972a. Nonspherical perturbations of relativistic gravitational collapse: II. Integer-spin, zero-rest-mass fields. *Phys Rev.*, D **5**, 2439–54. [Section 5.4.3]

Rácz, I. and Wald, R. M. 1996. Global extensions of spacetimes describing asymptotic final states of black holes. *Class. Quantum Grav.*, **13**, 539–52. [Section 5.5.2]

Randall, L. and Sundrum, R. 1999a. Large mass hierarchy from a small extra dimension. *Phys. Rev. Lett.*, **83**, 3370–73. [Section 3.12]

Randall, L. and Sundrum, R. 1999b. An alternative to compactification. *Phys. Rev. Lett.*, **83**, 4690–93. [Section 3.12]

Reissner, H. 1916. Über die Eigengravitation des elektrischen Feldes nach des Einsteinschen Theorie. *Ann. Phys. (Germany)*, **50**, 106–20. [Section 5.2.1]

Robertson, H. P. 1935. Kinematics and world structure. *Astrophys. J.*, **82**, 248–301. [Section 2.6, Problem 2; Sections 3.6.2 and 3.8]

    1936. Kinematics and world structure. *Astrophys. J.*, **83**, 187–201. [Section 2.6, Problem 2; Sections 3.6.2 and 3.8]

Robinson, D. C. 1975. Uniqueness of the Kerr black hole. *Phys. Rev. Lett.*, **34**, 905–6. [Section 5.4.3]

Sachs, R. K. 1962. Gravitational waves in general relativity: VIII. Waves in asymptotically fiat space-time. *Proc. Roy. Soc. London*, **A270**, 103–26. [Section 4.4]

Schutz, B. F. 1980. *Geometrical Methods of Mathematical Physics*. Cambridge: Cambridge University Press. [Section 1.12]

1985. *A First Course in General Relativity.* Cambridge: Cambridge University Press. [Preface]

Schwarzschild, K. 1916. Über das Gravitationsfeld eines Massenpunktes nach der Einsteinschen Theorie. *Sitzber. Deut. Akad. Wiss. Berlin, Kl. Math.-Phys. Tech.*, 189–96. (For an English translation, see http://xxx.lanl.gov/abs/physics/9905030. [Section 5.1.1]

Smarr, L. 1973. Mass formula for Kerr black holes. *Phys. Rev. Lett.*, **30**, 71–3. [Section 5.3.12]

Sorkin, R. D. 1998. The statistical mechanics of black hole theormodynamics. In *Black Holes and Relativistic Stars*, ed. R. M. Wald, pp. 117–94. Chicago: University of Chicago Press, [Section 5.6]

Sudarsky, D. and Wald, R. M. 1992. Extrema of mass, stationarity, and staticity, and solutions to the Einstein–Yang–Mills equations. *Phys. Rev.*, D **46**, 1453–74. [Section 4.4]

Sullivan, B. T. and Israel, W. 1980. The third law of black hole mechanics: What is it? *Phys. Lett.*, **79A**, 371–2. [Section 5.6]

Synge, J. L. 1960. *Relativity: The General Theory*. Amsterdam: North-Holland. [Section 1.12]

Szekeres, G. 1960. On the singularities of a Riemannian manifold. *Publ. Mat. Debrecen*, **7**, 285–301. [Section 5.1.2]

Thirring, H. and Lense, J. 1918. Über den Einfluss der Eigenrotation der Zentralkörper auf die Bewegung der Planeten und Monde nach des Einsteinschen Gravitationstheorie. *Phys. Z.*, **19**, 156–63. [Section 3.10]

Thorne, K. S., Price, R. H. and Macdonald, D. A. 1986. *Black Holes: The Membrane Paradigm*. New Haven: Yale University Press. [Section 5.6]

Vaidya, P. C. 1953. 'Newtonian time' in general relativity. Nature, **171**, 260–61. [Sections 4.3.5, 5.1.8, and 5.5.6]

Visser, M. 1995. *Lorentzian Wormholes: From Einstein to Hawking*. Woodbury: American Institute of Physics. [Section 2.5]

Wainwright, J. and Ellis, G. F. R. 1997. *Dynamical Systems in Cosmology*. Cambridge: Cambridge University Press. [Section 2.5]

Wald, R. M. 1984. *General Relativity*. Chicago: University of Chicago Press. [Preface; Sections 1.12, 2.5, 3.12, 4.4, and 5.6]

1992. Thermodynamics and black holes. In *Black Hole Physics: Proceedings of the NATO Advanced Study Institute (12th Course of the International School of Cosmology and Gravitation of the Ettore Majorana Centre for Scientific Culture)*, Erice, Italy, May 12–22, 1991, ed. V. de Sabbata and Z. Zhang, pp. 55–97. Dordrecht: Kluwer Academic. [Section 5.6]

Walker, A. G. 1936. On Milne's theory of world-structure. *Proc. London Math. Soc.*, **42**, 90–127. [Section 2.6, Problem 2; Sections 3.6.2 and 3.8]

Weinberg, S. 1972. *Gravitation and Cosmology*. New York: Wiley. [Section 1.12]

York, J. W. 1979. Kinematics and dynamics of general relativity. In *Sources of Gravitational Radiation*, ed. L. M. Smarr, pp. 83–126. Cambridge: Cambridge University Press. [Section 4.5, Problem 4]

Index

A

Acceleration, 17, 18, 92, 104, 105, 185
Accretion onto a black hole, ix, 109, 201, 223
Action functional
  for gravitation, x, 118, 121, 122, 124, 126, 127,
    136–9, 141, 145, 157, 158
  in general, 118–22, 125, 128, 129, 132, 138
ADM mass, *see* Mass, Arnowitt–Deser–Misner
Advanced time, 152, 161, 165, 173, 199, 204, 207,
  213, 214, 218
Affine parameter, 1, 7, 10, 16, 17, 20, 21, 45, 48,
  50–2, 56, 58, 61, 63, 66, 98, 104–6, 108, 110–13,
  171, 173, 186, 193, 194, 199, 204, 207
Angular momentum
  definition in terms of gravitational Hamiltonian, ix,
    x, 147, 156, 177
  definition in terms of Komar integral, 149, 151,
    177, 209
  flux vector, 155
  of a black hole, ix, 109, 163, 187, 190, 192, 200,
    201, 206, 209–11, 221–3
  of a particle, 10, 27, 155, 183, 188, 192
  of a rotating shell, 94, 109
  of an axisymmetric body, 148, 150, 151
  of an isolated body, ix, 118, 119, 146, 147, 205,
    210
  transfer across a hypersurface, 155, 156, 160–62
Angular velocity
  of a rotating black hole, 190, 191, 200, 204, 209,
    221
  of a rotating shell, 96, 97, 112
  of a rotating star, 82
  of an inertial frame, 94, 97
  of an observer, 27, 189, 190, 192
  of null generators, 110, 112, 191
  of ZAMOs, 188
Apparent horizon, 171–6, 178, 180, 190, 191, 203,
  214–7
Area theorem, *see* Black-hole mechanics, second law
Asymptotically-flat spacetime, x, 54, 118, 119, 126,
  127, 138, 146–8, 151, 157, 162, 180, 183, 187,
  198, 217

B

Basis vectors, 19, 22, 24, 29, 57, 66, 73, 75, 99, 101,
  103, 106–8, 110, 113, 130, 150, 207, 222
Bianchi identities, 15, 16, 127, 128
Bifurcation two-sphere, 176, 199, 200, 204, 208
Birkhoff's theorem, 163, 164
Black hole
  extreme, 178, 184, 190, 199, 200, 203, 213, 214,
    216
  general, viii, ix, 84, 151, 163, 201–4
  Kerr, 97, 109, 187–201, 205, 206, 208, 209, 211,
    213, 215, 218, 219, 221
  Kerr–Newman, 187, 221–3
  Reissner–Nordström, 176–187, 191, 194, 196–8,
    206, 213, 216, 217
  Schwarzschild, 55, 90, 109, 110, 163–176, 180,
    184, 205
  stationary, 204–6
Black-hole entropy, 214–6
Black-hole mechanics
  first law, viii, 192, 209, 211–2, 215, 216, 223
  second law, viii, 192, 201, 212–3, 215, 216
  third law, viii, 213–4, 216
  zeroth law, viii, 208–9, 214–6, 220, 221
Black-hole tunnel, 180, 181, 183, 184, 198, 217, 218
Bondi–Sachs mass, *see* Mass, Bondi–Sachs
Boundary, viii, 69, 90, 92, 117–9, 126, 132, 134, 136,
  141, 142, 148, 150, 151, 156, 163, 167, 169, 172,
  174, 196, 202, 203, 209, 218
Boundary terms, x, 118, 119, 121, 124, 137–42, 144,
  146, 156, 212
Boyer–Lindquist coordinates, 187, 192, 196
Brane worlds, 114

C

Canonical momentum, 118, 128, 129, 131, 133, 140
Carter constant, 192, 219
Casimir effect, 32

Cauchy horizon, 217, 218
Cauchy problem, 218
Causal structure, 163, 168, 170, 174, 194, 198, 202, 203
Caustic, 41, 51, 203, 212
Centrifugal force, 183
Chandrasekhar on the perfection of black holes, 206
Charge, 25, 176, 177, 184, 187, 206, 213, 216, 217, 221, 223
Christoffel symbols, 6, 7, 22, 23, 86, 87, 101, 122, 219
Cofactor, 65, 69, 131
Collapse, gravitational, *see* Gravitational collapse
Completeness relations, 29, 63, 67, 68, 78, 79, 99, 102, 107, 123, 134, 135, 207, 222
Conformal flatness, 84
Congruence
    of null geodesics, viii, 28, 45–54, 56–8, 105, 116, 163, 171–5, 186, 190, 192–4, 203, 205, 207, 212, 220, 221
    of timelike curves, 54, 129, 135
    of timelike geodesics, viii, 28, 36–45, 54, 55, 76, 85, 99, 100, 104, 114
Connection, 4–6, 8, 9, 11, 12, 22, 73, 74, 86
Conservation laws, 16, 71, 91, 116, 125, 153, 155, 185
Constraint equations, 79–84, 116, 145, 146
Coordinates
    Boyer–Lindquist, *see* Boyer–Lindquist coordinates
    Eddington–Finkelstein, *see* Eddington–Finkelstein coordinates
    Fermi normal, *see* Fermi normal coordinates
    Kerr–Schild, *see* Kerr–Schild coordinates
    Kruskal, *see* Kruskal coordinates
    Painlevé–Gullstrand, *see* Painlevé–Gullstrand coordinates
    Reimann normal, *see* Reimann normal coordinates
Cosmological constant, 117
Cosmology, ix, 41, 54, 80, 112–3
Covariant differentiation, x, 1, 4–6, 8, 9, 47, 73, 74, 76, 123, 139, 143, 159
Cross section, 42–5, 48, 51–3, 66, 105, 174, 200, 202, 207–9, 212
Curvature tensors
    Einstein, x, 2, 16, 78, 88, 89, 101, 142, 144, 152
    extrinsic, *see* Extrinsic curvature
    Ricci, x, 2, 16, 26, 31, 55, 78, 79, 82, 84, 88, 98, 115, 121, 137, 146, 207, 220
    Riemann, x, 2, 15–7, 19, 23, 24, 26, 59, 76–8, 86–9, 98, 101, 115, 116, 188, 204
    Weyl, x, 26, 55, 58, 98
Curve, 1–6, 8, 9, 17, 18, 21, 22, 36, 39, 46, 52, 53, 60, 62, 83, 111, 129, 130, 135, 160, 164, 166, 170, 171, 202, 217

D

de Sitter spacetime, 117
Delta function, *see* Dirac distribution
Density, *see* Mass density or Energy density
Derivative
    covariant, *see* Covariant differentiation
    Lie, *see* Lie differentiation
Determinant, 2, 12–4, 45, 53, 64–8, 71, 121, 122, 131, 135, 207

Deviation vector, viii, 16–8, 21, 23, 36, 37, 45–7
Deviation, geodesic, *see* Geodesic deviation equation
Dirac distribution, 85–9, 98, 101, 104, 113, 158
Divergence formula, 13, 70, 72, 133
Dragging of inertial frames, 97, 110, 188, 189
Dual vector, 1–3, 5, 9
Dust, 90, 91, 153, 155, 173, 175, 213, 216–8

E

Eddington–Finkelstein coordinates, 167, 168, 173
Einstein field equations, viii, ix, 2, 16, 26, 31, 40, 50, 59, 78, 80, 84, 86, 88, 90, 91, 101, 115, 117, 118, 122, 124–6, 145, 146, 151–4, 157, 158, 163, 164, 173, 177, 187, 201, 205, 207, 209, 220
Einstein tensor, *see* Curvature tensors, Einstein
Einstein–Hilbert priority dispute, 157
Einstein–Maxwell field equations, 176, 187, 217, 221
Electromagnetic field, 32, 152, 157, 176, 177, 214, 216, 222
Embedding, 25, 59, 67, 68, 76, 115, 127, 134, 135, 138, 147, 148, 153, 160
Energy
    binding, 93
    density, 29, 30, 32, 55, 113, 153, 156, 173, 214, 218
    electrostatic, 184
    flux vector, 155
    gravitational potential, 159
    internal, 214
    kinetic, 93, 159
    mass-energy, *see* Mass
    of a particle, 10, 27, 42, 92, 109, 155, 181, 191, 192
    rotational, 192
Energy conditions
    averaged, 32
    dominant, 32, 209, 220
    in general, ix, 28–33, 54, 173, 174, 214
    null, 31, 50, 174, 203–5, 212, 215
    strong, 31, 40, 41, 204
    violations, 32, 54
    weak, 30–1, 213, 214
Entropy, *see* Black-hole entropy
Equivalence principle, *see* Principle of equivalence
Ergosphere, 189, 191, 192
Euler–Lagrange equation, 6, 7, 120, 121, 128
Event horizon, viii, ix, 28, 163, 170–6, 178, 180, 185, 186, 189–92, 196, 197, 199, 200, 202–5, 207–9, 211–3, 215–7, 220–23
Expansion scalar, 28, 34, 35, 37, 40–3, 46, 48, 51, 52, 55, 56, 58, 76, 105, 113, 114, 116, 171–5, 190, 203, 212, 220
Extrinsic curvature, ix, 59, 75–7, 79–81, 83, 89, 92, 95, 102, 103, 114, 115, 118, 121, 124, 127, 134–6, 138–40, 145, 147–50, 153, 158, 160

F

Fermi normal coordinates, 2, 18–24, 26
Field equations

Einstein, *see* Einstein field equations
Einstein–Maxwell, *see* Einstein–Maxwell field equations
Klein–Gordon, *see* Klein–Gordon equation
Maxwell, *see* Maxwell field equations
Flat space, 25, 60, 70, 80, 82–4, 115, 138, 147, 149, 154, 159, 160, 168, 217
Flat spacetime, 17, 26, 32, 51, 56, 63, 65, 66, 83, 93, 95, 107, 115, 116, 126, 127, 138, 146, 153, 160, 165, 194, 195, 217, 219
Flatness
  asymptotic, *see* Asymptotically-flat spacetime
  conformal, *see* Conformal flatness
  local, *see* Local flatness theorem
Fluid, 30, 55, 90, 96, 97, 111–3, 117, 153, 155, 158, 173, 218
Focusing action of gravity, ix, 28, 109, 116, 175, 205
Focusing theorem
  for null geodesics, 28, 50–1, 175, 212
  for timelike geodesics, 28, 40–1, 43
Friedmann–Robertson–Walker spacetime, 54, 81, 90
Frobenius' theorem, 28, 38–40, 48–50, 186

G

Gauss' law, 177
Gauss' theorem, 69, 71, 120, 133, 209
Gauss–Codazzi equations, 76–9, 115
Gauss–Weingarten equation, 75
Generators
  of a null hypersurface, 49, 51, 61–3, 65, 66, 98, 104–13
  of an event horizon, viii, 174–6, 185, 186, 191, 199, 200, 203–5, 207, 208, 212, 220, 223
Geodesic deviation equation, 2, 16–8, 21, 23
Geodesic equation
  in affine parameter form, 7, 10, 17, 18, 20, 22, 25, 39, 52, 56, 92, 100, 155, 158, 173, 186, 192, 200, 204, 219
  in general form, 7, 25, 50, 61, 186, 204
Geodesic, defined, 6–8
Gravitational collapse
  in general, 167, 176, 200, 205, 209
  of a pressureless star, 90–2
  of a thin shell, 93, 107–9
  of null dust, 216, 217
Gravitational repulsion, 184
Gravitational waves, 152, 156, 164, 223
GRTensorII, x

H

Hamilton's equations, 128, 132, 133, 145–6
Hamiltonian
  density, 131, 132, 140
  formulation of general relativity, x, 118, 136–46, 156
  gravitational, ix, 118, 119, 136–47, 157
  in general, 118, 128, 129, 131, 132
Hawking evaporation, 206, 214–6
Hawking temperature, 214

Hawking–Hartle formula, 223
Hilbert action, 121, 122
Hilbert–Einstein priority dispute, *see* Einstein–Hilbert priority dispute
Horizon
  apparent, *see* Apparent horizon
  Cauchy, *see* Cauchy horizon
  event, *see* Event horizon
  Killing, *see* Killing horizon
  trapping, *see* Trapping horizon
Hypersurface orthogonal, 28, 38–40, 42, 43, 48–50, 52, 56, 57, 76, 82, 105, 129, 186, 190, 205
Hypersurface, defined, 60

I

Ideal gas, 159
Induced metric, ix, 59, 62–3, 68, 71, 73, 74, 76, 80, 82, 84, 86, 91, 94–6, 102, 103, 106, 108, 110, 113–5, 117, 118, 121, 122, 124, 126, 131, 134–6, 139, 145, 148, 153, 158, 160, 168, 200
Inertial frame, 94, 97, 110, 112
Infinity
  null, 152, 153, 161, 169, 170, 202, 206
  spatial, ix, 71, 97, 110, 119, 126, 147, 151, 153, 168–70, 185, 188, 191, 209
  timelike, 169, 170
Initial-value problem, ix, 79–84, 114, 145
Integration over a hypersurface, ix, 59, 64–9, 118, 150, 151, 155, 156, 163, 209–11

J

Jacobian, 12, 14, 36
Junction conditions, ix, 60, 84–90, 92

K

Kerr spacetime, 94, 109, 163, 187–201, 205, 206, 208, 211, 215, 218, 219, 221
Kerr–Newman spacetime, 187, 221–3
Kerr–Schild coordinates, 194–6
Kerr–Schild metric, 195, 219
Killing horizon, 175–6, 185, 191, 204
Killing tensor, 192, 219
Killing vector, x, 1, 10–11, 25–7, 82, 96, 149, 150, 176, 185–9, 191, 192, 204, 207, 208, 219, 220
Klein–Gordon equation, 121, 125, 133
Komar integral
  for angular momentum, *see* Angular momentum, definition in terms of Komar integral
  for mass, *see* Mass, definition in terms of Komar integral
Kruskal coordinates, 164–8, 171, 176, 178–80, 184, 186, 196, 197, 216
Kruskal diagram, 166–8, 170, 172, 176, 180

L

Lagrangian
  density, 118, 120–2, 125, 131, 133, 139, 157
  formulation of general relativity, x, 118, 121–8, 156
  in general, 6, 118–20, 124

Laplace's equation, 84, 217
Lapse function, 118, 119, 130, 131, 139, 145–7
Lense–Thirring effect, *see* Dragging of inertial frames
Levi-Civita tensor, 2, 13–5, 26, 64, 65
Lie differentiation, x, 1, 8–10, 17, 25, 44, 53, 76, 80, 103, 127, 130, 131, 139, 146, 158, 207
Lie transport, 9, 45, 75, 103, 160
Light cone, 51, 63–4, 66, 170
Local flatness theorem, 1, 11–2, 18, 19, 22
Local Lorentz frame, 2, 11, 12, 15, 37, 46, 123, 185
Lorentz transformation, 11

M

Mach's principle, 97
Majumdar–Papapetrou spacetime, 216, 217
Manifold, 1–3, 8, 12, 26, 59, 60, 68, 69, 76, 120, 126, 127, 138, 146, 166, 169, 178, 202
Mass
  -energy, *see* Energy
  Arnowitt–Deser–Misner, 147, 151–6, 158, 159, 177, 187, 217
  Bondi–Sachs, 151–6
  definition in terms of gravitational Hamiltonian, ix, x, 147, 156, 177
  definition in terms of Komar integral, 149, 177, 209
  density, 84, 90–3, 96, 97, 102, 104, 106, 109, 111–3, 116, 117, 155, 159, 164
  in spherical symmetry, 82, 115, 152, 158, 164, 173, 184, 217
  of a black hole, ix, 55, 109, 110, 163, 176, 177, 184, 187, 200, 201, 206, 209–11, 213, 221–3
  of a collapsing shell, 93, 94, 108, 109, 111
  of a particle, 10, 27, 42, 157, 181, 185
  of an isolated body, ix, 82, 91, 92, 116–9, 146, 148, 150, 164, 198, 210
  transfer across a hypersurface, 155, 156, 160, 161
Matrix, 11–4, 34, 35, 56–8, 65, 66
Matter distribution, 16, 26, 30, 59, 82, 88, 90, 93, 109, 121, 175, 205, 210, 211, 213
Maximal extension of a spacetime, 167, 176, 196–8
Maxwell field equations, 157, 176, 177
Membrane paradigm, 216
Metric determinant, *see* Determinant
Minkowski metric, 11, 19, 22, 23, 29, 56, 64, 95, 195, 219
Minkowski spacetime, *see* Flat spacetime
Moment of time symmetry, 81–4, 115
Momentum vector, 32, 191
Multipole moments, 206

N

Normal coordinates
  Fermi, *see*, Fermi normal coordinates
  Riemann, *see* Riemann normal coordinates
Normal vector, 38, 48, 60–1, 63, 65–9, 71, 75, 80–2, 85, 87, 92, 95, 99, 100, 103, 109, 114, 115, 123, 126, 129, 130, 134, 136, 147–9, 153, 160, 174, 185, 190
Numerical relativity, 84, 114, 146, 157

O

Oppenheimer–Snyder collapse, *see* Gravitational collapse of a pressureless star

P

Painlevé–Gullstrand coordinates, 168
Parallel transport, 1, 4, 6, 7, 19, 22, 24, 25, 37, 47, 57, 58
Penrose process, 191–2
Penrose's description of the event horizon, 203, 212, 215
Penrose–Carter diagram, 168–70, 180, 198, 216
Permutation symbol, 13
Perturbation of a black hole, 184, 201, 211, 212
Poisson's equation, 84
Pressure, 29, 55, 90, 91, 93, 96, 97, 102, 104–7, 109, 111–3, 116, 117, 153, 159, 173
Price's theorem, 206
Principal null congruences of the Kerr spacetime, 192–4
Principle of equivalence, 6, 11
Projection of a tensor, 24, 47, 56, 73, 74, 77, 96, 100, 101, 139
Proper distance, 7, 8, 19–21, 85, 115, 160, 185
Proper time, 7, 8, 19, 20, 24, 25, 36, 41, 43, 85, 91, 99, 100, 104, 116, 147, 158, 168, 181, 188, 192

R

Radiation, 112, 152–5, 160–62, 173, 185, 206, 213, 215, 218, 223
Raychaudhuri's equation, 28, 40–41, 50–51, 55, 57, 58, 105, 116, 205, 207, 212
Redshift, 185
Reissner–Nordström spacetime, 163, 176–87, 194, 196, 197, 217
Retarded time, 152, 153, 155, 160, 161, 165
Ricci tensor, *see* Curvature tensors, Ricci
Riemann normal coordinates, 26
Riemann tensor, *see* Curvature tensors, Riemann
Rigid rotation, 161
Rotation tensor, 28, 35–41, 48, 50, 51, 55–7, 205, 220
Rotation, gravitational effects, 97

S

Schwarzschild spacetime, 26, 42, 52, 55, 90, 93, 108, 114–7, 152, 158, 163–76, 205, 217
Shear tensor, 28, 34–7, 40, 41, 48, 50, 51, 55–8, 105, 205, 220, 223
Shift vector, 118, 119, 130, 139, 145–7
Singularity
  of a congruence, *see* Caustic
  of null-dust collapse, 217
  of the Boyer–Lindquist coordinates, 187, 192
  of the Eddington–Finkelstein coordinates, 182
  of the Kerr spacetime, 187, 190, 195–6, 198
  of the Kruskal coordinates, 179, 180, 197
  of the Reissner–Nordström spacetime, 178, 180
  of the Schwarzschild coordinates, 164, 165, 178

of the Schwarzschild spacetime, 169, 170
theorems, 28, 204, 215
Smarr's formula, 200, 209–11, 215, 222
Stokes' theorem, 69, 71, 150
Stress-energy tensor
decomposed in terms of density and principal
pressures, 29
in general, ix, 16, 29, 105, 113, 125, 155, 205, 207,
211, 213, 215
of a perfect fluid, 30, 96, 155, 158
of a point particle, 158
of a scalar field, 125, 161
of a surface layer, 87–90, 93, 95, 96, 98, 100–4,
107, 111, 113, 116
of an electromagnetic field, 157, 177
of null dust, 153, 173, 213, 214, 218
Surface area of a black hole, ix, 163, 200, 201, 209,
211–3, 222, 223
Surface element, ix, 59, 64–9, 72, 106, 133, 149, 207,
210, 211
Surface gravity of a black hole, 184–7, 198–200, 204,
208, 209, 213–5, 220, 222, 223
Surface integral, *see* Integration over hypersurfaces
Surface layer, ix, 59, 84–90, 93–114, 116

T

Tangent plane, 3
Tangent tensor field, 73, 88
Tangent vector, viii, 1–5, 7–10, 17, 19–21, 25, 36, 39,
41, 44–7, 49, 51–3, 55, 56, 61–3, 67, 68, 73–6,
95, 98–100, 103, 104, 106, 108–10, 112, 113,
115, 116, 129, 130, 134, 135, 153, 171, 173, 176,
185, 186, 191, 193, 194, 199, 204, 207, 220, 221,
223
Thermodynamics, 159, 206, 214, 215
Thin shell, *see* Surface layer

Three-plus-one decomposition, 118, 129–31, 136,
146, 157
Three-tensor, 1, 44, 45, 59, 62, 68, 73–6, 78–80, 82,
84, 88, 102, 103, 115, 130, 134, 142, 144, 158
Transverse curvature, 103, 104, 107, 108, 110, 111,
113
Transverse space of a congruence, viii, 36, 37, 39, 44,
46–8, 50–3, 56, 62, 66, 98, 105, 207, 208
Trapped surface, 172, 180, 203, 204, 215, 216
Trapping horizon, 172, 203, 215

U

Uniqueness theorems for black holes, 205, 206, 211,
215, 216
Universe, *see* Cosmology

V

Vaidya spacetime, 152–5, 172–5, 213, 217
Variational principle, 118–46, 157
Vector potential for electromagnetism, 25, 157, 216,
221, 222
Velocity vector, 10, 25, 27, 30, 32, 42, 55, 91, 93, 96,
99, 100, 111, 112, 116, 117, 147, 153, 155, 159,
168, 173, 181–3, 185, 188, 189, 214, 218
Virial theorem of Newtonian gravitational physics,
159
Volume element, 12, 45, 64, 159

W

Wheeler's no-hair statement, 206
Work done
by a null shell, 106
on a stationary particle, 185
World line, 26, 32, 41, 54, 91, 92, 157, 158, 170
Wormhole, 32, 54

Printed in the United States
By Bookmasters